Mobile Source Emissions
Including Polycyclic Organic Species

NATO ASI Series

Advanced Science Institutes Series

A series presenting the results of activities sponsored by the NATO Science Committee, which aims at the dissemation of advanced scientific and technological knowledge, with a view to strengthening links between scientific communities.

The series is published by an international board of publishers in conjunction with the NATO Scientific Affairs Division

A	Life Sciences	Plenum Publishing Corporation
B	Physics	London and New York
C	Mathematical and Physical Sciences	D. Reidel Publishing Company Dordrecht, Boston and Lancaster
D	Behavioural and Social Sciences	Martinus Nijhoff Publishers
E	Engineering and Materials Sciences	The Hague, Boston and Lancaster
F	Computer and Systems Sciences	Springer-Verlag
G	Ecological Sciences	Berlin, Heidelberg, New York and Tokyo

Series C: Mathematical and Physical Sciences Vol. 112

Mobile Source Emissions Including Polycyclic Organic Species

edited by

D. Rondia
University of Liège, Liège, Belgium

M. Cooke
Battelle Memorial Institute, Columbus, Ohio, U.S.A.

and

R. K. Haroz
Battelle Memorial Institute,
Geneva Research Center, Geneva, Switzerland

Sponsored by
NATO Scientific Affairs Office
Battelle Memorial Institute
Belgian Regional Ministry for the Environment

D. Reidel Publishing Company

Dordrecht / Boston / Lancaster

Published in cooperation with NATO Scientific Affairs Division

Proceedings of the NATO Advanced Study Institute on
Mobile Source Emissions Including Polycyclic Organic Species
Liège, Belgium
August 30-September 2, 1982

Library of Congress Cataloging in Publication Data

NATO Advanced Research Workshop on Mobile Source Emissions
       Including Polycyclic Organic Species (1982 : Liège, Belgium)
       Mobile source emissions including polycyclic organic species.

       (NATO ASI series. Series C, Mathematical and physical sciences ; no. 112)
       "Published in cooperation with NATO Scientific Affairs Division."
       "Proceedings of the NATO Advanced Research Workshop on Mobile Source
Emissions Including Polycyclic Organic Species, Liège, Belgium, August 30—September 2,
1982" — T.p. verso.
       Includes index.
       1. Automobiles—Motors—Exhaust gas—Congresses. 2. Diesel motor exhaust
gas—Congresses. 3. Polycyclic compounds—Environmental aspects—Congresses.
I. Rondia, D. (Désiré), 1928—   . II. Cooke, Marcus, 1943—   . III. Haroz,
R. K. (Richard K.), 1941—   . IV. North Atlantic Treaty Organization. Scientific
Affairs Division. V. Title. VI. Series.
TD886.5.N38   1982   628.5'32   83—11069
ISBN-13: 978-94-009-7199-8   e-ISBN-13: 978-94-009-7197-4
DOI: 10.1007/978-94-009-7197-4

Published by D. Reidel Publishing Company
P.O. Box 17, 3300 AA Dordrecht, Holland

Sold and distributed in the U.S.A. and Canada
by Kluwer Academic Publishers,
190 Old Derby Street, Hingham, MA 02043, U.S.A.

In all other countries, sold and distributed
by Kluwer Academic Publishers Group,
P.O. Box 322, 3300 AH Dordrecht, Holland

D. Reidel Publishing Company is a member of the Kluwer Academic Publishers Group

CONTENTS

PREFACE

    This book contains the combined papers from a NATO sponsored meeting entitled "Mobile Source Emissions Including Polycyclic Organic Species". This meeting was held in Liège, Belgium, from August 30 to September 2, 1982, as an Advanced Research Workshop (ARW). Specialists from 13 countries met to discuss the impacts of fuel type and engine operation on emissions from gasoline and diesel-powered vehicles. The meeting was organized into five sessions: basic studies, biological activity, analytical methods, data bases, engineering and international research programs.

    The five sessions were each followed by intensive discussions on the important conclusions to be drawn from the data presented. From these session discussions, subcommittee efforts and a final group discussion on all the research results presented, a meeting summary was prepared by the editors that presents the major findings of the conference and gives detailed recommendations for future research. This summary is included at the end of the book. The summary includes highlights of data presented, areas of research needs, recommendations for standardization of test parameters and species analyzed, suggested new methodology for characterizing mobile emissions, trends in fuel use including diesel, gasoline, anti-knock additives and blending agents, operation of catalysts, control strategies, implications from both in vivo and in vitro health effects studies, and a presentation of the scope and direction of several large international programs that are major studies of mobile source performance.

    Scientists and engineers from government, industry and academic institutions brought the perspectives of each institution and specialty to the meeting and the discussions which followed served to illuminate key problems and major developments in this field. From these discussions, recommendations for future research are given. It is hoped that program planners starting to work on engine operation and mobile emission problems can find this book a helpful guide in formulating new research programs.

ACKNOWLEDGEMENTS

The organizers of this meeting express their appreciation to the group of sponsors whose financial support made this meeting possible. Direct financial support was provided by: the North Atlantic Treaty Organization, Scientific Affairs Division; the Battelle Memorial Institute; the Regional Ministry for Water, Environment and Rural Life in Belgium; the University of Liège; the Perkin-Elmer Corporation (Van der Heyden S.A.) who provided writing materials and badges for participants; and the Fabrique National, who arranged a visit to the exquisite Museum of Arms in Liège and provided the reception that followed for the participants.

The secretarial assistance of Mrs. Domine at the University of Liège and Mrs. Karen Rush at Battelle (Columbus) is greatly appreciated.

THE EDITORS

D. Rondia
M. Cooke
R. Haroz

CONTRIBUTORS

G.J.K ACRES
Johnson Mathey Research
  Centre
Blounts Court Sonning Common
Reading RG49NH
ENGLAND

I. ALFHEIM
Central Institute for
  Industrial Research
Forskningsven 1
Blindern, Oslo 3
NORWAY

W. BAEYENS
Analytische Chimie
Vrije Universiteit Brussel
1040 Brussels
BELGIUM

R. BERBER
Department of Chemical
  Engineering
Faculty of Science
University of Ankara
Besevler Ankara
TURKEY

A. BJØRSETH
Deminex (Norway) A.S.
Wergelandsveien 7
Oslo 2
NORWAY

R.W. BRAATEN
Energy Mines and
  Resources Canada
555 Booth Street
Ottawa, Ontario K1A 0G1
CANADA

L.J. BRASSER
Instituut Voor Milieuhygiene
  en Gezondheidstechniek TNO
Schoemakerstraat 97/2600 AE DELFT
NETHERLANDS

A. CANDELI
Direttore Instituto di Igiene
  Della Facolta di Scienze
Universita di Perugia
06100 Perugia
ITALY

E.J. CLAR
Punta Chullera
KM 144 Estepona
SPAIN

A. CLARK
Public Health Engineering Laboratory
Department of Civil Engineering
Imperial College
London SW7 2BU
ENGLAND

M. COOKE
Battelle's Columbus Laboratories
505 King Avenue
Columbus, Ohio 43201
USA

I. ELSKENS
Analytische Chimie
Vrije Universiteit Brussel
1040 Brussels
BELGIUM

B.J. FRENCH
Ethyl Corporation
1600 West Eight Mile Road
Ferndale, Michigan 48220
USA

xi

C.A. HALL
Ethyl Corporation
1600 West Eight Mile Road
Ferndale, Michigan 48220
USA

C.M. HAMPTON
Ford Motor Company
Engineering and Research Staff
Box 2053 Building 53061
Dearborn, Michigan 48121
USA

R. HAROZ
Battelle Research Centers
Centre for Toxicology
 and Biosciences
7 Route de Drize
1227 Carouge (Geneva)
SWITZERLAND

B. HARRISON
Johnson Mathey Research
 Centre
Blounts Court Sonning Common
Reading RG49NH
ENGLAND

A. HARTUNG
Volkswagenwerk A G
 Abt. Messverfahren
D3180 Wolfsburg
FEDERAL REPUBLIC OF GERMANY

A.C.S. HAYDEN
Energy, Mines and
 Resources Canada
555 Booth Street
Ottawa, Ontario K1A 0G1
CANADA

ING HOK
UTAC Laboratoire
Autodrome de Linas Monthlery
Linas 91310 Monthlery
FRANCE

J. JACOB
Biochemisches Institut
 für Umweltcarcinogene
Sieker Landstraße 19
D2070 Ahrensburg Holst
FEDERAL REPUBLIC OF GERMANY

R.H. JUNGERS
U.S. Environmental Protection
 Agency
Environmental Monitoring
 Systems Laboratory
Research Triangle Park,
 North Carolina 27711
USA

W. KARCHER
Commission of the European
 Communities
Joint Research Centre
PO Box 2
1755 ZG Petten/NL
NETHERLANDS

J. KRAFT
Volkswagenwerk A G
 Abt. Messverfahren
D3180 Wolfsburg
FEDERAL REPUBLIC OF GERMANY

G.W. KUNZ, JR.
DuPont Company

Wilmington, Delaware
USA

A. LAFONTAINE
Institute of Hygiene
 and Epidemiology
14 Rue J. Wystman
1040 Bruxelles
BELGIUM

J. LAHAYE
Centre de Recherches sur la
 Physico-Chimie des Surfaces
Solids, C.N.R.S.
24 Avenue du President Kennedy
Mulhouse 68200
FRANCE

G. LEPPERHOFF
Forschungs Gesellschaft
 für Energietechnik
Augustinergasse 2
D5100 Aachen
FEDERAL REPUBLIC OF GERMANY

J.N. LESTER
Public Health Engineering
 Laboratory
Imperial College
London SW7 2BU
ENGLAND

J. LEWTAS
U.S. Environmental Protection
 Agency
Health Effects Research
 Laboratory
Research Triangle Park,
 North Carolina 27711
USA

K.H. LIES
Volkswagenwerk A G
 Abt. Messverfahren
D3180 Wolfsburg
FEDERAL REPUBLIC OF GERMANY

F. MAGDONELLE
CEC, Dir XI.
Rue de la Loi 200
1049 Bruxelles
BELGIUM

R. MARANO
Ford Motor Company
Engineering and Research Staff
Box 2053 Building 53061
Dearborn, Michigan 48121
USA

A.E. McINTYRE
Public Health Engineering
 Laboratory
Imperial College
London SW7 2BU
ENGLAND

G. MOROZZI
Direttore Instituto di Igiene
 Della Facolta di Scienze
Universita di Perugia
06100 Perugia
ITALY

K. MORTELMANS
SRI International
333 Ravenswood Avenue
Menlo Park, California 94025
USA

M.G. NISHIOKA
Battelle's Columbus Laboratories
505 King Avenue
Columbus, Ohio 43201
USA

F. NUNNEMANN
Daimler-Benz AG
ABT Vang
Postfach 202
D7000 Stuttgart Unterturkheim 60
FEDERAL REPUBLIC OF GERMANY

I.K. O'NEILL
World Health Organization
150 Cours Albert-Thomas
F69372 Lyon Cedex 2
FRANCE

M. PAPUTA
Ford Motor Company
Engineering and Research Staff
Box 2053 Building 53061
Dearborn, Michigan 38121
USA

T.C. PEDERSON
General Motors Research
 Laboratories
Biomedical Science Department
Warren, Michigan 48090
USA

N. PELZ
ABT EGL
Daimler-Benz AG
D7000 Stuttgart 60
FEDERAL REPUBLIC OF GERMANY

H.J. PEPERSTRAETE
SCK-CEN Chemistry Department
Air Pollution Laboratories
Boeretang 200
2400 Mol
BELGIUM

R. PERRY
Public Health Engineering
 Laboratory
Department of Civil
 Engineering
Imperial College
London SW7 2BU
ENGLAND

B. PETERSEN
Battelle's Columbus Laboratories
505 King Avenue
Columbus, Ohio 43201
USA

G. PRADO
Centre de Recherches sur la
  Physico-Chimie des Surfaces
Solids, C.N.R.S.
24 Avenue du President Kennedy
Mulhouse 68200
FRANCE

T.J. PRATER
Ford Motor Company
Engineering and Research Staff
Box 2053 Building 53061
Dearborn, Michigan 48121
USA

R. PROKOPUK
Energy, Mines and
  Resources Canada
555 Booth Street
Ottawa, Ontario K1A 0G1
CANADA

T. RAMDAHL
Central Institute for
  Industrial Research
Forskningsven 1
Blindern, Oslo 3
NORWAY

T. RILEY
Ford Motor Company
Engineering and Research Staff
Box 2053 Building 53061
Dearborn, Michigan 48121
USA

D. RONDIA
Universite de Liege
Laborotoire de Toxicologie
  Industrielle BG
B4000 Liege
BELGIUM

I. SALMEEN
Ford Motor Company
Engineering and Research Staff
Box 2053 Building 53061
Dearborn, Michigan 48121
USA

S. SANTORO
Cogeneration Management
  Company
474 Brookline Avenue
Boston, Massachusetts 02215
USA

V. SCHAFER
EM 53 BMW A.G.
Petuelring 130
D8000 Munchen 40
FEDERAL REPUBLIC OF GERMANY

D. SCHUETZELE
Ford Motor Company
Engineering and Research Staff
Box 2053 Building 53061
Dearborn, Michigan 48121
USA

J. SCHULZE
Volkswagenwerk A G
  Abt. Messverfahren
D3180 Wolfsburg
FEDERAL REPUBLIC OF GERMANY

M.A. SHAPIRO
Graduate School of
  Public Health
University of Pittsburgh
Pittsburgh, Pennsylvania 15261
USA

L. SKEWES
Ford Motor Company
Engineering and Research Staff
Box 2053 Building 53061
Dearborn, Michigan 48121
USA

U. STENBERG
Department of Analytical
  Chemistry
University of Stockholm
S10691 Stockholm
SWEDEN

J.M. TIMS
Esso Petroleum Company
Esso Research Centre
Abingdon Oxfordshire OX13 6AE
ENGLAND

K. VAN CAUWENBERGHE
Chemistry Department
University of Antwerp (UIA)
Universitertsplein
B2610 Wilrijk
BELGIUM

L. VAN VAECK
Chemistry Department
University of Antwerp (UIA)
Universitertsplein
B2610 Wilrijk
BELGIUM

V.S. WILLOUGHBY
Ethyl Corporation
1600 West Eight Mile Road
Ferndale, Michigan 48220
USA

M. WYATT
Johnson Mathey Research
  Centre
Blounts Court Sonning Common
Reading RG49NH
ENGLAND

CATALYTIC CONTROL OF MOTOR VEHICLE EXHAUST EMISSIONS

G.J.K. Acres, B. Harrison and M. Wyatt

Johnson Matthey Research Centre, Sonning Common, Reading, Berkshire, England.

ABSTRACT

Legislation requiring strict control of emissions from motor vehicles is in force in the USA and Japan and is being introduced progressively in Europe. The regulations limit emissions of hydrocarbons, carbon monoxide and nitrogen oxides from gasoline and diesel engines and, in the USA, particulates from diesel engines. Although these regulations refer to 'total hydrocarbons', it is known that this encompasses a broad range of species from simple molecules such as methane to high molecular weight polynuclear aromatic hydrocarbons. The carcinogenic activity of certain polynuclear aromatic hydrocarbons is well characterised and has been a matter of concern in the USA where polynuclear aromatic emissions have been identified in the raw exhaust of gasoline vehicles running on 91 octane lead free fuel and also in the exhaust of diesel vehicles.

Since 1975 catalysts have been fitted to vehicles to control emissions, initially of hydrocarbons and carbon monoxide (oxidation catalysts), and latterly also of oxides of nitrogen (three way catalysts). This contribution will demonstrate the ability of precious metal based catalysts not only to control carbon monoxide, hydrocarbons and nitrogen oxides but also the polynuclear aromatic fraction from both gasoline and diesel fuelled vehicles. The data will include that from both fresh and aged catalyst systems and also those exposed to leaded gasoline.

1

D. Rondia et al. (eds.), Mobile Source Emissions Including Polycyclic Organic Species, 1–12.
© 1983 by D. Reidel Publishing Company.

1.   INTRODUCTION

Growing concern regarding the contribution of the car to
the pollution of the environment has resulted in increasingly
severe worldwide automobile emission legislation, with the
strictest controls being imposed in the USA and Japan.
Emission levels permitted in Europe are being lowered
progressively and present and proposed legislation for carbon
monoxide (CO) hydrocarbon (HC) and oxides of nitrogen (NOx) is
compared with US legislation in Fig. 1. Particulate emissions
from diesel vehicles are also controlled in the USA to a level
of 0.6 g/mile for 1982 and 0.2 g/mile from 1985.

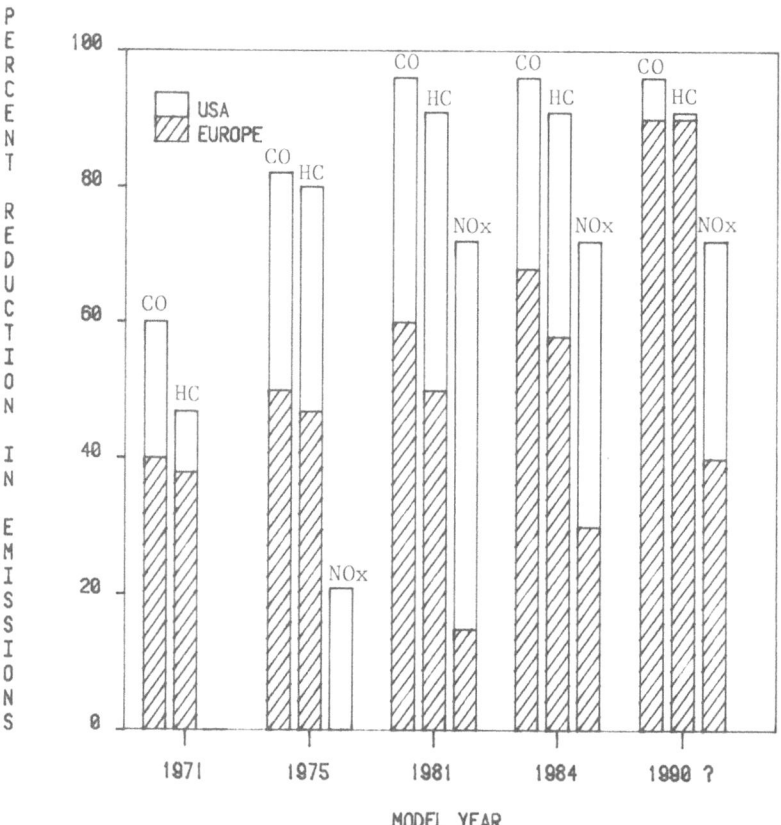

Figure 1 : USA and European Emission Regulations

Since 1975 catalysts have been fitted to vehicles in the USA to control emissions, initially of HC and CO (oxidation catalysts), and latterly also of NOx (three way catalysts). The mode of operation of these catalyst systems in the USA and Japan is now well characterised (1). The catalysts typically comprise the precious metals platinum, palladium and rhodium, either singly or in combination, together with base metal promoters or stabilisers, supported on alumina pellets or alumina coated ceramic monoliths. Catalysts for the US market are designed to withstand 50,000 miles of road use and must be operated in conjunction with lead free fuel since they are poisoned by lead.

As European emission legislation is strengthened, mechanical control techniques similar to those used in the USA until 1975 are available to car manufacturers and it is likely that most vehicles will be able to meet standards up to ECE 15-04 without catalysts, although, in some cases, a fuel economy benefit may be possible by re-tuning and using catalysts. However, a further tightening of legislation will necessitate catalytic control on the majority of cars and present evidence suggests that, although the lead level in gasoline will be reduced gradually from $0.4gl^{-1}$ to $0.15gl^{-1}$, lead free petrol is unlikely to be freely available. Thus, it follows that any catalysts fitted to cars in Europe should be lead tolerant. Recent studies have demonstrated the basic feasibility of lead tolerant catalysts and promising road durability trials have been completed (2).

The concern over emissions from diesel engines relates primarily to the problem of odour, which consists mainly of partially oxygenated hydrocarbon species, and smoke or particulate emissions. Catalysts capable of controlling diesel odour have been commercially available for some time (3) but particulate control has only recently been required by US legislation. Particulate control catalysts are designed to filter particulate matter, which contains adsorbed hydrocarbons, and then to catalytically convert these to carbon dioxide and water. These catalysts are capable not only of converting particulate and gaseous emissions from diesel engines but also of reducing the known mutagenic activity of the particulate matter. (4).

Current emission regulations for hydrocarbons refer to "total hydrocarbons", although it is known that this encompasses a broad range of species from simple molecules such as methane to high molecular weight polynuclear aromatic hydrocarbons (PAH). The carcinogenic activity of certain PAHs is well characterised and has been a matter of concern in the USA where PAH emissions have been identified in the raw exhaust

of gasoline vehicles running on 91 octane lead free fuel and
also in the exhaust of diesel vehicles. Several workers have
shown that PAH emissions are not directly related to the PAH
content of fuel (5-11), but that increasing the aromatic
content of the fuel does have a significant impact (7, 9,
11-14). The effect of lead antiknock components appears to be
slight (7, 9, 11-18), with some evidence that the use of
tetramethyl lead (TML) increases the lower PAHs (11).
Air/fuel ratio (A/F) is a significant factor with PAH emissions
decreasing as A/F increases (7, 8, 12, 16, 17, 19, 20) up to
the point of lean misfire where PAH emissions increase sharply
(7). Engine oil can affect PAH emissions in two ways, either
by accumulating and re-emitting PAH or by contributing to PAH
formation by oil pyrolysis (6-8, 10, 11, 17, 21). The major
mechanism of PAH formation is thought to be via fuel and oil
pyrolysis (10) where hydrocarbons are decomposed to acetylene
via a free radical chain reaction (14). Pyrolysis of acetylene
via a free radical mechanism in the absence of oxygen results
in the formation of PAH.

The control of vehicle emissions by the use of precious
metal catalysts is now well established in the USA and Japan
and there is some evidence that these catalysts are capable of
providing significant control of PAH emissions (9, 14, 22,
23). This contribution will demonstrate the ability of
precious metal based catalysts to control emissions of PAH from
both gasoline and diesel fuelled vehicles. The activity of
these catalysts, both fresh and aged and including those
exposed to leaded gasoline is presented and discussed.

2.   EXPERIMENTAL

2.1 Engine Test Procedures

Diesel emission results were obtained using a naturally
aspirated Perkins 6.354.4 direct injection diesel engine on a
test bed dynamometer under steady state conditions at 50% load
and peak torque speed (1300 r.p.m.). The cetane rating of the
fuel was 45. The catalyst used in the diesel tests comprised a
radially wound compressed stainless steel wire mesh, coated
with a high surface area alumina and catalytically active
precious metals (4).

Gasoline emission tests were conducted on a 1.6 l Ford
Escort car on a rolling road dynamometer. The vehicle was
modified to incorporate secondary air injection into the
exhaust to ensure an exhaust composition which was consistently
oxygen rich. The car was driven over the European ECE-15 test
cycle, and throughout the cycle exhaust gas was sampled at a

constant rate of 10 $lmin^{-1}$ through two cold traps.

Three gasoline fuels were employed during the tests: 97 octane lead free gasoline, the same base fuel doped with 0.15 gl^{-1} lead Motor Mix, and a standard 97 octane UK pump fuel containing $0.4gl^{-1}$ lead Motor Mix. Lead was included in the $0.15gl^{-1}$ fuel as tetramethyl lead (T.M.L) and "Motor Mix" refers to the scavenger package which includes ethylene dichloride (EDC) and ethylene dibromide (EDB). The quantity of scavengers added to the fuel is expressed in terms of the theoretical amount necessary to interact with all the lead and convert it to the corresponding dihalides. The most common additive package known as Motor Mix contains 1 Theory EDC and 1/2 Theory EDB, that is an excess of scavenger.

The catalysts employed for the gasoline studies were prototype monolith supported precious metal based lead tolerant systems (2). Two catalysts were used, one fresh, and one which had been aged on a 1.6 1 Ford Escort engine on leaded fuel $(0.4gl^{-1}$ Pb Motor Mix) for 500 hr over the CCMC cycle (approximately equivalent to 25000km).

2.2 PAH Collection and Analysis

The method of PAH collection in the case of the diesel tests involved drawing a known volume of diluted exhaust through a filter pad in order to collect the particulate matter. The pad underwent Soxhlet extraction with dichloromethane overnight.

In order to analyse gasoline engine exhaust, a constant volume was drawn through two 3 1 cold traps immersed in a dry ice/ethanol bath throughout the ECE-15 test cycle. The material collected in the cold traps and also in a downstream particulate filter was extracted into dichloromethane.

In the cases of both diesel and gasoline exhaust the extract was concentrated, chromatographically cleaned, filtered and separated by high pressure liquid chromatography (HPLC) using ultra-violet fluorescence detection.

Engine and vehicle testing and PAH analyses were performed at Ricardo Consulting Engineers plc.

3.   RESULTS AND DISCUSSION

3.1 Diesel Engine Tests

The baseline and catalyst test results obtained on the

Perkins Diesel engine under steady state conditions are shown in Table 1, where it can be seen that >90% of all PAHs were removed by the catalyst.

Table 1   –   PAH Collected in Diesel Exhaust

Polycyclic Aromatic Hydrocarbon	Emissions Without Catalyst (µg/hr)	Emissions With Catalyst (µg/hr)	Conversion (%)
Fluoranthene	376	17	96
Pyrene	1027	30	97
Chrysene	652	9	99
Benz(a)anthracene	56	3	94
Benzo(e)pyrene	753	26	97
Benzo(b)fluoranthene	76	4	95
Benzo(k)fluoranthene	15	1	91
Benzo(a)pyrene	29	1	97
Benzo(ghi)perylene	39	4	91
Coronene	6	NQ	–

NQ   –   Not quantifiable due to possible interference.

In addition, the catalyst removed 91% CO, 98% total HC and 66% of particulate emissions at a catalyst inlet temperature of 300° C.   The high level of PAH conversion is consistent with that reported by Enga et al. (4) who used a similar catalyst system, but is contrary to that reported by Williams and Swarin (22), although these authors used a pelleted catalyst system. The high levels of CO, HC and particulate conversions are also borne out in the results obtained on a similar wire mesh catalyst system on a 2 litre diesel engined car over the LA4 Cycle (cold start), as shown in Table 2, where it will be seen that a significant reduction in mutagenic activity (by Ames testing) is also recorded.

Table 2   –   Vehicle Diesel Catalyst Results (LA4 Cycle)

	Percentage reduction
Particulates (g/mile)	69
Hydrocarbon (g/mile)	78
Carbon monoxide (g/mile)	88
Mutagenic activity, (Ames Test), induced revertants/mile	88

3.2   Gasoline Engine Tests

Three gasoline fuels containing zero Pb, $0.15gl^{-1}$ Pb Motor Mix and $0.4gl^{-1}$ Pb Motor Mix were evaluated as well as two catalysts, one fresh and one aged under leaded conditions. It was thus possible to investigate the effect of lead on baseline PAH emissions as well as the effectiveness of fresh and aged catalysts in removing PAH. The results obtained are shown in Tables 3, 4 and 5.

A consideration of Tables 3 and 4 shows that the baseline emissions obtained without a catalyst were lower for leaded fuel than for unleaded fuel in all cases except benzo(b)fluoranthene where there was an increase on lead addition. The effect was relatively small, however, and the results are generally consistent with those reported by other workers (12, 14, 15), although Pederson (11) reported a slight increase in emissions of lower PAHs when TML was used as the lead additive. It is not possible to compare directly the results obtained at zero Pb and $0.15gl^{-1}$ Pb with those obtained on the pump $0.4gl^{-1}$ Pb fuel because two variables, i.e. fuel composition and lead level, have changed. However, PAH emissions observed with this fuel were very similar to those observed for the $0.15gl^{-1}$ Pb fuel. There was no significant difference in the total hydrocarbon emissions recorded with any of the fuels.

Table 3 – PAH Collected at Constant Rate
During ECE-15/04 Test

Polycyclic Aromatic Hydrocarbon	Gasoline Containing $0.0gl^{-1}$ Lead		
	Emissions without catalyst (µg)	Aged Catalyst	
		Emissions (µg)	Conversion (%)
Fluoranthene	1880	34	98
Pyrene	3460	76	98
Chrysene	1140	35	97
Benz(a)anthracene	220	2	99
Benzo(e)pyrene	1210	NQ	–
Benzo(b)fluoranthene	185	8	96
Benzo(k)fluoranthene	56	2	96
Benzo(a)pyrene	78	3	96
Benzo(ghi)perylene	380	9	98
Coronene	NQ	NQ	–

NQ:- Not quantifiable due to possible interferences

Table 4 – PAH Collected at Constant Rate

During ECE-15/04 Test

Polycyclic Aromatic Hydrocarbon	Emissions without catalyst (μg)	Gasoline Containing 0.15g l^{-1} Lead			
		Fresh Catalyst		Aged Catalyst	
		Emissions (μg)	Conversion (%)	Emissions (μg)	Conversion (%)
Fluoranthene	1400	ND	100	106	92
Pyrene	2440	6	100	198	92
Chrysene	700	3	100	63	91
Benz(a)anthracene	100	ND	100	4	96
Benzo(e)pyrene	460	NQ	–	NQ	–
Benzo(b)fluoranthene	240	0.3	100	17	93
Benzo(k)fluoranthene	37	0.3	99	4	89
Benzo(a)pyrene	48	0.7	99	6	88
Benzo(ghi)perylene	240	4	98	19	92
Coronene	NQ	NQ	–	NQ	–

NQ:- Not quantifiable due to possible interferences

ND:- Not detected

Table 5 - PAH Collected at Constant Rate

During ECE-15/04 Test

Polycyclic Aromatic Hydrocarbon	Gasoline Containing $0.4\mathrm{g}1^{-1}$ Lead				
	Emissions without catalyst (μg)	Fresh Catalyst		Aged Catalyst	
		Emissions (μg)	Conversion (%)	Emissions (μg)	Conversion (%)
Fluoranthene	1330	14	99	67	95
Pyrene	2310	14	99	114	95
Chrysene	596	5	99	44	93
Benz(a)anthracene	126	1	99	5	96
Benzo(e)pyrene	286	NQ	-	33	88
Benzo(b)fluoranthene	226	14	94	26	88
Benzo(k)fluoranthene	36	2	94	5	86
Benzo(a)pyrene	52	1	98	5	90
Benzo(ghi)perylene	215	4	98	32	85
Coronene	74	NQ	-	13	82

NQ:- Not quantifiable due to possible interferences

The effect of fresh, precious metal based, lead tolerant catalysts on PAH emissions can be seen in Tables 4 and 5, where,with the exception of two data points, PAH conversions are $>98\%$ in the case of tests on both 0.15 and $0.4g1^{-1}$ Pb fuel. The results obtained for the aged catalyst (Tables 3-5) also show high conversion levels, although slightly lower than those shown by the fresh catalyst at around 90%. Conversions of this magnitude have been reported previously for catalysts used with unleaded gasoline (9, 14, 22, 23) but no results have been reported for those aged in leaded gasoline. It is interesting to note that, when the aged catalyst was tested using lead free fuel, PAH conversions were generally $>96\%$. This may be due to a reactivation of the catalyst caused by running on lead free fuel or alternatively an effect of exhaust borne lead species on the ability of the catalyst to oxidise PAH. The total hydrocarbon conversion observed during the ECE-15 test on unleaded fuel was also slightly improved (70% compared to 65%).

4. CONCLUSIONS

The effect of supported precious metal catalysts on PAH emissions from diesel and gasoline engines has been evaluated with the following conclusions:

i) A radial flow, stainless steel wire supported diesel particulate catalyst removed $>90\%$ of all PAH emissions at 50% engine load and peak torque speed.

ii) The presence of lead in gasoline appeared to inhibit baseline PAH emissions from a vehicle in the ECE-15 test.

iii) A fresh catalyst was capable of removing approximately 98% of PAH emissions from a gasoline fuelled vehicle.

iv) A lead tolerant catalyst aged for 500hr (equivalent to 25000km) in leaded fuel was still capable of removing approximately 90% of PAH emissions. Some recovery of this catalyst was observed when it was tested under lead free conditions.

v) It has demonstrated that precious metal catalysts are capable of providing very high conversions of PAHs even to the extent where PAH conversions are higher than "total hydrocarbon" conversions. This is entirely in keeping with published data (24) which suggests that branched chain and aromatic hydrocarbons are simpler to convert by catalytic oxidation than simpler hydrocarbon molecules.

REFERENCES

1. Harrison, B., Cooper, B.J., and Wilkins, A.J.J.: 1981, Platinum Metals Rev. 25(1), pp 14-22.

2. Diwell, A.F., and Harrison, B.: 1981, Platinum Metals Rev. 25(4), pp 142-151.

3. Sercombe, E.J. : 1975, Platinum Metals Rev. 19(1), pp 2-11.

4. Enga, B.E., Buchman, M.F., and Lichtenstein, I.E. : 1982, SAE Paper No. 820184.

5. Begeman, C.R., and Colucci, J.M. : 1968, Science, 161, p 271.

6. Laity, J.L., Malbin, M.D., Haskell, W.W., and Doty, W.I. : 1973, SAE Paper No. 730835.

7. Gross, G.P. : 1972, SAE Paper No. 720210.

8. Janssen, O. : 1980, VDI-Berichte, 358, pp 69-79.

9. Padrta, F.G., Samson, P.C., Donohue, J.J., and Skala, H. : 1971, Am. Chem. Soc. Div. Petrol Chem. Prepr., 16(2), pp E13-E23.

10. Handa, T., Yamamura, T., Kato, Y., Saito, S., and Ishii, T.: 1979, J. Jpn. Soc. Air Pollut., 14 (11/12), pp 18-27.

11. Pederson, P.S., Ingwerson, J., Nielsen, T., and Larsen, E.: 1980, Environ. Sci. Technol., 14(1), pp 71-79.

12. Ingwerson, J., Pederson, P.S., Nielsen, T., Larsen, E., and Fenger, J. : 1980, Aerosols Sci., Med. Technol. - Biomed. Influence Aerosol - Conf., 7th pp 169-174.

13. Felt, A.E., and Kerley, R.V. : 1971, Automotive Eng., 79(3), pp 54-57.

14. Zaghini, N., Mangolini, S., Arteconi. M., and Sezzi, F.: 1973, SAE Paper No. 730836.

15. Candeli, A., Mastrandrea, V., Morozzi,G.,and Toccaceli, S.: 1974, Atmos. Environ. 8, pp 693-705.

16. Griffing, M.E., Maler, A.R., Borland, J.E., and Decker, R.R. : 1971. Am.Chem.Soc.Div.Petrol Chem. Prepr., 16(2), p E24.

17. Begeman, C.R., and Colucci, J.M. : 1970, SAE Paper No. 700469.

18. Rinehart, W.E., Gendermalik, S.A., and Gilbert, L.F. : 1970, Amer.Indust.Hygiene. Conf., Detroit, Michigan, 127, p 15.

19. Hoffman, C.S., Willis, R.L., Patterson, G.H., and Jacobs, E.S. : 1971, Am.Chem.Soc.Div.Petrol Chem. Prepr., 16(2), p E36.

20. Gofmekler, V.A., Maneta, M., Manusadshants, Z., and Stepanov, L. : 1963, Gigiena i Sanitariya, 28, pp 3-8.

21. Newhall, H.K., Jentoft, R.E., and Ballinger, P.R. : 1973, SAE Paper No. 730834.

22. Williams, R.L., and Swarin, S.J. : 1979, SAE Paper No. 790419.

23. Lee, F.S.C., Prater, T.J., and Ferris, F.: 1979, Polynucl. Aromat. Hydrocarbons, Int.Symp.Chem.Biol. - Carcinog.Mutagen., 3rd, pp 83-110.

24. Acres, G.J.K.: 1970, Platinum Metals Rev., 14(1), pp 2-10.

POLYCYCLIC ORGANIC MATTER IN THE ATMOSPHERE, POM CONCENTRATIONS[*] IN THE NETHERLANDS

L.J. Brasser [**]

TNO Research Institute for Environmental Hygiene

SUMMARY

After the discovery of large reserves of natural gas in The Netherlands, coal and oil were rapidly replaced by natural gas as a fuel
for production of electricity, heat, energy in industry etc. This
resulted in a change in the concentrations of air pollutants. Sulphur dioxide, smoke, grit and polycyclic aromatic hydrocarbons (as
far as related to the use of coal and oil as a fuel)went down in concentration.
PAH from traffic exhaust gases and from special sources, f.i. coke
ovens,did not diminish in quantity. Ratios between different PAH
components can be used as tracers for the different sources.

[*] Publication No.838 of the T N O Research Institute for Environ-
    mental Hygiene

[**] Prof. Ir. L.J. Brasser TNO Research Institute for Environmental
    Hygiene P.O. Box 214
                2600 AE Delft, The Netherlands
                Resp. University of Technology Eindhoven

1. INTRODUCTION

Starting in the winter of 1964-1965, the TNO Research Institute for
Environmental Hygiene (RIEH) has determined the PAH concentrations
of the atmosphere at several measuring stations in The Netherlands.
At the moment the measurements started, coal and at a smaller scale
oil, was used as a fuel for the heating of buildings and at more or
less equal rates for the production of energy. After the discovery
of large reserves of natural gas, the production of heat and electricity changed over to natural gas as a fuel. This resulted in a
much cleaner combustion. Concentrations of sulphur dioxide, black
suspended matter and PAH rapidly declined (1), (2), (3). More information on this subject will be presented in paragraph 3 dealing
with the trends in POM-concentrations.
During the same period motor traffic developed as a major source of
air pollution. Industry too was growing in size and quantity, old
installations were replaced by more modern ones. All this resulted
in a process of changing ratio's between different pollutants. This
process is still going on as in the meantime the come-back of coal
started; up to now mainly as a fuel in power stations. The process
has been studied by means of air quality measurements and it is the intention to proceed along this line of study.

2. MEASUREMENTS AVAILABLE FOR THIS STUDY

The measurements mentioned in the introduction were made at a large
number of stations. In the course of the years a selection was made, and
only stations representing a well described area were kept in operation. At all stations the following polycyclic aromatic hydrocarbons
were determined:

    pyrene (benzo [d, e, f] phenanthrene)
    fluoranthene (indryl)
    1,2-benzanthracene (benz [a] anthracene)
    chrysene   (1,2-benzophenanthrene)
    1,2-benzopyrene (benzo [e] pyrene)
    3,4-benzopyrene (benzo [a] pyrene)
    1,12-benzoperylene (benzo [g,h,i] perylene)
    anthanthrene
    coronene   (hexabenzobenzene)

It should be stressed that all measurements are determinations of
the air quality, producing data on PAH-concentrations in atmospheric air. This means that on all stations air pollutants emanating
from all kinds of sources can occur. However, the better defined
the areas around the stations are, the better fingerprint of special
sources is obtained.
Apart from these measurements at fixed stations, mobile measurements
were made under the plumes of typical sources. The data obtained in
this way give a better description about the pollution produced by
these typical sources (as seen against the background of general pollution in that area).

The samples were collected on filters and extracted for 16 hours in a
Soxhlet extractor. Using column chromatography the extract was divided
over 24 fractions. The absorbtion spectra of the fractions were measured
in the 270-400 nm region. Later in the study, when the PAH concentrations became lower and better laboratory techniques became available, determinations were made by HPLC.
As it is not well possible to reproduce here all data obtained, a
selection of the data was made. However,all data are presented in
reports of the TNO - RIEH (1), (2).
Several other institutes in The Netherlands conduct air quality studies: PAH concentrations are determined at a number of measuring
stations in regional networks. Data of one of these networks, where
one typical source (coke ovens) gives an important contribution to
the PAH concentrations,are used in paragraph 4 (4).

3. TREND OF THE PAH CONCENTRATIONS OVER THE YEARS

To prevent printing many long tables of concentrations a selection
was made of three measuring stations where long series of data were
available. As concentrations during the heating season are most important, especially for the years before the change over from coal
to gas as a fuel for domestic heating, only mean winter values are presented here (Winter = November through February). A selection of
four polycyclic aromatic hydrocarbons was made.
The three measuring stations are:

W 9 Naaldwijk   Country town of about 25000 inhabitants in the
                     centre of a large district covered with glass-
                     houses
D 12 Delft   Outskirts of a middle-sized town of about 85000
                     inhabitants, near residential areas and industry,
                     in the midst of university and research institu-
                     te buildings
V 10 Rotterdam   City centre of a major town. The centre is for
                     a large part heated by district heating with a
                     power station at several kilometres distance as
                     the major heat source.

The trend in the data presented in tables 1A and 1B is very clear.
The other PAHs that were measured (see paragraph 2) but not mentioned in the tables show the same trend,as do measurements on other
stations.
During the period covered in this study there are two important changes
in the possible sources of PAH. In the first place the fuels used
for domestic heating and for energyproduction changed. In 1965 coal
and oil were major sources of heat and energy in general. Natural
gas was just coming into use. About 1980 a very large part of the
production of heat and energy was covered by natural gas, oil was
still in use but it covered a far smaller part of the total energy

Table 1A. The trend for 1,2 benzantracene and chrysene in ng/m^3 in winter at three locations.

Winter ↓ station	1,2-benzantracene			Chrysene		
	W 9	D 12	V 10	W 9	D 12	V 10
1965 - 66	8	9	20	8	19	25
1966 - 67	9	7	10	15	7	11
1967 - 68	4	9	3	.	.	.
1968 - 69	19	14	13	22	14	20
1969 - 70	12	12	11	19	19	14
1970 - 71	9	9	7	13	15	14
1971 - 72	11	5	6	20	9	14
1972 - 73	7	17	7	9	13	9
1973 - 74	3	n.d.	n.d.	4	18	6
1974 - 75	.	.	2.0	.	.	6.7
1975 - 76	.	.	.	.	.	.
1976 - 77	3	n.d.	2.3	8	2	5
1977 - 78	1	1	1	1	2	1
1978 - 79	2	1	n.d.	3	n.d.	6
1979 - 80	n.d.	n.d.	n.d.	3.1	3.9	3.1
1980 - 81	n.d.	n.d.	n.d.	1.1	n.d.	n.d.

Table 1B. The trend for 1,2 benzopyrene and 3,4 benzopyrene in ng/m^3 in winter at three locations.

Winter ↓ station	1,2 benzopyrene			3,4 benzopyrene		
	W 9	D 12	V 10	W 9	D 12	V 10
1965 - 66	13	18	37	14	12	26
1966 - 67	15	13	22	11	6	17
1967 - 68	9	14	26	6	8	21
1968 - 69	26	22	24	19	10	22
1969 - 70	17	20	25	13	13	20
1970 - 71	11	12	17	5	13	15
1971 - 72	16	10	17	14	4	20
1972 - 73	8	12	16	13	18	17
1973 - 74	4	n.d.	2.4	9	10	5.3
1974 - 75	.	.	3.6	.	.	2.4
1975 - 76	.	.	.	.	.	3.6
1976 - 77	4	n.d.	3.2	n.d.	n.d.	1.3
1977 - 78	2	n.d.	n.d.	n.d.	n.d.	n.d.
1978 - 79	n.d.	n.d.	1	n.d.	n.d.	n.d.
1979 - 80	0.9	n.d.	2.3	n.d.	n.d.	n.d.
1980 - 81	1.1	n.d.	n.d.	n.d.	n.d.	n.d.

. no data available
n.d. not detectable

production than natural gas. Coal was only used in very small
quantities, mostly in large power stations.
The second change was the development of motor traffic, the total
number of motor vehicles grew from about 1 700 000 in 1965 to about
5 000 000 in 1980 in The Netherlands.
Comparing these data with the concentrations presented in tables
1A and 1B, brings us to the conclusion that changing coal for natural
gas as a fuel for heating and energy production(especially heating
in houses) brought down the PAH-concentrations at the measuring stations to very low values. The rapid growth of traffic density didnot interfere with this trend.
Ofcourse these conclusions are only true for the general air quality measuring stations like those mentioned in tables 1A and 1B. At
special stations, measuring air pollution from typical sources, the
trend for PAH may be different or less pronounced.

4. RATIOS BETWEEN PAH'S AS AFFECTED BY THEIR TYPE OF SOURCE

From already older literature (5), (6) it is known that the ratios
between the different polycyclic hydrocarbons depend very strongly
on the type of source by which they are produced. From this and other literature,table 2 was produced before (3), showing some of
these ratios.

Table 2.   Ratios between different PAH concentrations as measured
            in practice (data from older literature)

Ratio	Motor traffic	Burning of fuels
3,4-benzopyrene / coronene	< 0.4 - 1	> 1.7
3,4-benzopyrene / 1,12-benzoperylene	0.2 - 0.6	> 0.8
pyrene / 3,4-benzopyrene	2 - 6	0.3 - 0.8

As last years motor traffic engines have changed considerably, a
change in the ratios is probable too. Moreover sampling systems and
analysis techniques have been modernized and more reliable data are becoming available. The necessity, especially in stack and exhaust sampling, to collect both particulate and vapor phase PAH(7)is stressed.
Seen in this light and keeping in mind that most data are derived
from samples collected on filters, it is better to use only particulate PAH as a base for determining ratios between the different compounds. This means that for instance anthracene, fluoranthene and
pyrene are not well suited for fingerprinting types of sources,
whereas chrysene, benzopyrenes, benzoperylenes,

and coronene seem to be much better, as they can be sampled quite well
on filters.

As the measurements presented in paragraph 3 show very low concentrations, it is necessary to use data obtained in the vicinity of
typical sources of PAH or short time measurements made under the
plume of selected sources for the study of ratios between the different polycyclic hydrocarbons. This situation may be different in
other countries.

In recent literature (7),(8) some data on PAH emissions are presented.
These data can be used for the determination of ratios, this in addition to the field measurements available in The Netherlands.
It looks promising to determine the ratios between several frequently measured PAHs as chrysene, 1,2 benzopyrene, 3,4 benzopyrene,
1,12 benzoperylene and coronene. By taking the most commonly determined PAH-3,4-benzopyrene- as a standard by giving its concentration
the value of 100, the ratios for the other compounds can be expressed in relation to this value of 100.
The data from other literature as presented in table 2 produce the
following ratios:

Table 3.   Ratios calculated from data in table 2 taking 3,4 benzo-
            pyrene as a standard (= 100).

| Source | Ratio (3,4 b.p. = 100) | |
	coronene	1,12-benzoperylene
motortraffic	100--- >250	167----500
burning of fuels	< 59	< 125

Data derived from research by the author's institute (1),(2),(3)
and from measurements made by the Province of North Holland (4)
were combined into groups and arranged according to the type of
source.
Though all data are available, it does not seem useful to produce
them completely in this report as this would need several pages to
print. Therefore, a compilation was made in which the weighed mean
of the ratios and the range in which they fall are produced. These
ratios are presented in table 4.

Some information from recent literature, derived from measurements
at the source (7),(8) is presented in table 5.

Table 4.   Ratios derived from older research by the author's institute
            (A) and from measurements made by the Province of North
            Holland(B)

Kind of source	M/R	Ratio(3,4 b.p. = 100)			
		chr.	1,2 b.p.	1,12 b.p.	cor.
flowing traffic} (A)	mean	339	163	115	91
	range	50-up-wards	17-up-wards	50-up-wards	30-up-wards
halting traffic} (A)	mean	144	131	334	189 .
	range	57-up-wards	100-up-wards	79-up-wards	36-up-wards
traffic+back-ground } (A) in fog	mean	140	136	85	14
	range	59-217	91-200	55-200	8-50
towns+coke ovens}(B)	mean	83	101	105	21
	range	64-143	83-171	78-143	14-38
towns+traffic } (A) winter	mean	106	61	36	60
	range	106	56-64	6-68	27-67
town,no through traffic} (B)	mean	103	140	134	34
	range	86-118	129-175	100-186	25-41
coal heated towns (1965/66)} (A)	mean	100	131	65	12
	range	57-158	93-150	57-75	8-17
eel smoking (A)	mean	162	135	108	27
	range	133-183	117-200	67-200	17-50
background sta- (B) tion	mean	79	118	89	18
	range	78-150	94-200	50-100	17-25

upwards means extremely high ratio because of very low 3,4-benzo-
pyrene concentration.

Table 5.   Data derived from measurements at the source and presen-
           ted in recent literature

Literature	PAH				
	chr.	1,2 b.p.	3,4 b.p.	1,12 b.p.r	cor.
(7) Jones P.W. emission from coke oven doors	31.0	16.0 [1]		6.0 [2]	1.5
(8) Bennett R.L. et al. oil-fired powerplant PAH4	3.9 [3]	5.7	2.4	0.47	0.30

1) We assume that this total is more or less evenly divided over
   both compounds: 8 1,2-b.p.
                    8 3,4-b.p.
2) all benzoperylenes together; we assume 4 for 1,12-b.p.r.
3) together with 1,2-benzanthracene; we assume 3 for chr.

If our assumptions are right, we find the ratios given in table 6.
These are combined with ratios derived from table 2 and already presented in table 3.

Table 6.   Ratios derived from older (A) and recent (B) literature
            concerning measurements at different types of sources
            (making several assumptions)

Source	Ratio (3,4 b.p. = 100)			
	chr.	1,2-b.p.	1,12-b.p.	cor.
motor traffic (A)	.	.	167---500	100---> 250
burning of				
fuels (A)	.	.	< 125	< 59
coke oven (B)	390	100	50	19
oil fired				
powerplant (B)	125	240	20	13

It would be very welcome if more data obtained from measurements at
different types of sources could be made available. A search in the
literature, more complete than the quick scanning of literature
done by the author would be most welcome to give a better insight
into the different ratios between PAHs found in the sources and
into the reproducibility of these ratios from more sources of the
same kind.
This is especially true for The Netherlands where a change over of
fuels has been made from coal and oil to natural gas and where another changeover may be expected in future when coal will be rein-
troduced especially as a source of energy in larger sources as for
instance in power stations.
These and other related conclusions are presented in the next paragraph.

5. CONCLUSIONS

The study of the concentrations of polycyclic aromatic hydrocarbons
in The Netherlands leads to several conclusions, as:
-    During the period studied (1965-1981) the concentrations of PAH
     as a contribution to general air pollution in The Netherlands

decreased very strongly. The most important reason for this
trend is the replacement of coal as a fuel for domestic heating
by natural gas.
- The rapid growth of the density of the motor traffic in towns
  didnot interfere with the downward trend of PAH-concentrations
  in towns.
- The influence of the reintroduction of coal as a fuel in power
  stations cannot be estimated yet.
- Local sources of PAH-air pollution still contribute to local concentrations of PAH.
- The above mentioned conclusions are found under the special situation in The Netherlands: a flat country with high wind velo-
  cities, domestic heating with natural gas as a fuel, large contribution of natural gas as a fuel for energy production.

From this new study a second group of conclusions can be added to
the foregoing:
- The range of ratios between different PAHs and 3,4 benzopyrene
  is so large that further study is necessary before these ratios
  can be more than a faint indication of the type of source.
- In practice field measurements give information on a mixture of
  pollution emitted by different types of sources.
- More data from source measurements (modern literature and research) are necessary before ratios can be used in practice.
- Even then, practice has to prove the validity of this approach.
  It is however worth while to make an attempt in this direction,
  especially in a country as The Netherlands where a change in
  fuels has been made before (changeover to natural gas) and where
  another change may be expected in future (reintroduction of coal
  as a fuel).
- It may be expected that modernization of sources will not only
  lead to a lowering of the production of pollution, but also to
  other ratios between the quantities of the different compounds
  produced.

LITERATURE

(1).Several measuring reports of the TNO Research Institute for En-
    vironmental Hygiene, covering the period of 1965 - present.
    IMG-TNO Delft.

(2).Kommers, F.J.W.
    Polycyclic hydrocarbons
    Report G 689 of the TNO Research Institute for Environmental Hy-
    giene
    IMG-TNO, Delft, January 1976.

(3).Brasser, L.J.
Polycyclic Aromatic Hydrocarbon concentrations in The Netherlands
VDI-Berichte Nr. 358
VDI Düsseldorf 1980, p.171-180.

(4).Several periodical reports of the Service for Environmental
Hygiene of the Province of North Holland
Provinciale Waterstaat van Noord-Holland
Dienst voor de Milieuhygiëne
Verslagen Metingen Luchtverontreiniging

(5).Sawicki, E.
Analysis for airborne particulate hydrocarbons: their relative
proportions as affected by different types of pollution
Symposium Analysis of carcinogenic air pollutants
Cincinnati, August 1961
National Cancer Institute Monograph No. 9, p. 201-220.

(6).Lyons, M.J.
Comparison of aromatic polycyclic hydrocarbons from gasoline
engine and diesel engine exhaust, general atmospheric dust and
cigarette-smoke condensate.
Symposium analysis of carcinogenic air pollutants
Cincinnati, August 1961
National Cancer Institute Monograph No. 9, p.193-199.

(7). Jones, P.W.
Measurements of PAH emissions from stationary sources -
an overview
VDI-Berichte Nr. 358
VDI Düsseldorf 1980, p.23-38.

(8). Bennett, R.L. et al.
Measurement of polynuclear aromatic hydrocarbons and other
hazardous organic compounds in stack gases
In: Polynuclear aromatic hydrocarbons
Edited by P.W. Jones and P. Leber
Ann Arbor Science Publishers Inc.
Ann Arbor 1979, p. 419-428.

ADDENDUM TO
POLYCYCLIC ORGANIC MATTER IN THE ATMOSPHERE, POM CONCENTRATIONS
IN THE NETHERLANDS

In the above mentioned paper only summaries of the original data
are used to study the possibility of determining ratios between
the different POM constituents with the intention to use these ratios for description of the source of the original pollution. In
this addendum more information is presented on the original data
and on the ratios calculated from them.
To do so four tables giving a compilation of the data and four tables presenting the ratios calculated from them are available.
This material forms the basis for paragraph 4 of the original paper.

Table A 1.
Measurements made in a tunnel, along a motor road and at a border-
crossing of a motor road

Station	ng PAH/m^3				
	chr.	1,2 b.p.	3,4 b.p.	1,12 b.p.r.	cor.
Amsterdam,	11	10	14	16	7
Coentunnel 1971	2	1	.	4	2
	.	2	12	11	10
	46	49	2	10	4
	43	4	10	8	17
Roelof Arends-					
veen motorroad					
Wi'72	3.1	3.7	2.7	2.9	0.8
Su'72	1	1	2	1	n.d.
Denekamp'65	.	12	12	54	32
Bergh Su'69	3	7	1	27	16
Wi'69	.	15	14	11	5
Sp'70	6	5	1	16	7
Su'70	4	7	7	9	6

Table A 2.
Measurements made at a measuring station near to a motor road during
foggy days with low wind velocity

Weather conditions foggy days	ng PAH/m^3				
	chr.	1,2 b.p.	3,4 b.p.	1,12 b.p.r.	cor.
low wind velocity from Rotterdam + motor road	. 98 80	55 79 62	49 56 54	39 31 44	4 . 5
low wind velocity from heated glass-houses	66 29 18	60 17 16	60 17 12	48 12 12	6 3 3
low wind velocity from Rotterdam + motor road	46 33 55	38 74 88	39 56 50	31 37 39	5 6 5
low wind velocity along motor road	13	12	6	12	3
low wind velocity from Rotterdam + motor road	23	14	11	13	3
low wind velocity from The Hague + motor road	151 209 199	200 175 177	123 138 101	103 126 118	21 19 18
low wind velocity from Rotterdam + motor road	50	39	43	31	6

Table A 3.
Measurements made by the author's institute during routine studies
in several towns and near an eel smoking plant in a fishing village

Station	ng PAH/m^3				
	chr.	1,2 b.p.	3,4 b.p.	1,12 b.p.r.	cor.
W 9)	8	13	14	8	2
D 12) W 1965/66	19	18	12	9	2
V 10)	25	37	26	17	2
W 9)	3.1	0.9	n.d.	n.d.	n.d.
D 12) W 1979/80	3.9	n.d.	n.d.	n.d.	n.d.
V 10)	3.1	2.3	n.d.	n.d.	n.d.
W 9)	1.1	1.1	n.d.	n.d.	n.d.
D 12) W 1980/81	n.d.	n.d.	n.d.	n.d.	n.d.
V 10)	n.d.	n.d.	n.d.	n.d.	n.d.
The Hague W 1970	.	.	9	1	6
The Hague W 1971	17	9	16	1	16
Delft (centre) W 1969	.	14	22	15	6
Volendam eel smoking	11	7	6	6	1
	4	4	3	2	1
	3	3	n.d.	n.d.	n.d.
	3	4	2	4	1

W 9, D 12, V 10 : see chapter 3

W : Winter.

Table A 4.
Measurements made by the Service for Environmental Hygiene of the
Province of North Holland during routine studies in several towns

Station		ng PAH/m^3				
		chr.	1,2 b.p.	3,4 b.p.	1,12b.p.r	cor.
H 4	Summer 1979	0.4	0.7	0.4	0.4	0.1
H 9		1.0	1.2	0.7	1.0	0.2
H 10		2.1	4.3	3.3	3.6	0.5
H 16		1.5	2.3	1.3	1.5	0.5
H 17		0.3	0.4	0.2	0.1	n.d.
H 4	Winter1979/'80	2.0	2.4	1.7	1.7	0.7
H 9		3.1	3.0	2.8	2.7	0.5
H 10		8.5	7.8	8.0	8.0	1.1
H 16		3.0	3.9	4.5	3.5	0.8
H 17		1.4	1.7	1.8	1.6	0.3
H 4	Year 1980/'81	0.6	0.9	0.7	1.3	0.2
H 9		2.1	2.6	2.8	2.6	0.8
H 10		2.7	3.5	4.2	5.1	0.9
H 16		1.5	2.0	1.8	2.3	0.6
H 17		0.5	0.6	0.4	0.4	0.1

Table B 1.
Ratios calculated from table A 1

Station	Ratio (3,4 b.p. = 100)			
	chr.	1,2 b.p.	1,12 b.p.r.	cor.
tunnel	79	71	114	50
	3,4	b.p. data missing		
	.	17	92	83
	2300	2450	500	200
	430	40	80	170
motor road	115	137	107	30
	50	50	50	n.d.
bordercrossing	.	100	450	267
	300	700	2700	1600
	.	107	79	36
	600	500	1600	700
	57	100	129	86

Table B 2.
Ratios calculated from table A 2

Source during foggy days	Ratio (3,4 b.p. = 100)			
	chr.	1,2 b.p.	1,12 b.p.r.	cor.
motor road + Rotterdam	.	112	80	8
	175	141	55	.
	148	115	81	9
heated glasshouses	110	100	80	10
	171	100	71	18
	150	133	100	25
motor road + Rotterdam	118	97	79	13
	59	132	66	11
	110	176	78	10
motor road	217	200	200	50
motor road + Rotterdam	209	127	118	27
motor road + The Hague	123	163	84	17
	151	127	91	14
	197	175	117	18
motor road + Rotterdam	116	91	72	14

Table B 3·
Ratios calculated from table A 3

Station	Ratio (3,4 b.p. = 100)			
	chr.	1,2 b.p.	1,12 b.p.r.	cor.
W 9 ⎫ D 12 ⎬ W 1965/66 V 10 ⎭	57 158 96	93 150 142	57 75 65	14 17 8
W 9 ⎫ 1979 D 12 ⎬ 1980 V 10 ⎭ 1981	3,4 b.p. not detectable			
The Hague W 1970 The Hague W 1971 Delft(Centre) W1969	. 106 .	. 56 64	11 6 68	67 100 27
Volendam eel smoking	183 133	117 133	100 67	17 33
	3,4 b.p. not detectable			
	150	200	200	50

Table B 4.
Ratios calculated from table A 4

Station	Ratio (3,4 b.p. = 100)			
	chr.	1,2 b.p.	1,12 b.p.r.	cor.
H 4 ⎫ H 9 ⎪ H 10 ⎬ Summer H 16 ⎪ H 17 ⎭	100 143 64 115 150	175 171 130 177 200	100 143 109 115 50	25 29 15 38 n.d.
H 4 ⎫ H 9 ⎪ H 10 ⎬ Winter H 16 ⎪ H 17 ⎭	118 111 106 67 78	141 107 98 87 94	100 96 100 78 89	41 18 14 18 17
H 4 ⎫ H 9 ⎪ H 10 ⎬ Year H 16 ⎪ H 17 ⎭	86 75 64 83 125	129 93 83 111 150	186 93 121 127 100	29 29 21 33 25

PAH CONTENT OF EXHAUST GASES FROM FUELS WITH DIFFERENT AROMATIC
FRACTION

Adele Candeli^, Guido Morozzi^, and Maurice A. Shapiro^^

^Istituto di Igiene Università di Perugia (Italy)
^^Graduate School of Public Health University of
Pittsburgh (USA)

ABSTRACT

It has been ascertained that the quantity of aromatics in fuels is
an important factor affecting PAH emission. Many results however,
indicate the inadequacy of total fuel aromaticity as a general
predictor of PAH emission, especially for gasoline. In fact it
has been shown that the kind of aromatics exert a great influence,
and particularly that PAH emission increase with the increase of
the molecular complexity of aromatics in fuels. Aromatics with 9,
10 or more carbon atoms (C_9-C_{10+}) increase PAH emission much more
than does benzene or the methyl and ethyl derivatives of benzene.
    In the case of diesel fuels lowered aromatics reduce PAH
emissions and improve fuel quality. This dual advantage does not
apply to gasoline which when its aromatics content is reduced,
is lowered in quality. This can be avoided by cutting out the high
carbon atom number aromatic fraction which in addition is associated with high fuel PAH content.

INTRODUCTION

    In the epidemiological literature related to cancer,one finds
a continuing debate about the influence on cancer mortality rates,
particularly on lung cancer of what has been termed the "urban
factor" that is a sum of socio-economic conditions and environmental
agents, present in urban but not in rural areas. (7, 4, 1, 40, 29,46,
41, 37, 39, 15). Nevertheless, up to now, there is no agreement as
to the existance of a real excess of lung cancer mortality in

29

D. Rondia et al. (eds.), Mobile Source Emissions Including Polycyclic Organic Species, 29–47.
© 1983 by D. Reidel Publishing Company.

urban areas with respect to rural areas, and no agreement has yet
been reached on the role of polluted atmosphere of cities on this
debated difference (25).

Despite this lack of agreement, there is no doubt about the
fact that carcinogenic compounds, such as some polycyclic aromatic
hydrocarbons (PAH), exist in higher concentration in urban than in
rural environments (42, 3). It is also clear, as will be described
later, that a considerable portion of this PAH pollution derives
from gasoline and diesel engine emissions, especially because,
after the advent of Clean Air Acts and their regulations, the
PAHs derived from burning coal have been measurably reduced in
large cities such as London (11).

Furthermore it has long been demonstrated that automobile
exhaust condensates are carcinogenic in animals in skin painting
studies (47, 35). This has been ascertained while, until this
time, there has been no similar incontrovertible proof of the
carcinogenicity of the same exhaust gases to the respiratory system
when tested in chronic animal inhalation studies (32). This result
was obtained despite the fact that in other studies (30), cancer
of the respiratory tract was induced by intratracheal instillation
of benzo(a)pyrene (BaP) in Syrian Golden Hamsters. Moreover, it
is well known that the carcinogenic PAHs found in automotive exhaust
have also been found in cigarette smoke (for the comparison of the
two sources see Cuddihy et al. 12) and in the atmosphere of the
gas retort houses (33) and that for these sources, particularly
for cigarette smoke there is, besides experimental evidence, epidemiological evidence proving that they represent important lung
cancer risk factors (14, 29, 23, 24, 15).

In this context it is interesting to note that at least on
one point the literature seems to be in agreement, i.e. that the
lung cancer mortality rate among non smokers is extremely low,
irrespective of the area in which the non-smokers live, while the
effect of smoking a particular amount of cigarettes is greater in
urban than in rural areas, suggesting that tobacco smoke may
interact with carcinogens in urban air (1, 15, 16). This finding
is in agreement with the hypothesis according to which a large
proportion of human cancers may result from exposure to two or
more different agents (25).

In addition to the literature referred to above, we should
remember that recent studies have proved that the particulate
matter of polluted air(36) and of exhaust gases, both of gasoline
(44) and of diesel engines (10, 45), produce positive results when
subjected to the Ames test, indicating the presence of indirect
mutagens, such as the well known PAHs (44) and of direct mutagens
such as nitro-pyrene (43). Moreover since direct mutagenic activity
demonstrated in polluted air is correlated with its lead content,
it can be assumed that this is due to motor vehicle emissions (44)
Not insignificant, is the further observation according to which
the air-borne particles of respirable size, less than 3 um of
equivalent aerodynamic diameter (36), show mutagenic activity and

that more than 90% of the mass of the heavier PAHs (e.g. BaP)
found in exhaust gases particulate matter, is adsorbed on particles
below 1 um diameter (38).

All the data referred to up to this point leave no doubt
about the fact that automotive vehicle exhaust gases are rich in
mutagenic and carcinogenic compounds. Therefore, until further
epidemiological studies lead to general agreement that their presence in the atmosphere is not harmful for human health, we must
reduce their concentration in exhaust emissions as much as possible. The ways in which the carcinogenic burden in motor vehicle
emissions can be reduced are numerous as are the variables which
influence the emission.

In this paper we concentrate upon fuel characteristics giving
particular attention to the aromatic fraction of fuels used in
gasoline and diesel engines. A principal objective is to find, if
possible, which type of aromatic could be excluded or added in
order to lower PAH emissions and consequentely reduce carcinogenicity while maintaining the beneficial properties of such aromatics.
To accomplish this objective we will deal separately with the
major motive power engines namely, gasoline and diesel.

PAH EMISSION IN RELATION TO THE AROMATIC FRACTION IN GASOLINE FUELS

Because of its considerable effect on PAH emissions, since
the early 1960s, aromatics content in gasoline has received much
attention (31, 6). More recent data (Fig. 1) have confirmed the

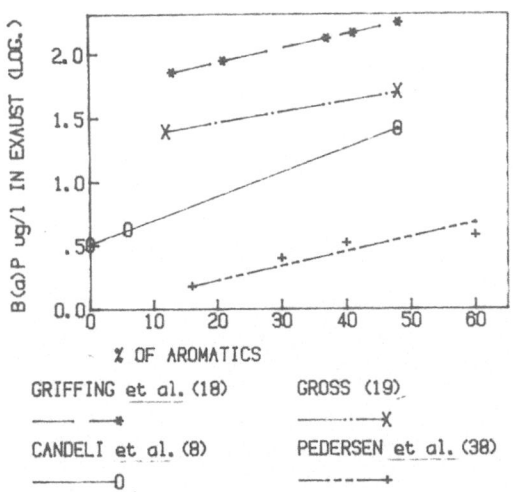

FIG. 1- EFFECT OF AROMATICS CONTENT IN GASOLINE ON B(a)P EMISSION

earlier findings. From the original data reported by the authors
quoted in Fig. 1, we have calculated the percentage change of
BaP emissions as a result of an increase in the aromatic content
of gasoline. From Griffing's et al. data (18) we find that a 1%
increase in aromatic content resulted in a 2.8% increase in BaP
emission. For the same increase of 1% aromatics, Gross (19) obtained a BaP increase of 9.8, Candeli et al. 14% (8) and Pedersen et
al. 2.8% (38).

      Further research has demonstrated however, that not all
aromatics have the same effect, this is so because the kind of
aromatics may greatly influence PAH emission. The low boiling
aromatics, have less capability of forming PAHs during combustion
than do the high boiling ones (8,48). Candeli et al. (8) comparing
the emission produced from a paraffinic fuel with a fuel containing
48% benzene, found that BaP emission increased from 3.2 to 6.7 ugs
per liter of fuel burnt. This is an increase of 109%. Moreover,
when benzene was substituted by a mixture of ethyl-benzene and
xylene (C_8 aromatics), an increase of about 137% in BaP, with
respect to the paraffinic fuel, was observed. When all the thir-
teen determined PAHs were taken into account an increase in total
PAH emission of 155% was obtained upon substituting benzene by a
mixture of ethyl-benzene and xylene (8). These data are in good
agreement with the more recent data obtained by Pedersen et al.
(38) which have demonstrated (Fig. 2) that benzene derivatives are

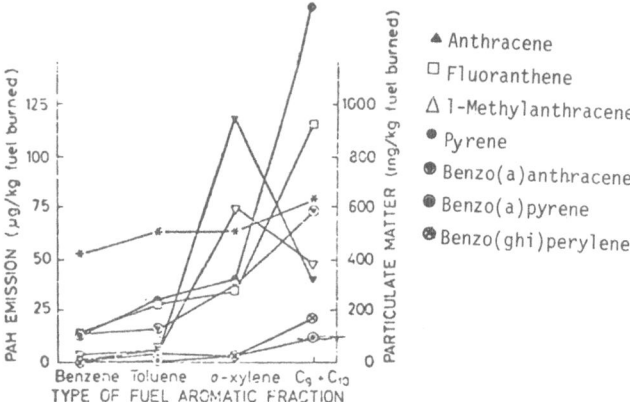

Figure 2. Effect on PAH emission (filter samples) of the type of
          aromatic hydrocarbon fraction at the constant aromatic
          content of 40 vol %. Fuels used were test fuels A3, A4,
          A5 and A6.

(Pedersen et al.)   (38)

better precursors of BaP than benzene itself. From the same figure
2 it is also evident that the remaining six PAHs, other than BaP,
increased in even greater amount.
      The described varying effect of the several classes of aromatics

may have important economic and technical consequences. In fact,
the aromatic fraction (produced by catalytic reforming) generally
contains a broad spectrum of aromatic compounds, ranging from C_6
to C_{10} or C_{10+}, although the heaviest compounds (C_9-C_{10+}) are present in a low percentage. This is demostrated by the composition
of the aromatic fraction of a fuel having a distillation curve
characteristic of a commercial fuel and whose aromatic composition
is a typical blend of catalytic reformate (Table 1).

DISTILLATION CURVE		COMPOSITION OF AROMATIC FRACTION % VOL		AROMATIC CARBON ATOMS No.
ASTM D 36				
IBP°C	39	Benzene	3.04	C_6
(% vol. evap.)		Toluene	12.46	C_7
5	53	Ethyl-Benzene	2.59	C_8
10	58	P-Xylene	3.76	C_8
20	67	M-Xylene	6.66	C_8
30	74	O-Xylene	3.57	C_8
40	87	Iso-Propyl-Benzene	0.12	C_9
50	101	N-Propyl-Benzene	0.65	C_9
60	118	1:3 + 1:4 Ethylbenzene	4.09	C_9
70	133	1:3:5 Trimethyl-Benzene	1.53	C_9
80	144	1:2 Methyl-Ethyl-Benzene	0.84	C_9
90	158	1:2:4 Trimethyl-Benzene	4.44	C_9
95	169	1:2:3 Trimethyl-Benzene	1.47	C_9
FBP°C	188	Aromatics C_{10+}	2.47	C_{10}
(% vol.)				
Recovery	98	Aromatic Fraction	48%	
Residue	1	Aliphatic Fraction	52%	
Loss	1			

Candeli et al. 1974 (8)
TABLE 1 - DISTILLATION CURVE AND COMPOSITION OF AROMATIC FRACTION
OF A TYPICAL COMMERCIAL FUEL

From Table 1 it is evident that 15.6% of the whole aromatic fraction
are C_9-C_{10+} aromatics. The results obtained by Candeli et al. (8),
using this typical catalytic refined commercial fuel, demonstrated
a significant increase in BaP emissions, both with respect to a
fuel containing only benzene (288%) and with respect to a fuel
containing ethyl-benzene and xylene (242%) as aromatic fraction.
These results suggest the hypothesis that increase in PAH emission
is dependent upon the complexity of the structure of aromatics
present in gasoline. To test this hypothesis high boiling aromatics,
namely the C_9-C_{10} (38), C_9-C_{10+} (9), C_{10}-C_{13}, and C_{10}-C_{14} (21) were
investigated. The results were compared with those obtained by
burning a base fuel which either did not contain any or contained
only a very low amount of high boiling aromatics. The fuels rich
in high boiling aromatics were prepared in three different ways:
    a) by use of a high boiling point fraction of catalytic

reformate,
    b) by adding to a base fuel different fractions of the bottom
of the catalytic reformate which was obtained by previous atmospheric distillation of the same catalytic reformate (Table 2), and

AROMATIC CARBON No	"A"	"B"
10	9.2	35
11	5.1	32
12	2.5	19
13	1.5	4.7
14	0.8	0.8
15	0.4	0.4
16	0.2	0.2

"A"- 20 g/gallon of boiling fractions (195°C-229°C + 229°C-305°C)
    of vacuum distillate of the bottom of the catalytic reformate.
"B"- 20% volume of heavy summer catalytic naphta.

Gross, 1972 (20)

TABLE 2 - GRAMS OF AROMATICS OF INDICATED CARBON NUMBER PER
         GALLON OF BLEND OBTAINED.

    c) by adding to a base fuel heavy summer catalytic naphta
(Table 2).
    The results obtained by use of different heavy aromatics
added to the "base fuel" are summarized in Table 3.

"BASE FUEL" (references)	% INCREASE OF BaP WITH RESPECT TO "BASE FUELS"			
	Kind of high boiling aromatics added			
	C_9-C_{10}	C_9-C_{10+}	$C_{10}-C_{13}$	$C_{10}-C_{14}$
A (Gross) (22)			50°	53°
B (Candeli et al.) (9)		228		
C (Pedersen et al.) (38)	250			

A-Unleaded fuel containing 47.2% of aromatics with low amount (?)
   of the C_{10+} fraction with BaP = 14 ug/l
B-Unleaded fuel containing 48 % of aromatics with low amount (2.47%)
   of the C_{10+} fraction with BaP = 270 ug/l
C-Leaded fuel with 40% of aromatic mixture $C_6:C_7:C_8:C_9+C_{10}=8:30:$
   40:20 BaP = 120 ug/l
°Tests carried out with stabilized deposits

TABLE 3 - THE EFFECT OF HIGH BOILING AROMATICS ADDITION ON BaP
          EMISSION - PERCENT INCREASE OF BaP WITH RESPECT TO
          "BASE FUELS"

The gasoline rich in heavy aromatics (C_9-C_{10} and C_9-C_{10+}) cause an appreciable increase in BaP emission with respect to the base fuel. This base fuel contains the same amount of aromatics but only a very low amount of C_9-C_{10+} aromatics. The data obtained by Gross (22) however, differ from those of Candeli et al. (9) and from those of Pedersen et al. (38). The higher values obtained by these two last groups of authors can be explained by the fact that in their experimental fuels there is a contemporary increase of C_9-C_{10+} hydrocarbons and of PAHs as can be seen in Figure 3 describing

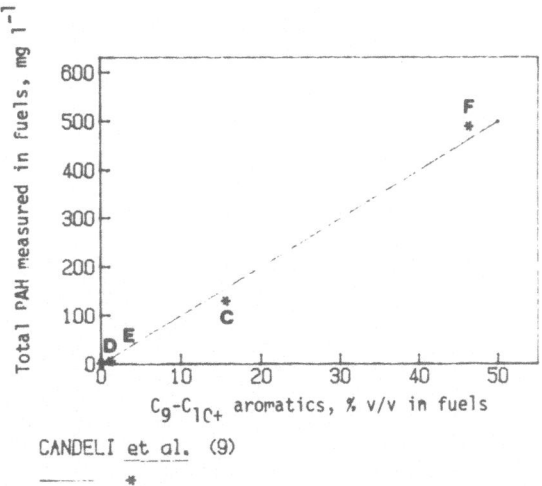

CANDELI et al. (9)

Fig.3 - Relationship between total PAH concentration ($mg\ l^{-1}$) and C_9-C_{10+} aromatic percentage in fuels D,E,C,F

the data of Candeli et al. (9). In Gross' experiments (22) this is avoided by adding to the "near zero PAH" base fuel, either the fractions with a boiling point range of 195-229°C plus 229-305°C obtained by vacuum distillation of catalytic reformate bottom or by adding heavy summer catalytic naphta (Table 3). However, as seen in Table 2, in this last case the amount of different C_{10}-C_{13} hydrocarbons is higher than in the first case. In pratice it is very difficult to separate the effect of PAH in gasoline on PAH emission, from that of the high boiling aromatics because, generally, the high boiling fractions of catalytic reformate contain both high boiling aromatic C_9-C_{10} hydrocarbons and PAHs (Fig. 3). In experimental fuels however, it was possible to see the effect of PAH fuel content on PAH emission by burning fuels containg large amounts of PAH. This was determined by adding, to reference "near zero" PAH fuels, the highest boiling fraction of the vacuum distillate of the bottom of catalytic reformate which contains only C_{16} and C_{16+} PAHs but not the other high boiling aromatics which are listed in Table 4 (20).

No OF CARBON ATOMS	PROBABLE COMPOUNDS	WEIGHT OF COMPOUNDS OF INDICATED CARBON NUMBER PRESENT IN DIFFERENT FRACTIONS					
		181°-183°C	183°-188	188°-195	195°-229	229°-305	>305°
6	Benzene	0.21	0.94	0.78	0.19	0.10	
7	Toluene	2.93	1.15	0.74	0.54	0.15	
8	Xylenes, Ethyl-Benzene	1.88	17.56	5.63	1.88	0.45	
9	Trimethyl Benzenes	42.58	78.99	88.49	66.97	3.66	
10	Naphtalene	51.99	1.33	4.37	27.21	21.53	0.17
11	C_4-Benzenes	0.32			2.88	32.16	
12	Acenaphtene	0.03			0.26	21.30	0.06
13							
14	Anthracene / Phenantrene					12.03	2.07
15						5.59	25.16
16	Pyrene					2.34	32.22
17						0.36	18.29
18	BaA; Crysene					0.03	8.77
19							5.91
20	BaP; BeP						2.76

Adapted from Gross, 1972 (20)

TABLE 4 – CHARACTERISTICS OF CATALYTIC REFORMATE FRACTIONS OBTAINED BY VACCUM DISTILLATION OF THE "CATALYTIC REFORMATE BOTTOM" OF PREVIOUS ATMOSPHERIC DISTILLATION OF THE SAME REFORMATE.

The results show a very large increase in PAH emission when the content of gasoline PAH increased (Table 5). The increase seems to be proportional to the amount of PAH present in the gasoline as reported by Gross (21). However on the influence of PAH in gasoline on PAH emission a complete agreement with the data here referred has not been reached (34, 38).

TEST FUELS	UNLEADED-FIELD PREMIUM GASOLINE BaP = 7 ug/l [°1]	REFERENCE FUELS LEADED GASOLINE BaP = 120 ug/l [°2]	REFERENCE FUELS LEADED GASOLINE BaP = 0.0 ug/l [°3]	REFERENCES
Unleaded field premium gasoline BaP = 320 ug/l [°1]	177▲			Gross (1973) (21)
Unleaded field premium gasoline BaP = 2150 ug/l [°1*]	285▲			Gross (1973) (21)
Leaded gasoline BaP = 2800 ug/l [°2]		35C●, 520●		Pedersen (1980) (38)
Leaded gasoline BaP = 960 ug/l [°3]			148□	Stichting Concawe (1974) (42)

[°1] Aromatic content 37.4%   [°1] Aromatic content 40.3%
[°1*] Aromatic content 37.4%   [°1*] Aromatic content 32.4%
[°2] Aromatic content 40.0%   [°2] Aromatic content 40.0%
[°3] Aromatic content 43.0%   [°3] Aromatic content 44.0%

▲ Fresh deposits are present
● Stabilized deposits are present
□ Calculated from data obtained in engines without thermal or catalytic reactor.

TABLE 5 - PERCENT INCREASE IN BaP EMISSION USING "RICH PAH" FUELS WITH RESPECT TO "NEAR ZERO PAH" FUELS (REFERENCE FUELS)

An unanswered question is whether PAHs emitted are those which "survive" the combustion process or are produced by a rearrangment of PAH present in gasoline. Begeman's data (2) indicate that 0.1 to 0.2%, and those of Handa et al. (26) that 1.0 to 2.0% of fuel BaP survives the combustion process and is recovered in the exhaust. However Pedersen's et al. (38) recent data demonstrated that when 2,800 ug/l of BaP only are added to a fuel already containing 120 ug/l of the same compound the BaP emission increased in two separate tests by 350 and 520% respectively, while other PAH increases were much lower. See Figure 5 in Pedersen et al. (38).

In Tables 3 and 5 the presence of soot deposits in the engine combustion chamber was considered because this seems to be an omportant factor in PAH emission, especially if deposits are formed by combustion of PAH rich fuels. Griffing et al. (18) observed that the use of Indolene, a fuel whith high BaP content, caused a very large amount of BaP emission. This occurred, not only in the test in which Indolene was used, but also in successive tests carried out with fuel of a low PAH content. This "carry over" effect has been carefully investigated by Gross (21) who reached the conclusion that when a test is carried out with PAH rich fuels there is first a storage of PAH in the engine deposits and, in successive tests, a re-emission of the same PAH at a progressively lower rate. After a number of tests, when deposits are stabilized. the PAH emissions remain constant at a lower level than that obtained in a test carried out immediately after the one in which PAH rich fuel was used. Therefore, according to Gross (22) "the critical factor was apparently the PAH in the deposits rather than in the test fuel being used".

PAH EMISSIONS IN RELATION TO THE AROMATIC FRACTION IN DIESEL FUELS

The influence of the total aromatic content of diesel fuels on PAH and particularly on BaP emissions has been studied by different investigators (27, 28, 13, 5, 17) using different diesel engines (heavy duty and light duty) and different operating conditions.In Table 6 we have summarized some of the data obtained using fuels with different percentage of aromatics, chosing among the numerous data those we thought more representative of the spectrum of fuels used by the various authors.

The information given in Table 6 was plotted both as aromatics, in the fuels tested by the quoted authors,versus BaP, and as aromatics versus cetane number. A regression line was calculated for each plot (see Figures 4 and 5). The data by Hare et al. (27), (the authors quoted in Figure 5) have not been referred to in Figure 4 as the BaP concentration in the emission was expressed in ug/Kg of fuel burnt, instead of ug/Km. As can be seen from these figures there is positive correlation between the percent of aromatics in the fuels and BaP emission (Fig. 4) and a negative correlation between the same percent of aromatics in the fuels and their cetane numbers (Figure 5).

FUEL CODES	% AROMATICS	CETANE NUMBER	BaP EMISSION ug/Km	REFERENCE
EM-242-F (premium)	12.4	53.0	1.8	
EM-240-F	13.0	47.4	2.0	Hare and Bainess
EM-239-F	21.6	48.7	2.3	1979 (28)
EM-239-F	29.8	48.6	1.9	b
EM-241-F (minimum quality)	34.6	41.8	7.0	
EM-401-F	2.7	56.0	0.4	
EM-400-F	10.5	49.6	0.5	Dietzman
EM-242-F	16.9	52.1	0.4	et al.1980
EM-329-F	21.3	42.0	0.8	(13) a
EM-241-F	35.8	42.0	1.1	
EM-395-F	5.8	62	0.47	
EM-438-F	6.8	64	0.21	Bykowski
EM-430-F	8.8	63	0.96	et al.1981
EM-463-F	30.8	60	0.43	(5) b
EM-434-F (+HAN)	31.5	44	0.37	
EM-460-F (+ individual aromatics)	32.1	46	0.73	
Reference	30.5	47.8	3.10	Gabele et
Oil Shale	34.3	57.0	11.8	al. 1982
Amoco	39.9	46.0	15.5	(17) b

a   Heavy duty vehicle
b   Light duty vehicle

TABLE 6 - EFFECT OF AROMATIC CONTENT IN DIESEL FUEL ON BaP EMISSION

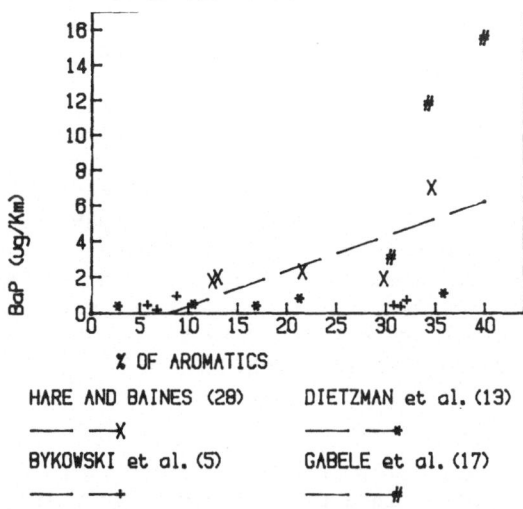

FIG. 4 EFFECT OF AROMATIC CONTENT IN DIESEL FUELS ON BaP EMISSIONS

and 0.69 and 0.15 for the Oldsmobile".

CONCLUSIONS

The data analyzed in this paper are concerned with fuel
characteristics, both of gasoline and diesel, as they relate
to PAH content of automotive exhaust. A major objective of the
analysis has been to find those data which, when applied, would
lead to a reduction in the PAH content in the exhaust and consequently in the general environment. For the sake of clarity the
collected information has been described separately for gasoline
and for diesel fuel.

Gasoline

1) The amount of aromatics present in gasoline has a definite
influence on PAH emission. This is so because as the aromatic
content in fuel is increased the PAH concentration in the exhaust
gases increases. Therefore in order to reduce the PAH emissions,
the aromatic content of the gasoline must be lowered. This
requirement, however, is in conflict with fuel quality needs,
since to maintain an antiknock capacity, fuel with a trend toward
a lowered lead content, has to have its aromatic content increased.
Consequently it is the other characteristics, namely quality and
not quantity, of aromatics which require control.
2) The kind of aromatics in gasoline has been found to have
an important influence on PAH emission. Therefore, keeping constant the quantity of aromatics, the aim of reducing PAH emission
can be achieved by reducing the level of high carbon atom number
aromatics (C_9-C_{10+}). Maintaining a desired concentration of this
class of compounds can be accomplished by preparing a blend containing, besides paraffinic hydrocarbons, benzene and its derivatives with low carbon atom numbers (C_7-C_8), such as toluene,
ethyl-benzene and xylene. This is theoretically possible, but
because of economic factors, may be difficult to realize in pratice.
3) PAH content of gasoline also has been claimed to have a
positive influence on PAH emission. However, since it has been
found that when a fraction contains a high concentration of C_9-
C_{10+} aromatics it also contains a large amount of PAH, reducing
the C_9-C_{10+} aromatic content reduces the PAH content as well.
4) Another aspect related to fuel composition is the
employment of catalytic converters to reduce automotive pollutant
emission which requires the use of lead free fuels. This has the
added advantage of reducing PAH emission.

Diesel Fuels

1) The amount of aromatics present in diesel fuel correlates
positively with PAH emission in a similar manner to that exhibited

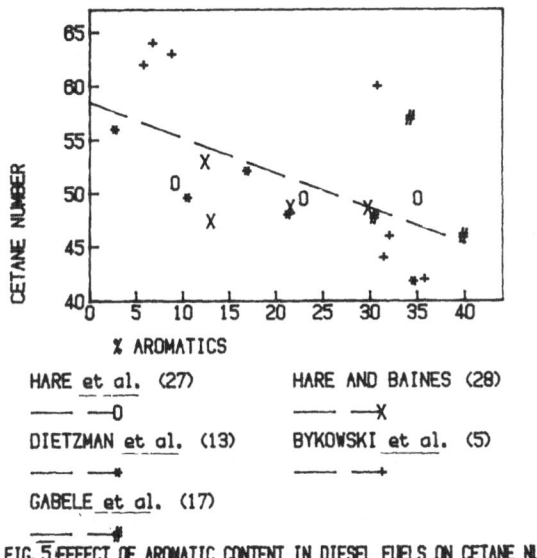

X AROMATICS

HARE et al. (27)   HARE AND BAINES (28)
——— —0   ——— —X
DIETZMAN et al. (13)   BYKOWSKI et al. (5)
——— —•   ——— —•

GABELE et al. (17)

——— —#

FIG. 5 EFFECT OF AROMATIC CONTENT IN DIESEL FUELS ON CETANE NUMBER

These results provide an important indication that if the aromatic content of the fuels is lowered the cetane number increases, while the BaP in the exhaust gases decreases.

Very few data have been found in the literature concerning the influence of the kind of aromatics in fuels on PAH emission. As far as we know only Bycowsky et al. (5) made an attempt of facing the problem of studying the influence of a mixture of individual pure mono-aromatics ranging from C_8 (ethyl-benzene) to C_{14} (octyl-benzene). This mixture was added to a high quality petroleum base fuel containing 5.8% of aromatics, to raise the aromatic content to about 30%. The second fuel was prepared by blending enough heavy aromatic naphta (HAN) into 25 liters of base fuel to raise the aromatic content to approximately 30%. The results obtained, shown in Table 6, reveal an opposite influence of these two kinds of aromatics on BaP emission. In the case of the fuel with the pure mono-aromatic mixture there was an increase of 55% of BaP emission with respect to the base fuel while when HAN was used to obtain nearly the same percentage of aromatics, the result was a decrease of 21.28% BaP in the emission.

The influence of PAH content in diesel fuel on PAH emission was studied by Gabele et al. (17). They studied four different fuels and measured the concentration of eleven PAH ranging from phenantrene to coronene while in the exhaust only pyrene and BaP (plus the direct mutagen: nitro-pyrene) were quantified. As the same authors say "linear regression analysis indicated that the exhaust SOF (soluble organic fraction) PAH level was not a strong function of the fuel PAH level: coefficients of determination for BaP and pyrene were respectively 0.11 and 0.36 for Volkswagen, by gasoline. In constrast to gasoline, in diesel fuel the quality (represented by the cetane number) is improved when the percentage of aromatics is lowered. The consequence of this is that lowering the aromatic content of the fuel results in a dual advantage: a) improving their quality and   b) lowering the PAH content of the emission.

2) In diesel fuels less attention, as compared to gasoline, has been paid to the kind of aromatics and to their PAH content, so that any precise conclusion about the influence of PAH emissions of these two fuel characteristics can not be drawn. On the other hand, in the case of diesel fuels, this problem is less important than for gasoline since in diesel fuels the whole aromatic fraction can be lowered significantly without worsening its quality. On the contrary, it is improved.

Finally, as a general conclusion we wish to point out that the research carried out in the last three decades to determine the quality and the quantity of carcinogenic substances, particularly PAHs present in exhaust gases, has produced significant results. This research has resulted in proposed pratical means of reducing the carcinogenic burden in exhaust gases and consequently in the general environment. This is important, even though the debate concerning the relationship between polluted atmosphere of cities and lung cancer incidence has not yet been resolved, because, as Hammond and Garfinkel (25) point out "if no efforts were made to control air pollution, then at some future date it might increase to a level such that it would result in a significant increase in the risk of lung cancer."

REFERENCES

1. Albert, R.E., and Burns, F.J. Carcinogenic atmospheric
pollutants and the nature of low-level risks: 1977, in Origins
of Human Cancer, Hiatt, H.H., Watson, J.D., and Winsten, J.A.
eds.,pp. 289-292, Cold Spring Harbor Laboratory.

2. Begeman, C.R., and Colucci, J.M. Benzo(a)pyrene in gasoline
partially persists in automobile exhaust: 1968, Science, 161,
p. 271.

3. Björseth, A., and Lunde, G. Long-range transport of polycyclic
aromatic hydrocarbons: 1979, Atmospheric Environment 13, pp.
45-53.

4. Blot, W.J., and Fraumeni, Jr. J.F. Geographic patterns of
lung cancer: industrial correlations: 1976, Amer. J. Epidem. 103,
pp. 539-550.

5. Bykowski,B.B., Hare, C.T., and Baines, T.M. Effects of a
narrow-cut no. 1 fuel, and variation in its properties, on light-
-duty diesel emissions:1981, Paper No 811193 presented at SAE
Fuels Lubricants Meeting, Tulsa, Oklahoma pp. 37-64.

6. Candeli, A. Ricerche sugli idrocarburi aromatici eterociclici
ad azione cancerigena nei gas di scarico degli autoveicoli: 1966,
Riv. ital. Igiene 26, pp. 217-230.

7. Candeli A., Mastrandrea, V. Fattore urbano e cancro polmonare:
1971, G. Ig. Med. prev. 12, pp. 89-120.

8. Candeli, A., Mastrandrea, V., Morozzi, G., and Toccaceli, S.
Carcinogenic air pollutants in the exhaust from a European car
operating on various fuels: 1974, Atmospheric Environment 8,
pp. 693-705.

9. Candeli, A. Morozzi, G., Paolacci, A., and Zoccolillo, L.
Analysis using thin layer and gas-liquid chromatography of
polycyclic aromatic hydrocarbons in the exhaust products from a
Euroepan car running on fuels containing a range of concentrations
of these hydrocarbons: 1975, Atmospheric Environment 9, pp.
843-849.

10. Chan, T.L., Lee, P.S., and Siak, J.S. Diesel-particulate
collection for biological testing. Comparison of electrostatic
precipitation and filtration: 1981, Environ. Sci. Technol. 15,
pp. 89-93.

11. Commins, B.T., and Hampton, L. Changing pattern in concentrations of polycyclic aromatic hydrocarbons in the air of central
London: 1976, Atmospheric Environment 10, pp. 561-562.

12. Cuddihy, R.G., Griffith, W.C., Clark, C.R., and McClellan, R.O. Potential health and environmental effects of light duty diesel vehicles: 1981, Lovelace Biomedical and Environmental Research Institute, Inhalation toxicology Research Institute Report LMF 89, pp. 1-107.

13. Dietzmann, H.E., Parness, M.A., and Bradow, R.L. Emissions from trucks by chassis version of 1983 transient procedure: 1980, Paper No 801371 presented at SAE Fuels & Lubricants, Baltimore, Maryland, pp. 1-16.

14. Doll.,R., and Peto, R. Mortality in relation to smoking: 20 years' observations on male British doctors: 1976, Brit. Med. J. 2, pp. 1525-1536.

15. Doll., R., and Peto, R. The causes of cancer; quantitative estimates of avoidable risks of cancer in the United States today: 1981, J. Nat. Cancer Inst. 66, pp. 1193-1309.

16. Fraumeni, Jr. J.F. Epidemiological approaches to cancer etiology: 1982 in Ann. Rev. Publ. Health 3, pp. 85-100 Annual Reviews inc., Palo Alto, California USA.

17. Gabele. P.A., Zweidinger, R.,and Black, F. Passenger car exhuast emission patterns; petroleum and oil shale derived diesel fuels: 1982,Environmental Protection Agency, Research Triangle Park, N.C. 27711 pp. 1-37.

18. Griffing, M.E. Maler, A.R., Borland, J.E.,and Decker, R.R. Applying a new method for measuring benzo(a)pyrene in vehicle exhaust to the study of fuel factors: 1971, presented at the Symposium on Current Approaches to Automative Emission Control, American Chemical Society, Los Angeles, California. Ethyl Corporation Research Laboratories, Detroit, Michigan, pp. 1-13.

19. Gross, G.P. The effect of fuel and vehicle variables on polynuclear aromatic hydrocarbon and phenol emissions: 1972 Paper No 720210, presented at SAE Automative Engineering Congress Detroit, Michigan, pp. 1-20.

20. Gross, G.P. Gasoline composition and vehicle exhaust gas polynuclear aromatic content: 1972, Third annual report CRC-APRAC Project No CAPE-6-68 Esso Research and Engineering Company Products Research Division, Linden, New Jersey, pp. 1-106.

21. Gross, G.P. Gasoline composition and vehicle exhaust gas polynuclear aromatic content: 1973, Fourth annual final report CRC-APRAC Project No CAPE-6-68. Esso Research and Engineering Company Products Research Division, Linden, New Jersey, pp. 1-135.

22. Gross, G.P. Automotive emissions of polynuclear aromatic hydrocarbons: 1974, Paper No 740564, presented at SAE National Combined Farm, Construction & Industrial Machinery and Fuels and Lubricants Meeting Milwaukee, Wis., pp. 1-22.

23. Hammond, E.C., Garfinkel, L., Seidman, H., and Lew, E.A. Some recent findings concerning cigarette smoking: 1977, in Origins of Human Cancer pp. 101-112, Cold Spring Harbor Laboratory.

24. Hammond, E.C., and Seidman, H. Smoking and cancer in the United States: 1980 Prev. Med. 9, pp. 169-173.

25. Hammond, E.C., and Garfinkel L. General air pollution and cancer in the United States: 1980, Prev. Med. 9, pp. 206-211.

26. Handa, T., Yamamura, T., Kato, Y., Saito, S., and Ishii, T. Determination of predominant factors of polynuclear aromatic hydrocarbon emissions from gasoline engine vehicles in ordinary city service: 1979, Taiki Osen Gakkaishi 14 pp. 464-473.

27. Hare, C.T., Springer, K.J., and Bradow, R.L. Fuel and additive effects on diesel particulate-development and demonstration of methodology: 1976, Paper No 760130 presented at SAE Automative Engineering Congress and Exposition, Detroit, Michigan, pp. 527-555.

28. Hare, C.T., and Baines, T.M. Characterization of particulate and gaseous emissions from two diesel automobiles as functions of fuel and driving cycle: 1979, Paper No 790424 presented at SAE Congress and Exposition Cobo Hall, Detroit, Michigan, pp. 1-42.

29. Higginson, J. and Jensen, O.M. Epidemiological review of lung cancer in man: 1977, in Davis, W., Mohr, U., Schmaell, D., and Tomatis, L. eds. Air Pollution and Cancer in Man. IARC, Lyon France, pp. 167-189.

30. Hilfrich, J., Bresch H., Misfeld, J., and Mohr, U. Investigation on the carcinogenic burden by air pollution in man. V. Tumours of the respiratory tract in Syrian Golden Hamsters after intratracheal instillation of benzo(a)pyrene: 1973, Zbl. Bakt. Hyg., I. Abt. Orig. 158, pp. 59-61.

31. Hoffmann, D., Theisz, E., and Wynder, E.L. Studies an the carcinogenicity of gasoline exhaust: 1965, J. Air Pollut. Control. Assoc. 15, pp. 162-165.

32. Kaplan, H.L., Springer, K.J., MacKenzie, W.F., Schreck, R.M., and Vostal, J.J. A subchronic study of the effects of exposure of three species of rodents to diesel exhaust: 1981, EPA Diesel

Emission Symposium, Raleigh, N.C., Oct. 5-7, pp. 1-20.

33. Lawther, P.J., Commins, B.T., and Waller R.E. A study of the concentrations of polycyclic aromatic hydrocarbons in gas works retort houses: 1965, British Journal of Industrial Medicine 22, pp. 13-20.

34. Meyer, J.P., and Grimmer, G. Einflüsse PAH-haltiger und PAH freier Kraftstoffe auf die Emission von polycyclischen aromatischen Kohlenwasserstoffen eines Fahrzeugs mit Ottomotor im Europa-Test; 1974 DGMK-Forschungsbericht 4547, II, Hamburg.

35. Misfeld, J., and Timm, J. Investigations on the carcinogenic burden by air pollution in man. I. Mathematical planning of experiments: 1973 Zbl. Bakt. Hyg., I Abt. Orig. 158, pp. 4-21.

36. Møller, A. Alfheim, I., Larssen, S., and Mikalsen, A. Mutagenicity of airborne particles in relation to traffic and air pollution parameters: 1982, Environ. Sci. Technol. 16, pp. 221-225.

37. Nasca, P.C., Burnett, S.W., Greenwald, P., Brennan, K., Wolfgang, P., and Carlton, K. Population density as an indicator of urban-rural differences in cancer incidence, upstate New York, 1968-1972: 1980, Amer. J. Epidem. 112, pp. 362-375.

38. Pedersen, P.S., Ingwersen, J., Nielsen, T., Larsen, E. and Fenger, J. Effects on pollution of a reduction or removal of lead addition to engine fuel. 1978. Publication No. Miljø-Projekter 15, Copenhagen.

39. Robertson, L.S. Environmental correlates of intercity variation in age-adjusted cancer mortality rates: 1980, Envir. Health Perspectives 36, pp. 197-203.

40. Saracci, R. Epidemiology of lung cancer in Italy: 1977, IARC Scient. Publ. No. 16, pp. 205-215.

41. Saxen, E., Teppo, L., and Hakulinen, T. Epidemiology of lung cancer in Scandinavia: 1977, IARC Scient. Publ. No. 16, pp.217-228.

42. Stichting Concawe. Effect of gasoline aromatic content on polynuclear aromatic exhaust emission: 1974, Report No. 6/74, September Stichting Concawe the Hague, pp. 1-63.

43. Tejada, S.B., Zweidinger, R.B., and Sigsby Jr., J.E. Analysis of nitroaromatics in diesel and gasoline car emission: 1982, Environmental Sciences Research Laboratory U.S. Environmental Protection Agency, Research Triangle Park N.C., pp. 1-28.

44. Wang, Y.Y., Rappaport, S.M., Sawyer, R.F., Talcott, R.E., and Wei, E.T. Direct-acting mutangens in automobile exhaust: 1978 Cancer Letters 5, pp. 39-47.

45. Wei, E.T., Wang, Y.Y., and Rappaport, S.M. Diesel emissions and the Ames test: a commentary; 1980, J. Air Pollut. Control Ass. 30, pp. 267-271.

46. Weiss, W. Lung cancer mortality and urban air pollution: 1978, Amer. J. Publ. Hlth. 68 pp. 773-775.

47. Wynder, E.L., and Hoffmann, D. A study of air pollution carcinogenesis; III. Carcinogenic activity of gasoline engine exhaust condensate: 1962, Cancer 15, pp. 103-108.

48. Zaghini, N., Mangolini, S., Arteconi, M., and Sezzi, F. Polynuclear aromatic hydrocarbons in vehicle exhaust gas: 1973, Paper No. 730836 presented at SAE, National Combined Farm, Construction & Industrial Machinery and Fuels and Lubricants Meetings, Milwaukee, Wis., pp. 1-10 .

THE AROMATIC SEXTET

Eric CLAR

University of Glasgow, Scotland.

ABSTRACT

The use of different spectra (UV, visible and NMR spectra) is
demonstrated for the identification of polycyclic hydrocarbons.

Soon after Kekule published his formula of benzene it became
apparent that there should be 2 isomers (Fig. 1). This could
not be found and Kekulé assumed that the three double bonds change
their place so fast that an isolation of these isomers would
be impossible.

Fig.1.   Fig.2.

Later a circle was used to symbolize aromaticity without a
clear definition (Fig.2). Pauling assumed that the isomers did
indeed not exist, but that the real state of benzene was the
result of "resonanze" between the two isomers.

E. Hückel used a mathematical method which gave molecular orbi-
tals which had no double bonds at all. He subdivided the
6 Π electrons in groups of 4 Π electrons and 2 Π electrons.
Both methods could be expressed by a circle (Fig.3)

49

D. Rondia et al. (eds.), Mobile Source Emissions Including Polycyclic Organic Species, 49–58.
© *1983 by D. Reidel Publishing Company.*

Fig.3.

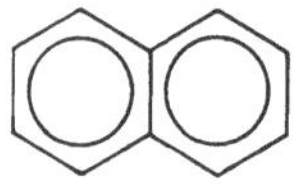

The difficulties of this method became obvious when more rings
were fused together. Thus in naphthalene 10 carbons have to share
8 π electrons only. In this way two formulations of naphthalene
are possible (Fig.4).

Fig.4

Before going into details of this method, one has to keep in
mind, that the huge success of Kekulé is based on his strict
adherence to the principle of the 4 valences of the C atom
(π electrons). Applying this principle to naphthalene in Fig. 3
is impossible and illogical because in this formula 10 C atoms
have only 8 π electrons and not 10 as required in Fig. 4. However,
the Fig. 5 is correct because the sextet circle can migrate from
one ring to the others as indicated by the arrows.

Fig.5.

This can be tested experimentally. Only 2 π electrons have to
migrate between the rings, in order to restore logic and bring
the theory in harmony with the experiments as shown below.

Using the double bond formula for anthracene, the situation
becomes even more unreal (Fig.6). Ring 1 should then be different
from ring 2 which is not the case.

Fig.6.

The sextet formula for tetracene, for pentacene and for hexacene
are respectively depicted in Fig.7; these structures correspond
to the properties of the hydrocarbons. Anthracene is colourless
but rather reactive and easily adds maleic anhydride; tetracene
is orange yellow and more reactive; pentacene is violet and so
reactive that it must be prepared under CO_2.

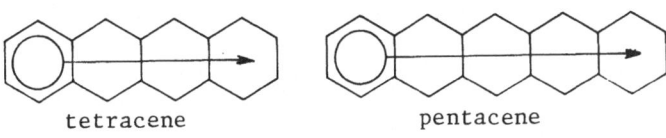

tetracene   pentacene

Fig.7.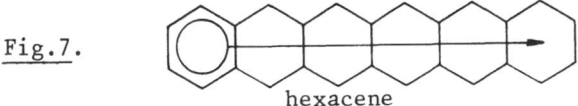

hexacene

These reactivities are considerably reduced if angular rings are
fused (Fig.8). These bring a second sextet. The following series
show that the reactivity is strongly reduced with the second
sextet. It becomes clear that the stability of benzene is best
symbolized by a circle.

Benzanthracene is still colourless and less reactive than
anthracene. Benzotetracene is pale yellow and benzopentacene is
violet red. These hydrocarbons are all less reactive than the
parent hydrocarbons. The benzenoidity must be inherent in the
circle. In the angle where the two arrows meet, a new sextet,
an induced one, is formed by the contribution of 2 Π electrons
each from the neighbouring rings and the inherent double bond of
the central ring.

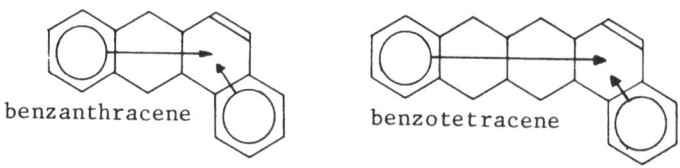

benzanthracene   benzotetracene

Fig.8

benzopentacene

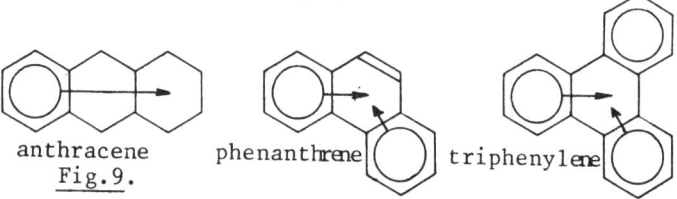

These changes become still more pronounced if a third ring is
fused. Anthracene is still colourless, but adds maleic anhydride.
The isomeric phenanthrene does not but can be oxidized only with
chromic acid. The third ring in triphenylene makes it the least
reactive hydrocarbon of the isomers.

This shows that hydrocarbons consisting of aromatic sextets only,
must always be the most stable of all isomers. The following
series demontrate this (Fig.9).

anthracene   phenanthrene   triphenylene
   Fig.9.

There are more properties typical of hydrocarbons of this kind.
It was found that a benzene ring fused to the fixed double bond
in phenanthrene has very little influence on the spectrum of
triphenylene.

A similar observation can be made in going from anthracene to
benzanthracene. A double bond is fixed at the terminal ring
(marked). Further rings fused to the position have a small
influence. In the tetracene series a similar change takes place.
This is even more pronounced in going from pentacene to benzopentacene and dibenzopentacene. This is shown by the three series
below : the number of aromatic sextets determines the different

properties in isomeric hydrocarbons and in particular their
stability (Fig. 10).

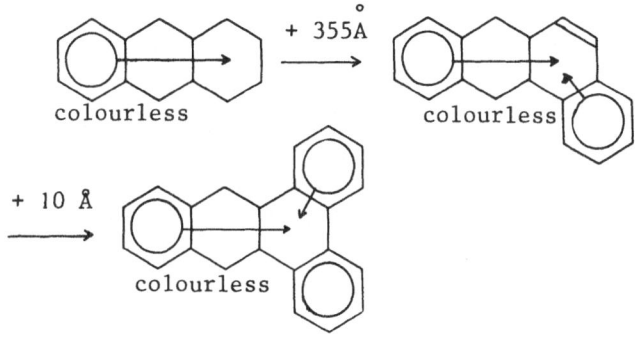

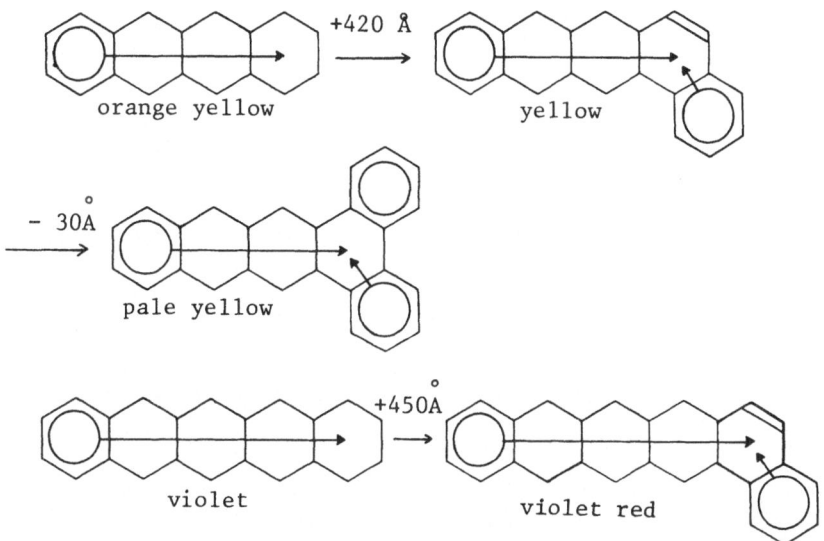

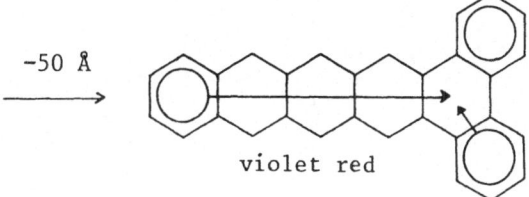

Fig. 10.

The number of sextets is most important as the following group
shows (Fig. 11) :

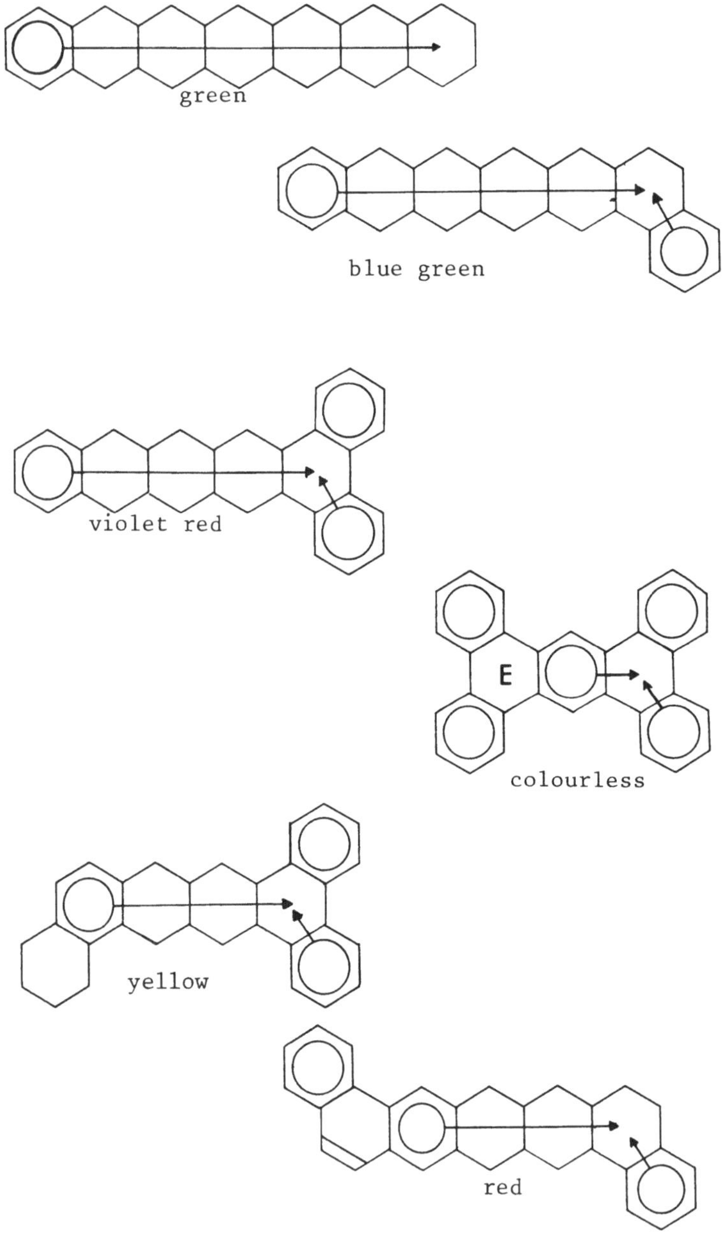

green

blue green

violet red

colourless

yellow

red

Fig. 11.

The number of sextets determines the colour and the reactivity, both going parallel. These effects are much bigger than
between straight and branched polyenes.

Fig. 12.

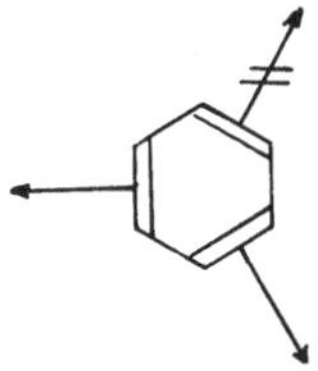

The sextet enforces the fixation of a double bond which would
otherwise not be possible. The annellation of two rings in the
direction shown by the arrows (Fig.12) has almost the same effect
as indicated by ⟶ . However, annellation to the fixed double
bond ⫫⟶ has very little influence on the intensive bands. One
could say the fixed double bond cuts off the aromatic conjugation.
Hydrocarbons are formed which are called "Starphenes".

The strong asymmetry is shown by the asymmetric annellation as
shown below :beginning with naphthalene a shift towards the red
of + 650 Å is recorded by fusion of a diphenyl system. The
second annellation brings + 395 Å only. An empty ring is formed
which has no sextet and is marked "E" (Fig. 13)

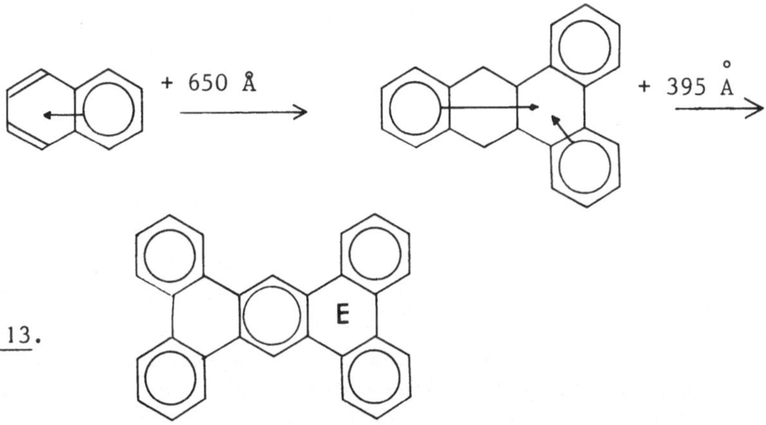

+ 650 Å

+ 395 Å

Fig. 13.

In the anthracene series a similar asymmetry results (Fig.14)

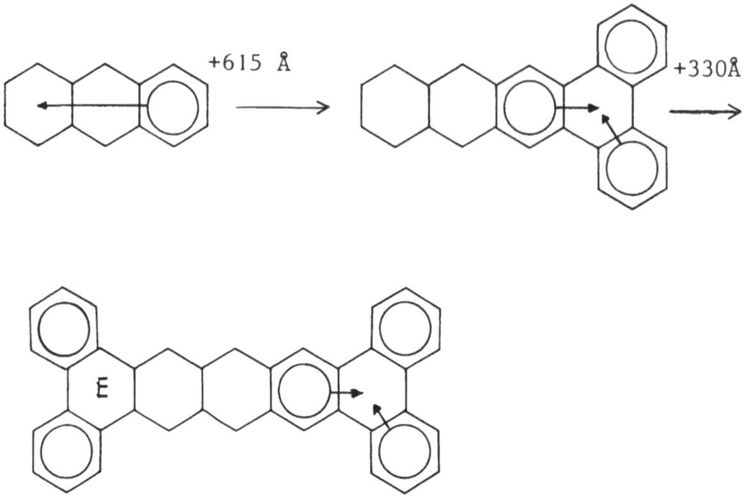

+615 Å   +330Å

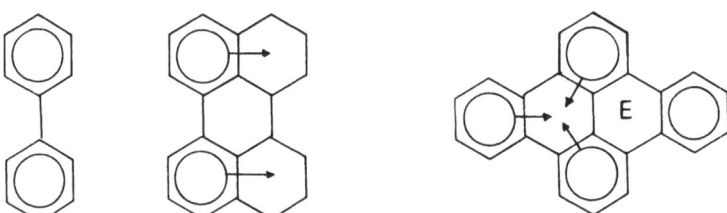

<u>Fig.14.</u>

The strong asymmetry can be only the result of the aromatic
sextet.

The sextet rule is also valid in condensed hydrocarbons like
diphenyl, perylene, pyrene and their benzologues (Fig.15).

<u>Fig.15.</u>

The asymmetry appears clearly in pyrene, benzopyrene and diben-
zopyrene. Moreover the annellation shows a particular interesting asymmetry in the pyrene series above. Only where the
arrows are, is an aromatic conjugation possible. This proves that
only 2 Π electrons can migrate, the other ones must stay in the
ring; an empty ring must be formed, marked "E" which causes no
annellation of a diphenyl system.

This is shown by the fact that only the first annellation,
forming the triphenylene system, produces a strong shift in the
fluorescence spectrum which is not observable after the second
annellation of a diphenyl system.

If there are any double bonds in aromatics, then the ones which
are most related to aliphatics are the 9.10 double bonds in
phenanthrene. The lowest degree in reactivity is reached in
tetrabenzonaphthalene. This is the only benzologue which is
reactive in the centre as indicated by the arrows. Chromic acid
gives the cyclic ketone and lithium a double addition.

The subdivision of the aromatic sextet.

The NMR method is particularly suitable to see whether double
bonds are in an aromatic sextet. The NMR spectrum of 1-methyl-
naphtalene is rather complicated as a result of multiple coupling.
This can be simplified by the introduction of bromine in position
4. The CH_3 signal is thus transformed into a doublet with a
separation of 0.05 Hz. If one sextet is in the first ring, then
it appears that there is only one double bond adjacent to the
CH_3. The CH_3 group separation in the central ring of phenanthrene
has about the same value 1.0 Hz. It appears that in an aromatic
sextet, there is only one double bond which comes very near to
an aliphatic double bond. This is confirmed in many other
examples. E. Hückel's rule is brought in harmony with the sextet,
if the 6 Π electrons are subdivided into 2 Π electrons which can
migrate through a linear branch of cata-annellated rings, whilst
the other 4 Π electrons cannot leave the ring with the aromatic
sextet (Fig.16)

Fig.16.

Toluene has a very complicated CH_3 signal. In 4-bromotoluene,
this is transformed into a septet, the result of an overlapping
triplet of triplets. In 2.4.5-tribromotoluene there is only a
doublet left. This can be simplified by decoupling at the 3-
position. This clears the spectrum into a doublet with the
separation of 0.95 Hz (Fig.17).

Since in 9-methylphenanthrene the CH_3 separation is 1.0 Hz, which
is the maximum in aromatics, there can be only one double bond,
2 free Π electrons mobile within the sextet and 2 Π electrons
capable ot migrating from one ring to the other.

There are many more examples of this kind. They all exclude the
existence of a second double bond in a sextet.

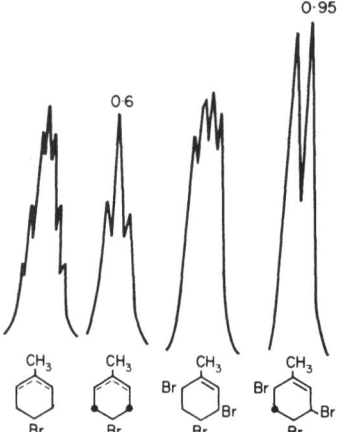

Fig. 17.

The above strict symbolic prevents grave errors and follows
Kedulé's ideas. I hope it will help to solve the problems in
which this conference is devoted.

EFFECT OF LEAD ANTIKNOCK REGULATIONS ON GASOLINE AROMATICS AND AROMATIC EXHAUST EMISSIONS

C. A. Hall, V. S. Willoughby, and B. J. French

Ethyl Corporation, Ferndale, Michigan

ABSTRACT

When lead antiknocks are restricted, refiners generally in-crease the aromatic content of motor gasolines to recover the lost octane numbers. This paper includes curves developed to show the relationship between aromatic content and lead content for leaded regular fuels in the United States and leaded regular and premium fuels in Europe. These curves show that lowering the lead content from 0.4 to 0.15 g Pb/litre increases aromatic con-tent by about 20% in U.S. gasolines and over 30% in European gasolines. This paper also summarizes the scientific literature on the effect of increased gasoline aromatics on emissions of polynuclear aromatic hydrocarbons and other aromatic pollutants from vehicles. Lowering the lead content from 0.4 to 0.15 g Pb/litre in European gasolines would increase benzo(a)pyrene emissions by about 25%.

INTRODUCTION

When the lead content of motor gasolines is restricted, re-finers must turn to other means of octane improvement in order to maintain the octane quality of their gasolines. The most widely used approach is to increase the amount of high-octane aromatics in gasoline. This will increase the polynuclear aro-matics (PNA--sometimes referred to as polynuclear aromatic hydrocarbons or PAH) and the reactive smog-forming constituents in vehicle exhaust emissions.

D. Rondia et al. (eds.), Mobile Source Emissions Including Polycyclic Organic Species, 59–76.
© 1983 by D. Reidel Publishing Company.

AROMATICS--A REPLACEMENT FOR LEAD ANTIKNOCKS

Since individual refineries vary greatly in processing equipment, available feedstocks, and product mix for their markets, we can only generalize as to how they would replace the octane quality lost if the use of lead antiknocks was restricted. The refiner can construct additional processing equipment, operate existing process units at higher severity, divert low-octane materials from gasoline to other uses, divert high-octane petrochemical aromatics to gasoline, and/or use high-octane blending agents such as alcohols and ethers. Construction of new process capabilities requires considerable time, but could provide a long-term solution. However, U.S. refiners are reluctant to commit large capital investments for octane processing equipment that may not be needed beyond the early 1980s. since gasoline demand is generally decreasing.

Certainly, aromatics are a significant replacement for lead antiknocks. This will have an impact on the availability and cost of aromatics now going to the petrochemical industry. The amount of additional aromatics required will depend on the volume and octane levels of the various grades of gasoline, the use of high-octane blending stocks from alkylation and isomerization processing units, and the use of alcohols and ethers.

U.S. GASOLINES

In the U.S., motor gasoline demand has decreased from the peak of 7.6 million barrels per day in 1978 to 6.6 in 1981, a decrease of 13 percent. In 1981, 10% of this gasoline was 96-RON unleaded premium, 39% was 92-RON unleaded regular, 50% 93-RON leaded regular, and only 1% 97-RON leaded premium. The government lead phasedown regulation as of October 1, 1980 requires all large refiners to meet a pool lead limit of 0.5 gram per U.S. gallon (0.13 g/L). In contrast to European lead antiknock restrictions, which are based on maximum allowable lead content per gallon, the U.S. restrictions are based on the maximum allowable pool lead (grams per gallon) for all motor gasoline produced by each individual refinery. Small refiners are allowed a more relaxed phasedown schedule.

To evaluate the effect of lead antiknock restrictions, we have developed a general relationship of aromatic content to Research octane number for both unleaded gasolines and the clear base stocks of leaded regular gasolines, as shown in Figure 1. These curves are the same as in our earlier paper (1). The lower unleaded-fuel curve was developed from data on commercial gasolines. Although the relationship is expressed by a line, in actuality it is a rather broad band. There are, of

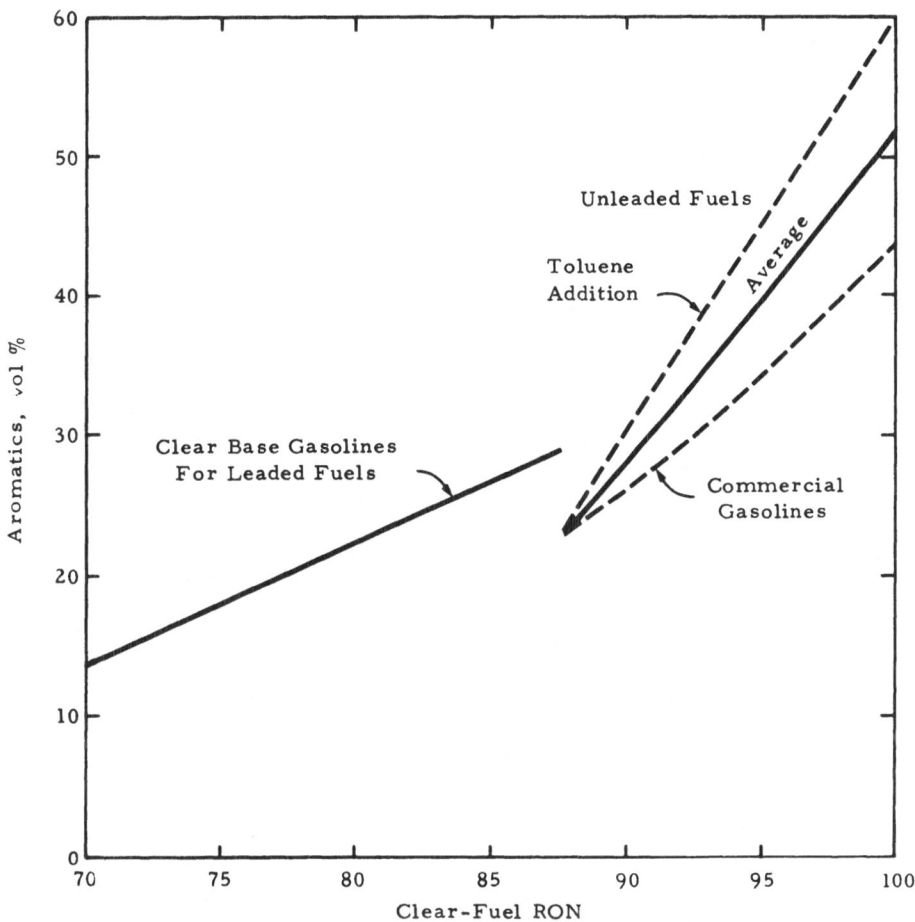

Figure 1. Aromatic-Octane Relationship for
U. S. Leaded and Unleaded Fuels

course, a wide variety of commercial unleaded gasolines, varying
considerably in the proportions of blending stocks (isomerate,
alkylate, catalytically cracked, and low-severity and high-
severity reformates) used to obtain the desired octane quality.
Thus, this curve does not represent the direct effect of aromatics on octane number, but rather the average aromatic concentration being used, along with other refinery blending stocks,
to achieve the desired octane quality.

The upper unleaded-fuel curve shows the octane numbers obtained by adding various amounts of toluene to an 87.5-RON (estimated U.S. clear pool) base gasoline containing 23% aromatics.

This curve represents a situation where a refiner would use aromatics alone to achieve the desired increase in antiknock quality. The solid line is the average of the two curves and may be more realistic of future operations of the refining industry.

A similar relationship for the clear base of leaded regular-grade gasolines was developed from more than 200 commercial gasolines obtained from service stations. Knowing the lead content of each fuel and using our updated lead susceptibility charts (2), the clear RON can be estimated. Here again, the data result in a wide band. The best-fit line has been determined, and its equation is:

$$\% \text{ Aromatics} = -47.3 + 0.872 \text{ (Estimated RON Clear)}$$

The slope of this best-fit line for the clear base gasolines for the commercial leaded fuels is noticeably less than that for the commercial unleaded fuels. There are several reasons. To make clear fuels below 90 RON, the refiner generally has available several base stocks in addition to reformate that can provide octane numbers. These include alkylate, isomerate, and catalytically cracked stocks containing olefins. Furthermore, these stocks plus straightrun have excellent lead susceptibility compared to aromatics. Therefore, one would expect to see less aromatics in leaded fuels and the aromatic increment for each increase in RON would be less. For unleaded fuels above 90 RON, reformate, which consists mostly of aromatics, is the most generally used blending stock to provide the required octane numbers.

Some refiners operate BTX units, which extract benzene, toluene, and xylene from catalytic reformate. Most of these aromatics go to the petrochemical industry. In the future, we would expect operation of catalytic reformers and BTX units to be tailored to keep more toluene and xylene going into unleaded gasoline.

Effect on Gasoline Aromatic Content

The lead phasedown regulation and the increasing amount of unleaded gasoline have both caused substantial increases in the average aromatic content of the different grades of gasoline. We have calculated the effect of the lead regulation on additional aromatics required in 1981 93-RON leaded gasolines. If there were no lead regulation, we believe refiners would use an optimum level of 0.62 g/litre (2.35 g/gal.) in their leaded gasoline. With 51% of the gasoline leaded in 1981 and assuming all of it regular grade (leaded premium was only 1% of total), the overall pool lead content would be 0.32 g/litre (1.2 g/gal.). Using our lead susceptibility charts, we find that the clear base gasoline would be 80.3 RON. Using this clear RON value, we can determine the aromatic content of this fuel to be 22.7% from the leaded-fuel curve in Figure 1. We can obtain estimates for the pool lead standard in the same manner. This is summarized in Table 1 and shows that the lead regulation has resulted in an additional 5 percentage points of aromatics in leaded regular gasolines. Surveys of aromatic contents of marketed gasolines agree quite well with the calculated effect of the lead restrictions. The additional large amount of aromatics required in gasoline affects the petrochemical industry, which must compete with the refining industry for available aromatics.

TABLE 1

Effect of Pool Lead on Aromatic Content of
93-RON Leaded Regular for 1981

Overall Pool Lead, g/L	Leaded Gasoline Vol %	Pb, g/L	RON of Clear	Aromatics Percent	Aromatics Δ Percentage Points	Aromatics Δ Billion Gal./Yr.
0.32*	51	0.62	80.3	22.7	-	-
0.13†	51	0.25	86.0	27.7	5.0	2.57**

* Optimum lead usage with no lead regulation.
† Present U.S. standard for major refiners.
** 6,586 M B/D x 0.51 x 0.05 x 42 x 365 = 2.57 billion gallons per year.

To show the relationship between percent aromatics and lead content, we have calculated the curve in Figure 2 using the leaded regular-fuel curve in Figure 1 and a lead susceptibility chart.

Effect on Crude Oil Usage

In addition to the lead phasedown regulation resulting in gasolines with higher aromatic content, the restriction on lead usage also requires more crude oil to produce the same volume of gasoline. In 1983, when only 41 percent of U.S. gasoline will be leaded, we now estimate that an additional 77,000 barrels of crude oil per day will be required because of the phasedown regulation. This is a net energy loss and no additional products will be produced.

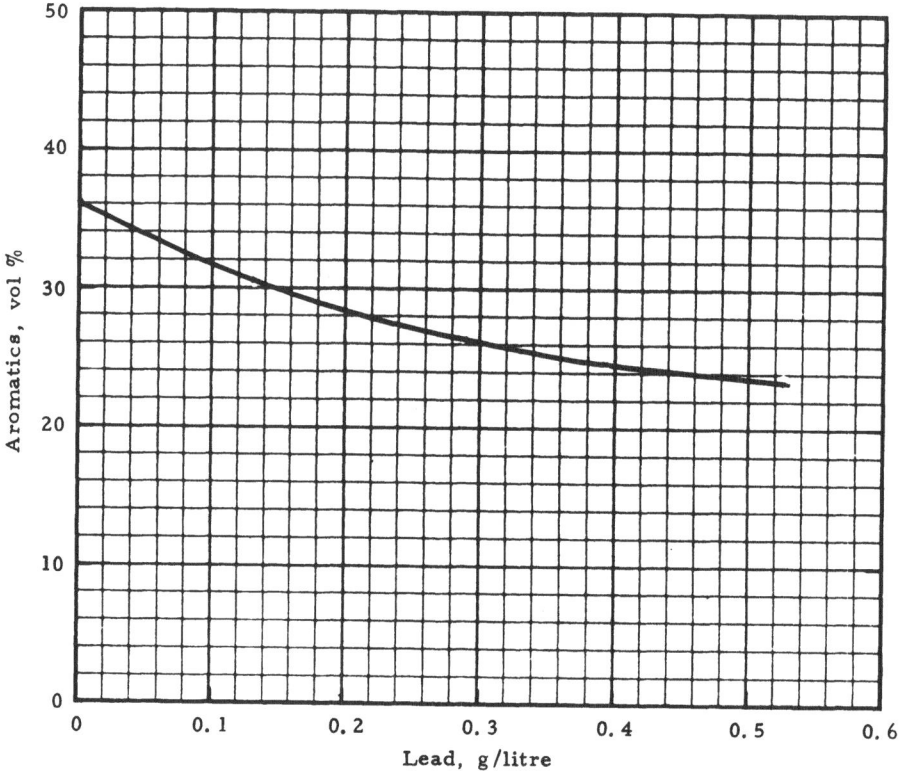

Figure 2. Aromatic Content vs. Lead Concentration
U.S. Regular Gasolines

EUROPEAN GASOLINES

   To determine the extent to which refiners in Europe have
replaced lead antiknocks with aromatic octanes, we analyzed the
aromatic content vs lead content for premium and regular-grade
gasolines. The data used were from surveys conducted by Ethyl
S.A. of service station samples collected in selected European
cities during three sampling periods from April 1979 to
September 1980.

   For the analysis, we used all of the premium-grade samples,
which had octane numbers ranging from 97.2 RON to 100.8 RON, with
most of the samples being in the range of 98 RON to 100 RON. For
the regular-grade samples, we restricted the analysis to those
samples in the range of 91.0 RON to 94.0 RON. As a result, the
study involved 93 premium gasolines and 54 regular gasolines.
Most of the gasolines containing not more than 0.15 g Pb/litre
were in West Germany.

The data for the 147 gasolines are plotted in Figure 3, along with the calculated best-fit lines for premium and regular gasolines. The equations for the two lines are:

Premium: % Aromatics = 43.30-37.32 (g Pb/litre)

Regular: % Aromatics = 26.72-26.45 (g Pb/litre)

The effect of reducing lead concentration from 0.4 g Pb/litre to 0.15 g Pb/litre is to increase aromatic concentration by about one-third, as shown in Table 2.

TABLE 2

Effect of Lead Restriction on Aromatic Content

	Aromatics, vol %		Δ Percentage Points	Percent Increase
	0.4 g Pb/Litre	0.15 g Pb/Litre		
Premium	28.4	37.7	9.3	32.7
Regular	16.1	22.7	6.6	41.0

This large effect of lead antiknock concentration on aromatic content of European gasolines is explained in a CONCAWE report (3), which stated:

"The average benzene content of European gasolines is higher than for USA gasolines (i.e. 3.5 vol. % vs 1.5 vol. %). In the USA high gasoline requirements on crude have resulted in the use of cracking processes on a much wider scale than in Europe. Consequently European gasolines have less cracked products and a higher proportion of benzene-containing reformates. Moreover the use of naphtha as feedstock for chemicals manufacture in Europe (rather than gas as in the USA) results in higher production and use of high benzene pyrolysis gasolines."

As previously noted, most of the gasolines containing 0.15 g Pb/litre or less were in West Germany. West German gasolines traditionally have been high in aromatics because of their extensive use of hydroskimming--catalytic reforming of the amount of the gasoline-range fraction that can be obtained by distilling a given amount of crude oil.

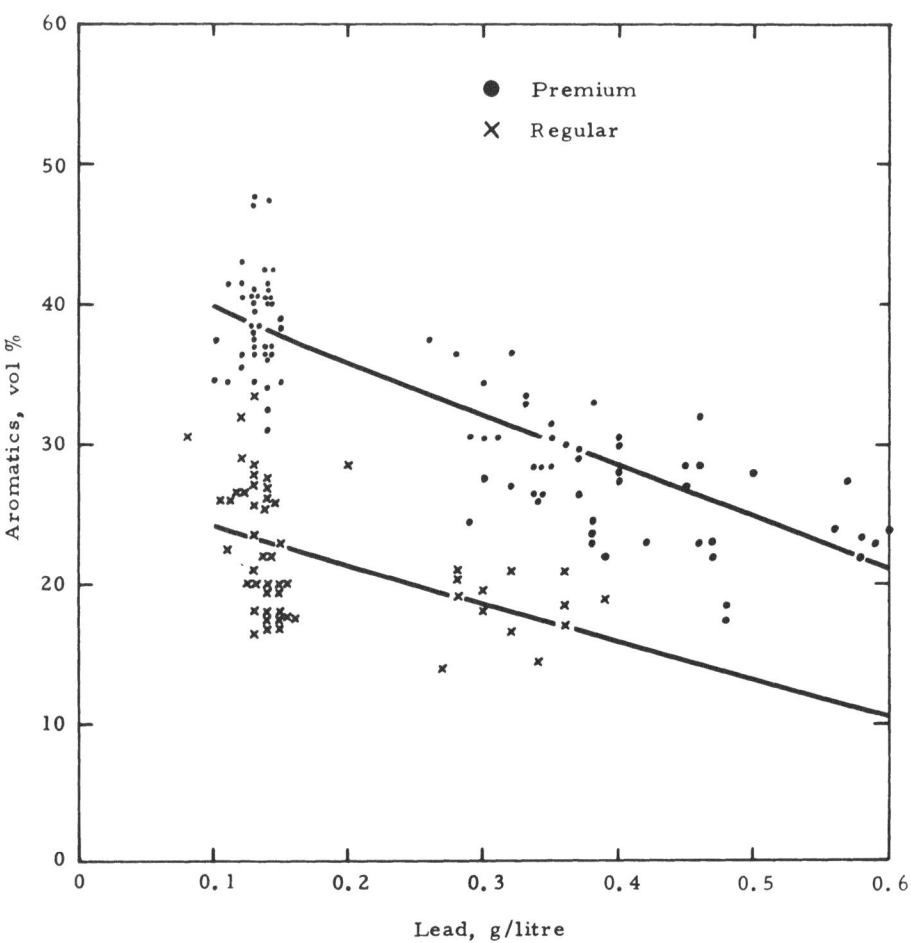

Figure 3. Aromatic Content vs. Lead Concentration
European Gasolines

1979-80

        To determine how much the West German gasolines affect the
analysis of the impact of lead restrictions on aromatic content,
the West German gasolines were deleted from the data for premium gasolines, and the slope and intercept were recalculated.
The results in Table 3 show that this deletion had relatively
little effect, with the remaining gasolines showing over a 25%
increase in aromatics as lead concentration is reduced from 0.4
g Pb/litre to 0.15 g Pb/litre.

TABLE 3

Effect of Lead Restriction on Aromatic Content

European Premium Gasolines Excluding West Germany

Aromatics, vol %		Δ Percentage Points	Percent Increase
0.4 g Pb/Litre	0.15 g Pb/Litre		
28.0	35.1	7.1	25.4

The impact of lead restrictions on the aromatic volumes required for gasolines can be illustrated by considering two countries--France and Italy. Table 4 shows the gasoline demand in these countries for 1980. If the average lead content of gasolines in these countries had been lowered from 0.4 g Pb/litre to 0.15 g Pb/litre and the aromatic content had increased by the values resulting from the best-fit lines in Figure 3, the results would have been added aromatic requirements of 1.59-million tonnes/year in France and 1.11-million tonnes/year in Italy.

TABLE 4

1980 Gasoline Demand

	Gasoline Demand, 10^6 tonnes/year	Increased Aromatics for 0.4 to 0.15 g Pb/L, 10^6 tonnes/year
France		
Normal (Regular)	3.1	0.21
Super (Premium)	14.8	1.38
Total	17.9	1.59
Italy		
Normal (Regular)	0.6	0.04
Super (Premium)	11.5	1.07
Total	12.1	1.11

The foregoing shows that (1) the aromatic content of European gasolines will increase significantly if lead antiknock content is lowered by government restrictions, and (2) the resulting increase in aromatic demand for gasolines would have a large impact on the availability of aromatics for petrochemicals.

PNA EMISSIONS FROM GASOLINE ENGINES

A considerable amount of research on this subject was conducted during the late 1960s and early 1970s in the U.S.,
Japan, and the German Federal Republic. Exhaust sampling techniques, critical to this investigation, were developed and engine operating variables were studied along with fuel composition factors. Throughout the 1970s, the DGMK and VDI in
Germany carried out rather comprehensive research in this area.
Their studies closely paralleled those done in the U.S. under the
auspices of CRC-APRAC. Japanese researchers have concentrated
on the effects of auto mileage and oil consumption on PNA emissions. Most of this work has been done by a group from ·the
Science University of Tokyo.

Most of the earlier work was carried out using benzo(a)-
pyrene (BaP) as the prime indicator of PNA emissions. More recently, analytical procedures have improved so that a variety
of PNA in the engine exhaust could be identified along with their
concentrations. Janssen, in his review of the DGMK work (4),
states that the early published work was done measuring only
BaP and that "data based solely upon one of a multitude of PAH
(PNA) are not necessarily a reliable base for conclusions of
general validity and applicability." The present ability to
determine 15 or more PNA using gas chromatography and mass
spectrometry is an improvement and helps to characterize specific
carcinogens. However, a close examination of the DGMK work indicates that, although a PNA profile of a test gives more complete information, BaP is a good indicator for the higher-
molecular-weight carcinogenic PNA. Previous literature may be
"limited," but BaP measurements alone will still indicate effects
of changing variables (e.g., fuel composition) on PNA emissions.

Effect of Air-Fuel Ratio

Of the various engine operating variables affecting PNA
emissions, changes in air-fuel ratio produce the largest effects.
Researchers have consistently found that the amount of PNA in
engine exhaust decreases with increasing air-fuel ratio (leaner
mixtures). A recent study confirms the decided effect of air-
fuel ratio. Pischinger et al.(5) extensively studied the effect
of air-fuel ratio on PNA emission and found that leaner mixtures
reduced PNA mass fraction in HC emissions, PNA emissions, and
mass fraction of the carcinogenic components of total PNA emissions. Since leaner mixtures supply excess oxygen, more complete combustion occurs. This results in much lower emissions
of both total hydrocarbons and PNA. Pischinger concluded that
there is "a severe reduction of the carcinogenic components with
engines which are operated only in the lean range."

In view of the large effect of changes in air-fuel ratio on PNA emissions, care must be taken in studying the effects of other variables, such as fuel composition, to make sure that variation in air-fuel ratio does not mask the prime variable under investigation.

Effect of Gasoline Aromatic Content

Studies in the early 1970's done in the U.S. by Begeman (6), Hoffman et al. (7), Griffing (8), and Gross (9) showed that increasing the aromatic content of gasoline increases PNA emissions as indicated by BaP. Table 5 summarizes the results of three of these studies.

TABLE 5

Effect of Increasing Gasoline Aromatic Content on BaP Emitted

Study	Range of Fuel Aromatic Content, vol %	Δ Percentage Points	Percent Increase in Fuel Aromatics	Increase in BaP Emitted*	
				%	Multiple
Hoffman(7)	15-44	29	193	(1) 152	2.5
				(2) 141	2.4
Griffing(8)	12-48†	36	300	120	2.2
Gross(9)	11-46	35	318	154	2.5

* Tests run in cars with no emission controls, except for Hoffman's 2nd series which were noncatalyst systems.
† Mole %.

TABLE 6

Effect of Increasing Gasoline Aromatic Content on BaP Emitted

Study	Range of Fuel Aromatic Content, vol %	Δ Percentage Points	Percent Increase in Fuel Aromatics	Increase in BaP Emitted	
				%	Multiple
Grimmer(10)	0-48	48	-	378	4.8
	26-42	16	62.0	111	2.1

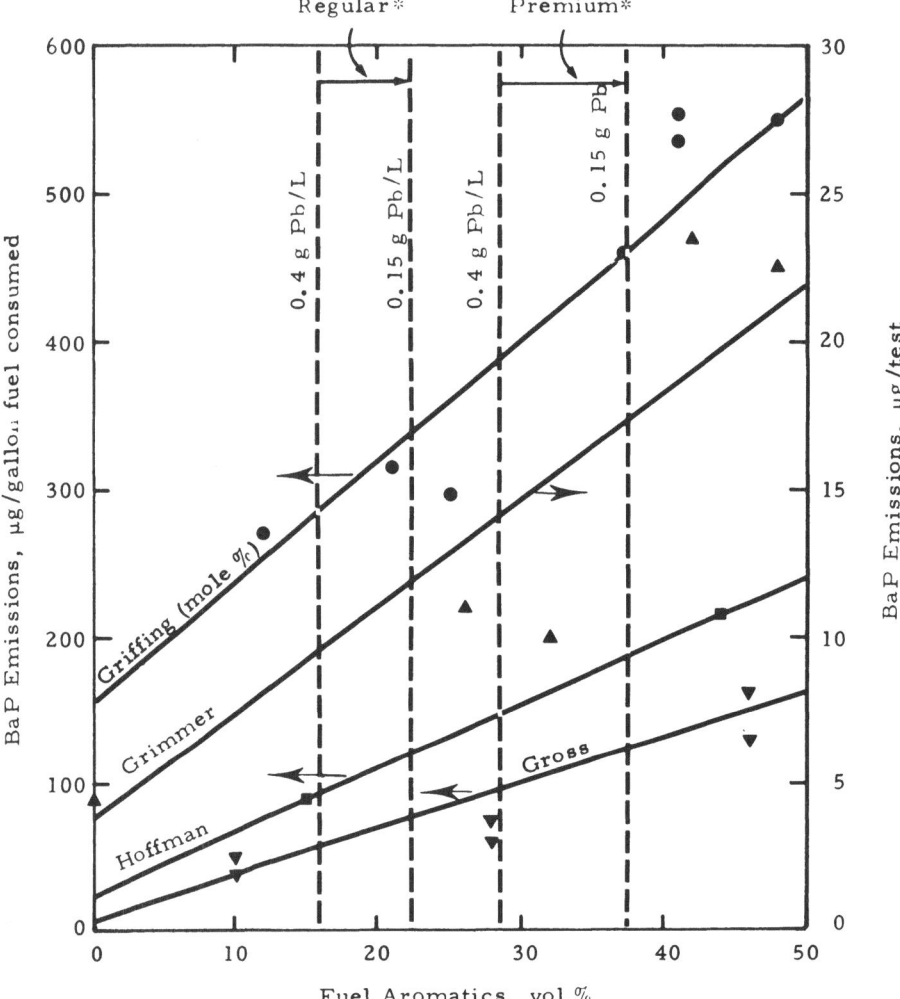

Figure 4. Relation of BaP Emissions to
Gasoline Aromatic Content

*Average increase in aromatic content due to restricting
lead from 0.4 to 0.15 g Pb/litre

As shown in Table 6, Grimmer (10) in West Germany found that
increasing aromatics from 0 to 48 vol % produced a 4.8-fold increase in BaP emissions. For intermediate fuel blends varying in
aromatic content from 26 to 42 vol %, a more likely range of commercial fuels, a 2.1-fold increase in BaP emissions occurred. A
similar finding occurred based on total PNA emissions.

The data on BaP emissions vs fuel aromatics obtained by Griffing, Grimmer, Hoffman, and Gross are plotted in Figure 4, and an approximate best-fit line is drawn for each set of data. The four vertical dotted lines represent the average aromatic content of the two grades of European fuels with 0.4 and 0.15 g Pb/litre. The BaP emission data at the intercepts of these dotted lines with the best-fit lines are presented in Table 7 for the four test programs. The average results indicate that restricting lead antiknocks in European gasolines from 0.4 to 0.15 g Pb/litre would result in an approximate 24% to 29% increase in BaP and other PNA emissions.

TABLE 7

Effect of Restricting Lead Antiknocks on BaP Emissions

	BaP Emissions, µg/gal. Fuel Consumed Premium Grade		Increase in BaP Emissions
Pb Content	0.4 g/L	0.15 g/L	
Aromatic Content	16%	22.5%	
Griffing(8)	285	340	19%
Grimmer(11)	9.3*	12*	29%
Hoffman(7)	92	120	30%
Gross(9)	56	77	38%
		Average	29%

	Regular Grade		
Pb Content	0.4 g/L	0.15 g/L	
Aromatic Content	28.5%	37.5%	
Griffing(8)	390	460	18%
Grimmer(11)	14.1*	17.4*	23%
Hoffman(7)	148	186	26%
Gross(9)	95	123	29%
		Average	24%

*µg BaP per test.

Effect of Gasoline PNA Content

Most gasolines contain small amounts of PNA. In general, the more high-boiling aromatics in the gasoline, the higher the PNA content. In the early 1970s, Begeman (6), Hoffman (7), Griffing (8), and Gross (9) all reported that high BaP emissions resulted from fuels containing high concentrations of BaP. In contrast, the German DGMK reports concluded that there was no effect of fuel PNA on PNA emissions.

The failure of the German studies to find a fuel PNA effect
is undoubtedly due to their testing sequence. Griffing was first
to find, later substantiated by Gross, that fuel PNA are stored
in the combustion chamber deposits with subsequent emission during severe operation or with a low-PNA fuel. The DGMK tests
were always run with a high-PNA fuel first followed by the low-
PNA fuel without "conditioning" (high-speed operation with the
low-PNA fuel) the deposits between the two tests.

OTHER EMISSIONS FROM AROMATICS IN GASOLINE

In addition to PNA, other deleterious compounds are produced when aromatics are burned in an engine. These include
phenols, which promote PNA carcinogenicity, and aldehydes, which
are eye irritants and participants in photochemical reactions
that lead to even more potent eye irritants. Also, aromatics
contribute to photochemical smog formation. Our review conducted several years ago disclosed the following information.

Phenols

Phenols are known promoting agents that strongly accelerate
tumor formation in laboratory animals, even for weak carcinogens.
Increasing the aromatic content of gasoline will proportionally
increase the phenolic content of the exhaust gas (12-14).

Aldehydes

Aldehydes, both aliphatic and aromatic, are powerful air
pollutants. They are present in exhaust gas and are also formed
from exhaust gas in smog-forming reactions in the air. They have
been reported to be eye irritants and irritants to mucous mem-
branes, plant toxicants, and participants in photochemical smog
reactions.

Aldehydes are present in significant quantities in exhaust
gas as it leaves the tailpipe. Increasing the aromatic content
of gasoline, while not affecting the total aldehyde content of
the exhaust gas, increases the aromatic aldehyde content proportionally. When increased aromatics are accompanied by lead
removal, the emission of aromatic aldehydes is further increased.
For example, in a 3-car test, changing from a U.S. leaded premium (30.8% aromatics) to an unleaded premium (46.6% aromatics)
increased the average emission of aromatic aldehydes by 123% (14).

General Motors has reported that benzaldehyde, the major
aromatic aldehyde in exhaust gas, is a necessary intermediate
in photochemical reactions leading to the formation of peroxybenzoyl nitrate (PBzN) (15). A major portion of eye irritation

resulting from Los Angeles smog is contributed by peroxybenzoyl nitrate (PBzN) (15), which is about 200 times as potent an eye irritant as formaldehyde.

Photochemical Smog

Reactivity is a measure of the smog-forming potential of hydrocarbons. Aromatics and olefins are the exhaust hydrocarbons that react most readily with nitric oxide (NO) in the atmosphere in the presence of sunlight to form nitrogen dioxide (NO_2), a key initiator in the formation of ozone, oxidants, and eye irritants.

Many investigations have shown a direct relationship between the aromatic content of gasoline and the photochemical reactivity of the exhaust. Therefore, any increase in aromatic content to compensate for the loss of octane quality by lead restrictions would increase the smog-forming potential of exhaust gas.

The most extensive and well-designed studies of these effects were made by the U.S. Bureau of Mines (16-19). They found, in actual smog-chamber measurements of the rate of photooxidation of NO to NO_2, that the exhaust gas from high-aromatic unleaded fuels is as much as 38% more photochemically reactive than the exhaust gas from conventional leaded fuels. The study involved prototype unleaded fuels with hydrocarbon composition modified to provide octane quality equivalent to that of leaded fuels.

Studies by General Motors Research (20) and Midwest Research Institute (21) have shown that lead compounds in car exhaust do not contribute to or enter into photochemical smog reactions.

REFERENCES

1. Hall, C. A., "Effect of Government Antiknock Regulations on Demand for Aromatics in U.S. Gasolines," presented at the 1979 Annual Fall Conference of the European Chemical Marketing Association, Munich, West Germany, October 16, 1979.

2. Unzelman, G. H. and Michalski, G. W., "Octane Improvement Economics - Antiknocks and Alternatives," NPRA Paper AM-79-46, NPRA Annual Meeting, San Antonio, Texas, March 25-27, 1979.

3. "Exposure to Atmospheric Benzene Vapor Associated with
    Motor Gasoline," Report No. 2/81, CONCAWE, Den Haag,
    February 1981.

4. Janssen, O., "Brief Review of Investigations on PAH in
    Vehicle Exhaust Round Robin Tests on Profile Analysis,
    Role of Fuels and Lubricants, Field Studies," VDI-Berichte,
    358, 69-79, 1980.

5. Pischinger, F. and Lepperhoff, G., "Influence of Air-Fuel
    Ratio on the PAH-Emission in the Case of Otto Carburetor
    Engines," VDI Reports, No. 358, 59-63, 1980.

6. Begeman, C. R. and Colucci, J. M., "Polynuclear Aromatic
    Hydrocarbon Emissions from Automotive Engines," presented
    to Society of Automotive Engineers, Detroit, Michigan,
    May 18-22, 1970.

7. Hoffman, C. S., Jacobs, E. S., Brandt, P. J., Patterson,
    G. H., and Willis, R. L., "Polynuclear Aromatic Hydrocarbon
    Emissions from Vehicles," presented at the Symposium on
    Current Approaches to Automotive Emission Control, American
    Chemical Society, Los Angeles, CA, March 31, 1971.

8. Griffing, M. E., Maler, A. R., Borland, J. E., and Decker,
    R. R., "Applying a New Method for Measuring Benzo(a)pyrene
    in Vehicle Exhaust to the Study of Fuel Factors," presented
    at the Symposium on Current Approaches to Automotive Emis-
    sion Control, American Chemical Society, Los Angeles, CA,
    March 31, 1971.

9. 2nd Annual Report on Gasoline Composition and Vehicle
    Exhaust Gas Polynuclear Aromatic Content," period ending
    April 15, 1971, CRC-APRAC Project No. CAPE 6-68, APCO/EPA
    Contract CPA-70-104.

10. Grimmer, G. and Voightsberger, P., "Influence of the Aro-
    matic Content of Fuel on the Emission of PAH from Passenger
    Car Otto Engines," Erdöl und Kohle - Erdgas - Petrochemie,
    33, 226, 1980.

11. Meyer, J. P. and Grimmer, G., "Optimization and Testing of a Collecting Method for Polycyclic Aromatic Hydrocarbons in Vehicle Exhaust Gas under Conditions of the European Driving Cycle," BMI-DGMK Community Project 4547, Part I, July 1, 1974.

12. Hoffmann, D. and Wynder, E. L., "Studies on Gasoline Engine Exhaust," J. Air Pollution Control Assoc., 13, 322, 1963.

13. Barber, E. D., Sawicki, E., and McPherson, S. P., "Separation and Identification of Phenols in Automobile Exhaust by Paper and Gas Liquid Chromatography," Analytical Chemistry, 36, 2442, 1964.

14. Hinkamp, J. B., Griffing, M. E., and Zutaut, D. W., "Aromatic Aldehydes and Phenols in the Exhaust from Leaded and Unleaded Fuels," Symposium on Current Approaches to Automotive Emission Control, American Chemical Society, Los Angeles, CA, March 31, 1971.

15. Agnew, W. G., "Future Emission Controlled Spark-Ignition Engines and Their Fuels," API Preprint No. 15-69, May 1969.

16. Dimitriades, B., Eccleston, B. H., and Hurn, R. W., "An Evaluation of the Fuel Factor Through Direct Measurements of Photochemical Reactivity of Emissions," presented at the 62nd Annual Meeting, Air Pollution Control Assoc., New York, June 22-26, 1969; J. Air Pollution Control Assoc., 20, 150, 1970.

17. Hurn, R. W., "Fuel: A Factor in Internal Combustion Engine Emissions," ASME Paper No. 69-WA/APC-8, ASME Winter Annual Meeting, Los Angeles, CA, November 16-20, 1969.

18. Eccleston, B. H. and Hurn, R. W., "Comparative Emissions from Some Leaded and Prototype Lead-Free Automobile Fuels," Bureau of Mines Report of Investigations-7390, May 1970.

19. Hurn, R. W., "Auto Exhaust Reactivity--Assessment and Trends," presented at 5th Technical Meeting, West Coast Section, Air Pollution Control Assoc., San Francisco, CA, Oct. 8-9, 1970.

20. Heuss, J. M., Nebel, G. J., and D'Alleva, B. A., "Effects of Gasoline Aromatic and Lead Content on Exhaust Hydrocarbon Reactivity," J. Environ. Sci. Tech., 8 (7), 641, 1974.

21. Morriss, F. V., Bolze, C., and Goodwin, J. T., Jr., "Smog Chamber Studies of Unleaded vs. Leaded Fuels," Ind. & Eng. Chem., 50 (4), 673, 1958.

THE TOXICOLOGY OF POLYCYCLIC ORGANIC MATTER FROM EXHAUST GASES

Richard Haroz

Center for Toxicology and Biosciences, BATTELLE
RESEARCH CENTERS, 1227 Geneva, Switzerland

ABSTRACT

The assessment of the potential toxic effects of polycyclic
organic matter from exhaust gases can be approached in a variety
of ways using short-term in vitro studies, whole animal studies
testing total exhaust emissions or fractions thereof, and
epidemiologic studies. Each of these approaches has its own
problems of interpretation and extrapolation. Many of the
experimental methods available today are claimed to yield results related to or predictive of human carcinogenesis. These
same methods are not claimed to be predictors of non-neoplastic
disease. These test systems are often used to evaluate fractions of the exhaust emissions or pure compounds which have been
identified in the total exhaust. Caution is necessary in the
interpretation of such results as many examples are known where
results differ when compounds are tested alone or as components
of a mixture. The majority of the in vitro tests monitor the
initiation phase of carcinogenesis, whereas we know that cancer
is a multistage process with a multifactorial etiology. The
efforts of dissecting the emissions with the intention of identification and removal of toxic components to produce a higher
quality emission must be weighed against simply reducing the
quantity of total emissions.

INTRODUCTION

It has been pointed out that the composition of participants for

D. Rondia et al. (eds.), Mobile Source Emissions Including Polycyclic Organic Species, 77–86.
© 1983 by D. Reidel Publishing Company.

this workshop is interdisciplinary. Therefore, my presentation will be of a general nature which, I hope, is not geared too low for this group. I also am afraid it will raise more questions than it will answer, but that is why we are here. There are several recent publications which review in greater detail some of the topics upon which I intend to touch. These are referenced in several other papers in this volume and I shall not here repeat these listings. I shall focus mainly on the toxic effects related to the carcinogenic process, and to a lesser extent on mutagenic and other genotoxic considerations which are covered in greater detail in the presentations by Joellen Lewtas, Tom Pederson and Dennis Schuetzle. The relationship of non-neoplastic diseases resulting from exposure to polycyclic organic matter or exhaust gases has not been extensively studied and I shall only say a few words about this, keeping in mind that such effects have also to be taken into account when assessing the overall health impact of exhaust emissions. I shall not cover such things as production methods, sampling procedures or extraction techniques which can introduce artifacts and effect the subsequent biological activity of the material to be tested. These topics are discussed elsewhere in the workshop.

The assessment of the potential toxic effects of polycyclic organic matter (POM) can be approached in a variety of ways. The toxic effects of any material can today be evaluated by the application of test methods ranging from short-term in vitro studies to extremely complex long-term inhalation studies. Additionally, epidemiology allows us to look back and attempt to relate present human health conditions to previous experiences. Appreciating the difficulties of extrapolating from laboratory animal experiments to man, epidemiologic data obtained in an ideal environment would identify the effects which occur in man and might even answer questions about threshold values. Unfortunately, we do not live in a perfect world and it is often impossible to isolate single causal factors from a multitude of potential contributing factors and to identify the resulting event where there already exists a background incidence of the disease in the population.

Polycyclic organic matter, derived from the total exhaust emission, is an extremely complex mixture. It includes a large number of compounds such as polynuclear aromatic hydrocarbons (PAH), derivations of PAH such as nitro-PAH and amino-PAH, oxygenated PAH such as phenols and quinones, and heterocyclic aromatic compounds containing sulfur and oxygen. In order to assist in the identification of classes of toxic compounds it is possible to fractionate the exhaust emissions into vapor and

particulate phases. The particulate phase can be further fractionated by, for example, solvent extraction. These fractions
can be tested, for such effects as mutagenicity. Individual
pure components can also be identified and tested for various
biological activities. Test systems such as many of the in vitro
short-term tests and mouse skin painting (MSP) usually examine
the effect of particulate matter or fractions extracted
therefrom.

One problem with such approaches is that one cannot a priori
equate the behavior of a single component or a separate fraction
in isolation from its behavior as part of a complex mixture.
There are many examples where, following fractionation procedures, the sum of the parts does not equal the whole. There are
other examples where the combination of components will multiply
or accentuate the effect of one component in isolation. Such
examples demonstrate co-mutagenic or co-carcinogenic effects.

It is known that many chemicals, e.g. polynuclear aromatic hydrocarbons require enzyme modification before becoming biologically active. The production of such so-called "ultimate mutagens or carcinogens" is dependent on a series of inducible
enzymes. The induction of these enzymes may increase or decrease
the ultimate effect of the parent molecule. These inducible
enzyme systems are also under genetic control, raising the
question of the existence of susceptible individuals in the
general population.

On the other hand, direct acting mutagens such as nitro-substituted polycyclic aromatic hydrocarbons have been detected in
diesel exhaust emissions and may be responsible for a major
portion of the mutagenic activity. However, in view of the presence of a variety of detoxification enzymes in human cells, the
distinction between direct acting and activatable mutagens may
not be justified for the assessment of health hazards.

Furthermore, there is still a great deal we do not know about
the process called cancer.

Distinct steps in the tumorigenic process have been characterized in the mouse skin painting system and today can be demonstrated in many other tissues and species. The process, which
can simplistically be divided into at least two stages, tumor
initiation and tumor promotion, may be ubiquitous. Although
initiation may contain a mutagenic event, this may not be sufficient to transform the cell into a stable, modified state which
is subsequently promotable to a tumor.

Experimental evidence suggests that completion of the initiation
phase is very rapid and irreversible. By this definition, and
knowing the variety of potential initiators around us, we are
probably all initiated. Therefore, the more important event in
the cancer process may be tumor promotion, which is reversible,
and initiators at inadequate doses may not be particularly important. Epidemiology resulting from studies of ex-heavy smokers
would also suggest that the risk of developing lung cancer is
reversible, consistent with an important role for promoters.
Taking the experimental animal data and the human epidemiology
together, one is tempted to conclude that tumor promotion or
enhancement is the more important event in the carcinogenic
process.

Test systems such as mouse skin painting and many of the in
vitro short-term assays usually only test a part of the exhaust
emissions, the particulate phase.

It would seem that, ideally, the toxic effects of exhaust gases
should be assessed by exposure to the whole material. This
implies exposure by inhalation, which is the predominant route
of exposure for humans. This also suggests the need for animal
models which are capable of developing lesions related to those
diseases, principally but not exclusively pulmonary, which occur
in man. Much work is still needed here and the development
costs are considerable.

So far, we have talked mostly about neoplastic diseases. What
about the effects of POM on non-neoplastic diseases? Here very
little information is available, although there is clear evidence that inhalation of particles can effect the macrophage
population and induce processes which may be important in the
production of emphysema and other chronic obstructive lung diseases. Immunological consequences of POM exposure have hardly
been studied and are little understood today.

Let us now consider some of these topics in greater detail, and
point out some of the problems and limitations of the test
methods currently available to us.

Short-Term Bioassays

A long list of potentially useful short-term tests have been
proposed as predictors of genetic and carcinogenic damage. The
basic similarity of genetic material in all cell types is the
scientific justification for employing such tests and making
extrapolations to higher organisms and more complex systems. A

few of these tests have undergone extensive validation to the point that their limitations are relatively well understood. Many others have not been subjected to systematic validation and their predictive value and limitations are much less understood. Short-term tests can be used for screening where additional animal studies may be suggested, for biological-activity directed fractionation and to compare relative effects of similar emission types.

Assuming that the tests have a predictive value, they could be extremely valuable as results are obtained in a short period of time and at a relatively low cost. Assuming successful development and employment, their value may lie both in reducing the cancer risk for the present generation and reducing mutation frequency in germ cells to the benefit of future generations.

Almost all of the proposed tests have to do with changes at the DNA or chromosome level. These DNA alterations (mutations, covalent adduct formation, etc.) or chromosomal modifications (sister chromatid exchange, micronucleus test, etc.) are proposed to be predictive of carcinogens. It is not claimed that they have any predictive value for non-neoplastic diseases.

In all of these test systems, as also in experimentation with animals, the problem of extrapolation to man is always present. In the case of short-term tests, we are generally extrapolating from an effect on more or less naked DNA or, in some cases, a single cell system devoid of the controls and detoxification mechanisms one would find in the whole animal.

As an example, one can cite some of the short-comings which may be attributed to a typical test system for mutagenicity, the so-called Ames Test.

This test uses specially constructed bacteria to detect reverse mutations. Each tester strain has enhanced sensitivity and selectivity for the classes of compounds it will pick up. Therefore, several tester strains are generally used to maximize the possibility of detecting a mutagenic compound. It is well known now that many compounds, for example polycyclic aromatic hydrocarbons, must undergo some kind of enzymatic or chemical modification before becoming reactive. Therefore, in many cases an enzyme preparation (commonly called S9) is added to the chemical in order to ensure conversion to a chemically-reactive species.

Many examples exist where both the quantity and quality of added S9 can modify the results. Therefore, it is sometimes possible

to manipulate the results. As was stated earlier, a positive
result in these tests does not establish proof of a carcinogenic
danger. What these tests demonstrate is a change in a sub-
cellular element, a cell, or tissue which may be correlated
with, in some cases, carcinogenic results in animal studies or,
in a few cases, cancer in man.

One other problem with such studies arises when mixtures of compounds are studied. Such mixtures may consist of only two
compounds, but the results can be completely different from when
either compound is tested alone. What then can be said about
the biological activities observed when all of the polycyclic
organic matter of an exhaust emission are examined as a whole or
as fractionated or purified parts. It is clear that the potential for completely misleading results exists.

Nitrated polynuclear aromatic compounds can account for a large
percentage of the mutagenic activity of diesel particle extracts
not requiring microsomal enzyme activation. However, it is
possible that the relevance to man is questionable as the endo-
genous bacterial nitroreductase enzymes may exaggerate the
potency. Certainly, more work is required in non-bacterial
systems including whole animal studies in order to better understand the potential health effect risk to man of this class of
compounds. Also, with gasoline particle extracts the opposite
appears to be true; the greatest mutagenic activity is seen when
an exogenous metabolic activation system is added, suggesting
that the different sources produce a different chemical mixture
with respectively different biological activity, at least in a
mutagenicity assay.

Although reasonably good correlation between mutagenic and carcinogenic potency may be found for certain classes of chemicals,
a general correlation may not exist for all chemicals.

Whole Animal Studies

Animal studies of engine exhaust emissions have generally taken
two forms: the first which uses only the particulate matter to
which condensable volatile matter adheres and, the second, which
uses the total exhaust emission including the particulate and
the gaseous or vapor phase.

The particulate portion of emissions has been studied by intratracheal instillation, sub-cutaneous injection and most of all
by mouse skin painting. Dr. Jacob will be describing his
experiments on mouse skin painting later in this meeting.

These experimental methods generally exclude consideration of
any contribution from the gases present in emissions.   There is
evidence that certain gases can produce an irritative effect and
may be carcinogenic in themselves or at least promote tumor formation. Formaldehyde is an example. These approaches, especially the extensively studied mouse skin painting test systems, do
have as their end-point a tumor. This is an important difference
from those test systems which only indicate a singular genetic
change. However, once again, mouse skin painting and the other
approaches are specific tests for predicting potential human
carcinogenicity.   Non-neoplastic diseases are not investigated
by such approaches.

By mouse skin painting it is also possible to examine the biological activities of fractions of the total particulate material.
This can be done in a complete test system, i.e. looking only at
the material of interest applied repeatedly to the skin, or in a
test system where the material of interest is looked at either
as a tumor initiator or tumor promotor.   Tumor initiation and
tumor promotion assessment offer some promise as relatively
short-term in vivo tests especially when a tumor susceptible
mouse like SENCAR is utilized.

It is known that tumor initiators are generally complete carcinogens at higher doses.   It is also true that tumor promotion
has been most extensively studied in the mouse skin.   As we
noted earlier, the promotion process may be the more important
aspect of tumorigenesis, thus deserving more study than has
occurred to date.

As with the earlier cited mutagenicity tests, the potential for
possible misleading results exists when working with pure compounds and with mixtures.   Just as co-mutagenesis has been
demonstrated in bacteria, so has co-carcinogenesis and even
anti-carcinogenesis been demonstrated in the mouse skin.   Just
as one can demonstrate either additive or antagonistic effects
with artificial simple mixtures of two compounds, one must be
aware of how a complex mixture is metabolized and handled in a
real life situation where, for example, genetic differences
exist and enzyme inducers can be found in the food we eat and
the cigarettes we smoke.

This takes us to the next topic - the most complex and the most
expensive, i.e., exposure of animals to the complete exhaust
emission by inhalation. Such a study mimics the chief route
of exposure of humans, takes into account any contribution of
the gaseous components and permits, in principle, an assessment

of the health risk not only related to cancer but also to other
non-neoplastic diseases which, when taken together, are much
more damaging to human health and our economy than cancer.

However, in order to hope to achieve a meaningful biological
endpoint, we must enter a large number of subjects (animals in
this case) into the experiment in order to be able to pick up
events which may only occur at a low level in the total population. To do this with large mammals is prohibitively expensive
and, anyway, may not tell us more than studies employing rodents.

Here again we have the problem of extrapolation but at least
from a whole animal. In the whole animal, as in man, the multiple processes of metabolic activation and inactivation, and
cellular interactions can occur and result in observable disease
which is not the case in isolated cells. Therefore, why have we
not been very successful in these types of studies? Firstly,
only a few life-time inhalation studies have been completed.
Secondly, there has been only a limited success in developing
animal models for human diseases, pulmonary and otherwise. We
are interested in those diseases which are most likely to be
associated with the inhalation of incompletely combusted plant
and/or animal material. These diseases include lung cancer,
emphysema, chronic bronchitis and cardiovascular disease.

In order to investigate some of the changes associated with the
development of these diseases, the rodent may not be the best
model. The mouse, for example, is almost too small an animal to
measure certain lung and heart functions, even in those laboratories which have perfected very sophisticated approaches.
Therefore, do we experiment with mice, realizing that solely the
pathological findings will be available to assess the affect
achieved?

What about lung cancer? Where is the animal model that reflects
the human situation? Does it exist? To date, we have been
unable to develop a fully satisfactory animal model for the
induction of lung cancer by cigarette smoke despite its apparent carcinogenic effect in people. Life-time studies with
rats exposed to exhaust emissions or to cigarette smoke have
indicated that these animals will not develop neoplastic lesions
in the lungs due to such exposure. What about a combined study
employing cigarette smoke and engine emissions? This type of
study has not yet been conducted. Our studies are generally
carried out with the healthiest specific pathogen free animal
strains that money can buy. Maybe this is wrong. Examples
exist where severe atmospheric conditions have created levels of

pollution which lead to many deaths. Generally speaking, the
individuals who died were already weakened by a prior history of
pulmonary or cardiovascular disease. Maybe then a better animal
model for that portion of the human population which may be at
high risk to exhaust emissions would be one with a pre-existing
health condition.

We are aware too that genetics can play a role in susceptibility. Should one therefore select a sensitive animal model or a
resistant one? For example, for skin carcinogenesis, the SENCAR
mouse is regarded as extremely sensitive to many carcinogens, in
particular polycyclic aromatic hydrocarbons. On the other hand,
C57 Black mice are insensitive to benzopyrene as an initiator.
If we are trying to assess the health risk to and protect the
most sensitive individuals in the population, we must at least
select sensitive animal models even if this results in a large
safety margin for the majority of the population.

Inhalation studies, even though expensive, offer certain advantages and are the sole means of investigating certain questions.
The route of administration simulates that of most concern to
human beings, the fate of the inhaled material can be studied in
a whole-body environment, and serial observation of the development of a disease is possible. The contribution from gaseous
phase components is also assessible, in contrast to most other
in vivo and in vitro approaches.

Only in such studies, many of which are being conducted by the
General Motors Research Laboratories, can the effectiveness of
defense systems such as ciliastasis, mucus clearance and phago-
cytosis be assessed. It is clear that such defense mechanisms
are operating. For example, macrophages recovered by bronchopul-
monary lavage and extracted with an organic solvent contain
mutagenic material immediately following exposure to diesel exhaust. However, within another day or two, the mutagenic material is no longer detectable. These kind of considerations must
be taken into account if one is trying to assess the potential
real health risk for man. Perhaps this can only be done experimentally by inhalation.

It is clear that, cigarette smoking aside, lung cancer occurs at
a very low incidence level in the population. This may be at
least as true in an animal population. Researchers at the Frau-
enhofer Institute have initiated or primed hamsters in order to
make them more likely to develop respiratory tract tumors. In
studies with diesel engine exhausts, they were able to demonstrate a significant increase of tumors due to diesel emission

exposures. However, the particle-free gaseous phase also augmented the tumorigenic response raising the question of carcinogenic or promoting effects due to gaseous phase components. These types of approaches require further study and cautious interpretation, but may indicate the need to study rather the complete exhaust emission including gases as opposed to fractions or even pure components thereof.

All that has been said raises more questions than it answers. It is not difficult to assume that the value of estimating cancer risk to man should be in the order epidemiologic studies > whole animal carcinogenesis assays > cell transformation and mutagenesis assays > chemical analysis alone. Each of these approaches has its own problems of interpretation and extrapolation, and requires many assumptions about how the human machine functions and lives.

Finally, the efforts being devoted to the dissection of exhaust gases for the purpose of identifying and removing toxic components, resulting in a higher quality emission, must be weighed against a simple reduction in the total quantity of exhaust emissions.

AUTOMOTIVE LEAD TRAPS: POTENTIAL UNDER CANADIAN CONDITIONS

A.C.S. Hayden and R.W. Braaten; G.W. Kunz, Jr.

Canadian Combustion Research Laboratory, EMR Canada;
DuPont USA

ABSTRACT

The information presented in this paper is directed to those
individuals concerned with the fuel consumption benefits associated
with the use of leaded gasoline and the reduction of lead emissions to the atmosphere through the use of automotive lead traps.
A two-year experimental program conducted by the Canadian
Combustion Research Laboratory of Energy, Mines and Resources
Canada under winter and summer driving conditions has demonstrated
that automotive exhaust lead traps can reduce lead emissions by
an average of 80%. Trapping efficiency was not affected by
ambient temperature, which varied from -20°C to 24°C, nor by city
or highway driving. With effective reduction in lead emissions
possible through the use of lead traps, energy savings associated
with the use of lead in gasoline can be realized, even in the
event that lead tailpipe emissions might be restricted by pollution considerations.

INTRODUCTION

Lead antiknocks have been used worldwide for more than fifty
years to improve gasoline octane quality. The widespread availability of high octane gasoline has enabled the automotive
industry to produce efficient, high compression ratio engines.
Without lead antiknocks, additional refining is required to
obtain the same high octane quality. This additional refining
uses more energy, reducing the amount of gasoline that can be
obtained from a given amount of crude oil. Thus, the use of lead
antiknocks both increases the amount of gasoline available and

87

D. Rondia et al. (eds.), Mobile Source Emissions Including Polycyclic Organic Species, 87–100.
© 1983 by D. Reidel Publishing Company.

saves energy in refining, resulting in more efficient use of
crude oil.

Since 1975, catalytic converters have been installed on a
majority of new Canadian automobiles. These converters require
unleaded gasoline and, as more converters are used, the energy
savings associated with leaded gasoline decreases each year in
Canada. However, many non-catalytic cars operating on leaded
gasoline have met the Canadian gaseous emission standards without
increased fuel consumption (1). The need for catalytic converters
on many new Canadian cars is questionable; it is possible that
the number of non-catalytic cars operated on leaded gasoline
could increase in the future.

Increased use of leaded gasoline in Canada may generate
pressures to control automotive exhaust lead emissions, although
there is no established health-based lead-in-air standard to
serve as the basis for such control. If reduction of automotive
lead emissions into the atmosphere should be required, controls
should be placed on the amount of lead emitted from the tailpipe,
similar to the manner by which gaseous emissions are controlled.
Such action is more energy efficient than reducing the amount of
lead used in gasoline. One effective way to control tailpipe
lead emissions is the use of automotive exhaust lead trap that
replaces the standard muffler (2).

This report describes the two-year road test program, supported by laboratory analysis, conducted by the Canadian Combustion
Research Laboratory of Energy, Mines and Resources Canada to
evaluate the effectiveness of automotive exhaust lead traps. The
exhaust traps were designed and built by the Du Pont Company.
Du Pont and other major lead antiknock manufacturers have been
developing automotive exhaust lead traps since the middle 1960s
(3). Other road tests have shown these traps to be effective
under benign climate conditions, but none have considered the
wide range found in Canada.

EXPERIMENTAL METHODS

Test Cars

The test fleet consisted of eight identically-equipped 1974
four-door Chevrolet Biscaynes operated by the Department of
National Defence, Canada. These cars were equipped with 5.74 litre
V-8 engines having air pumps to supply secondary air to the exhaust
manifolds to reduce gaseous emissions. Four vehicles were equipped
with new standard muffler and exhaust pipe systems and were used
as control cars. The standard mufflers and exhaust pipes of the
second group of four cars were replaced with lead trap systems.

New crossover pipes and tail pipes were installed on all test cars. The test vehicles had an average of 80,000 kilometres before the start of the lead emission trial. Prior to the commencement of the trial, each automobile was given a tune-up by an authorized Chevrolet dealer to meet manufacturer's specifications.

Lead Traps

The trap system consisted of two components, as shown in Figure 1. The forward component was a tube cooler, which replaced the standard muffler. A cutaway sketch of the tube cooler is shown in Figure 2. The primary purpose of the cooler was to maintain exhaust temperatures of 300°C to 400°C in the downstream trap. Without the cooler, inlet temperatures of the trap would have been about 500°C during highway operation. Most of the lead salts in the exhaust are lead halides, which are in the vapour state at temperatures about 400°C and mostly solids below 300°C.

The particle size of lead salts can be increased through agglomeration by promoting surface condensation. With the extended cooled surfaces of the tube cooler and the turbulence produced by its design, some surface condensation of lead halides takes place. The majority of the lead salt agglomeration occurs on the large surface area produced by the bed of alumina pellets in the trap, shown in Figure 3. Most of the agglomerated particles collected in the alumina bed are entrained into the exhaust gas flow entering the downstream dual parallel cyclones of the trap. The exhaust particles separated by the cyclones are retained in collection chambers that can be sized to hold all of the exhaust particulate matter collected over 150,000 kilometres, an average car life in Canada.

Mileage Accumulation

Test mileage was accumulated under in-use driving conditions. Mileage was obtained under routine taxi-like service in the transport pool of the Department of National Defence for seven of the vehicles. The eight automobile was used in the transport pool, but at one-half the rate of the other seven.

Six specific lead emission tests were performed on each test car over a two-year period. Tests were scheduled at approximately five-month intervals to cover the wide range of ambient temperatures and operating conditions experienced in Canada.

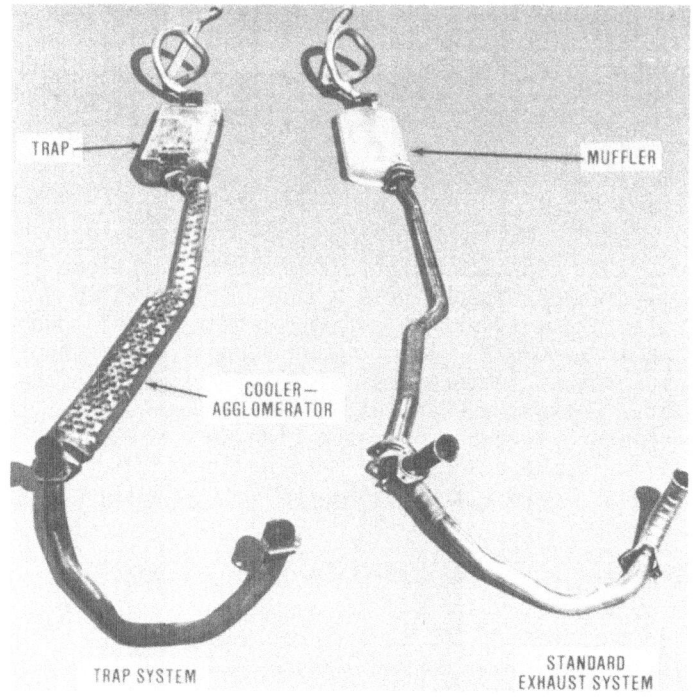

Figure 1. Lead trap system and standard exhaust system.

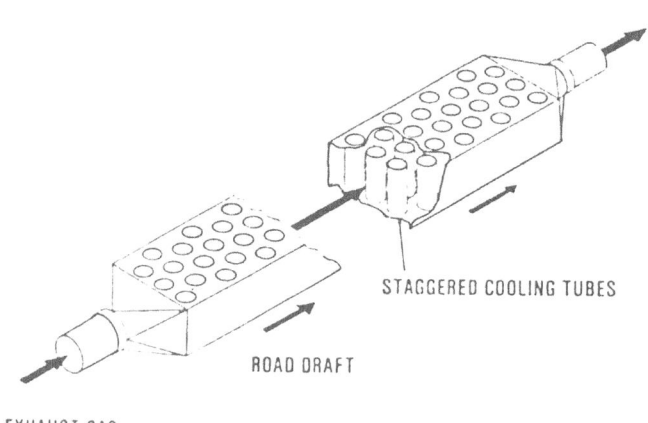

Figure 2. Cutaway view of the tube cooler. The tube cooler
           ensures that the trap inlet temperature is below
           400°C.

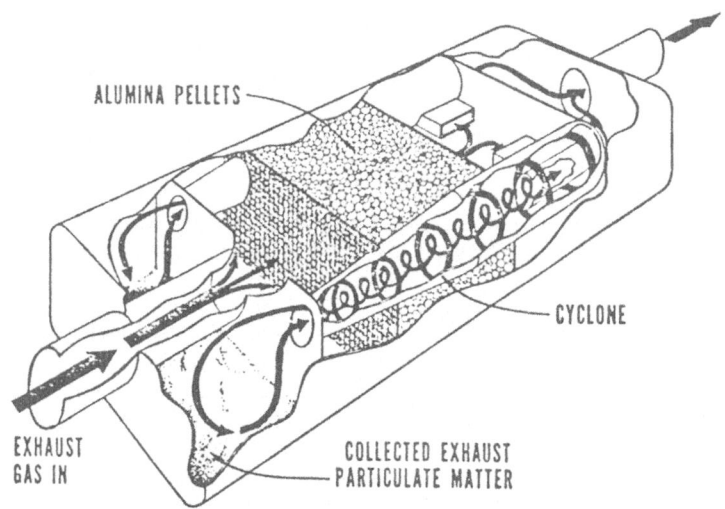

Figure 3.   Cutaway view of lead trap showing location of
             alumina pellets upstream of cyclone separators.
             The alumina pellets promote agglomeration and
             the dual parallel cyclones separate and retain
             the exhaust particulate matter in collection
             chambers.

Lead Emission Measurement

     Lead emissions were measured by passing all of the exhaust
from the tailpipe of the vehicle under test through an analytical
filter. The filter captured all of the lead passing from the car,
and was later analyzed to provide a quantitative measure of the
amount. The filter was contained in a "back-pack" attached on
top of the trunk, as shown in Figure 4. Installation and removal
of the back-pack and fuel meter on the test car was carried out
at the Canadian Combustion Research Laboratory. An extension
was used to connect the tailpipe with the back-pack. During the
winter months, the back-pack had to be insulated, to minimize
condensation of water vapour in the exhaust. The back-pack
material was fibreglass paper 0.041 cm thick and has a typical
efficiency of less than 0.001% penetration of dioctyl phthalate
monodispersed aerosol of 0.3 microns diameter, at a velocity of
3.2 m/min.

     After each test, the back-pack was disassembled and lead
deposited on the filter was extracted in hydrochloric acid. The
lead deposited on the tailpipe extensions and on internal parts
of the back-pack upstream of the filter was extracted with an
aqueous solution of tetrasodium salt of ethylenediamine tetra-
acetic acid. These solutions were analyzed by the Solid Fuels
Analysis Section, CANMET, Energy, Mines and Resources Canada.
Lead concentrations were determined by atomic absorption.
Separate sets of back-pack and tailpipe extensions were used for
the standard and for the trap cars. The amount of lead measured
in the washing of each set of back-pack and tailpipe extensions
after a series of lead emissions tests was prorated to the corresponding trap or standard car, based on the amount of lead
measured on the individual filters used for each car.

Fuel Consumption and Analysis

     Fuel consumption was measured at the Canadian Combustion
Research Laboratory, using a Fluidyne Instrumentation Model 1250
Precision Fuel Meter installed in the vehicle. When removing the
fuel meter, a sample of the fuel in the test car was taken and
analyzed for lead. Table 1 shows the lead concentration in the
test fuels as a function of test date. The minimum value was
0.3 grams of lead per litre of fuel during Test No. 2.1; the
highest value was 0.67 grams of lead per litre of fuel during
Test Nos. 5.5 to 5.8. The product of the amount of fuel consumed
and the lead content in the fuel gave the amount of lead input
during a lead emission test. Lead emissions were computed as a
percentage of the input lead.

Figure 4. Installation of back-pack filter and tailpipe extension used to measure exhaust lead emissions.

TABLE 1.   Lead concentration in test fuel as a
            function of test dates.

Test No.	Test Car No.	Test Date	Test Fuel (g Pb/l)	Test No.	Test Car No.	Test Date	Test Fuel (g Pb/l)
1.1	Std. 74	22/11/77	0.45	4.4	Trap 62	22/02/79	0.48
1.2	Trap 60	24/11/77	0.48	4.5	Std. 74	23/02/79	0.43
1.3	Std. 56	08/12/77	0.52	4.6	Trap 69	27/02/79	0.44
1.4	Trap 69	12/12/77	0.50	4.7	Trap 65	01/03/79	0.50
1.5	Trap 65	14/12/77	0.50	4.8	Std. 56	09/04/79	0.61
1.6	Trap 62	16/12/77	0.50				
1.7	Std. 70	20/12/77	0.49	5.1	Std. 70	21/08/79	0.63
1.8	Std. 77	22/12/77	0.52	5.2	Std. 77	22/08/79	0.64
				5.3	Trap 60	23/08/79	0.64
2.1	Std. 74	25/04/78	0.30	5.4	Std. 74	24/08/79	0.63
2.2	Trap 69	27/04/78	0.48	5.5	Trap 69	30/08/79	0.67
2.3	Std. 77	28/04/78	0.40	5.6	Trap 62	31/08/79	0.67
2.4	Trap 60	03/05/78	0.42	5.7	Trap 65	06/09/79	0.67
2.5	Std. 70	10/05/78	0.55	5.8	Std. 56	07/09/79	0.67
2.6	Std. 56	12/05/78	0.59				
2.7	Trap 62	17/05/78	0.42	5.9*	Trap 60	20/10/79	0.61
2.8	Trap 65	13/06/78	0.50	5.10*	Trap 60	21/10/79	0.63
				5.11*	Std. 56	03/11/79	0.63
3.1	Trap 69	29/08/78	0.61	5.12*	Std. 56	04/11/79	0.39
3.2	Std. 74	31/08/78	0.62				
3.3	Std. 77	05/09/78	0.62	6.1	Trap 69	14/12/79	0.51
3.4	Trap 62	08/09/78	0.62	6.2	Std. 70	18/12/79	0.45
3.5	Std. 70	12/09/78	0.61	6.3	Std. 74	19/12/79	0.43
3.6	Trap 60	14/09/78	0.61	6.4	Trap 60	27/12/79	0.43
3.7	Std. 56	18/09/78	0.61	6.5	Trap 65	28/12/79	0.44
3.8	Trap 65	20/09/78	0.62	6.6	Std. 77	12/01/80	0.49
				6.7	Std. 56	19/01/80	0.42
4.1	Std. 77	19/02/79	0.40	6.8	Trap 62	26/01/80	0.42
4.2	Std. 70	20/02/79	0.44				
4.3	Trap 60	21/02/79	0.47	*Urban and Highway Trials			

Test Cycle

    The test cycle used for the lead emission tests was developed
by Energy, Mines and Resources Canada to represent typical driving
patterns that might be expected in most Canadian cities. The test
cycle which was used had 9.2 kilometres of urban driving and
16.5 kilometres of highway driving, as shown in Figure 5. The
average speed of the test cycle was 51 km/h; the test course was
lapped twelve times during each lead emission test, for a total
of 309 kilometres. This distance was necessary to ensure a
sufficient amount of lead was emitted to permit accurate measurement.

EXPERIMENTAL RESULTS

Lead Emissions

    Mileage Effects. Lead emissions of the eight test cars as
a function of test kilometres are plotted in Figure 6. The
overall average lead emissions of the standard and trap cars are
summarized in Figure 7. The average standard car emitted 33% of
the lead burned during the emission test periods, while the
average trap car emitted 7% of the lead burned during the test
period. Hence, the average overall reduction in lead emissions
was 80% due to the use of the automotive exhaust lead traps.

    Ambient Temperature Effect. As indicated in Table 1, the
six lead emission test series were conducted during different
times of the year, ranging from mid-winter to summer, and covering
an ambient temperature range from -20°C to 24°C. In Figure 8,
the lead emissions of the eight test cars are plotted as a
function of the average daily ambient temperature of each lead
emission test. Wide scatter of the lead emissions of the standard
cars occurred during days of similar ambient temperatures, whereas
the low lead emissions of the trap cars were consistent over a
wide range of ambient temperatures. It is clear that the lead
traps are equally effective from winter to summer driving conditions.

    Urban and Highway Driving Effects. At the end of Test
Series No. 5, additional tests were conducted on standard car
No. 56 and trap car No. 60 in order to quantify the difference
in lead emissions between urban and highway operation. Results
are shown in Table 2. Lead emissions were measured separately
for urban and highway cycles, each consisting of continuous
operation of approximately 325 kilometres. The urban test on
each car was performed first. The trap system reduced lead
emissions by 68% during urban driving and by 92% during highway
driving. The lead emission of the standard car was relatively

Figure 5. Test course used to measure lead emissions. One lap
           has 9.2 kilometers of urban driving and 16.5 kilometers
           of highway driving. The run between CCRL and the
           start/stop point of the test course was only done at
           the beginning and end of the test.

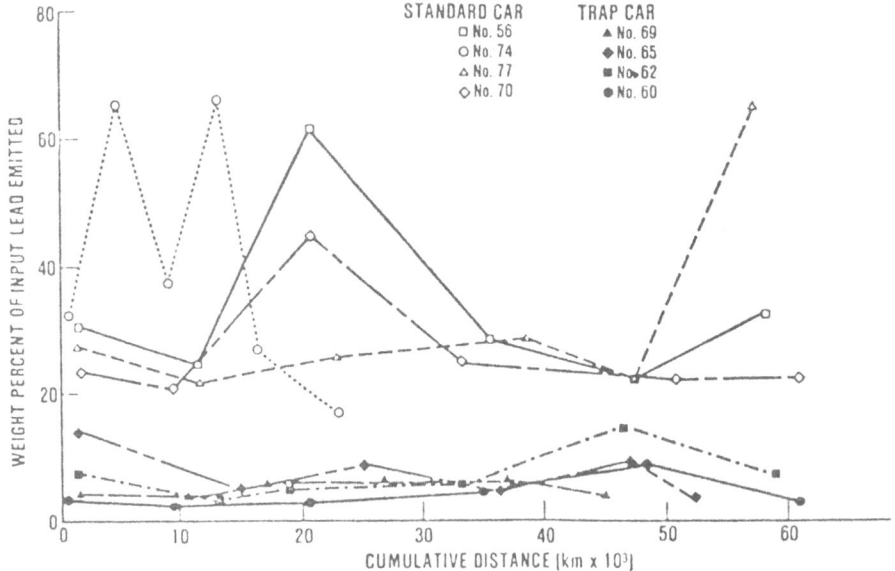

Figure 6. Lead emissions from standard and trap cars as a
           function of distance traveled.

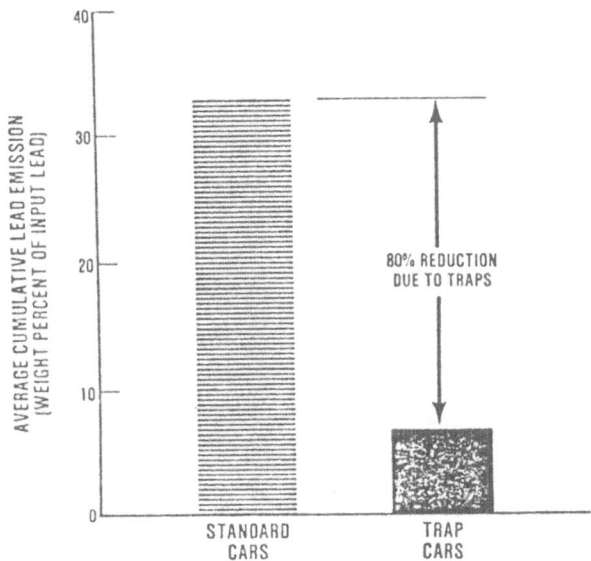

Figure 7. Comparison of average cumulative lead emission
          between standard and trap cars.

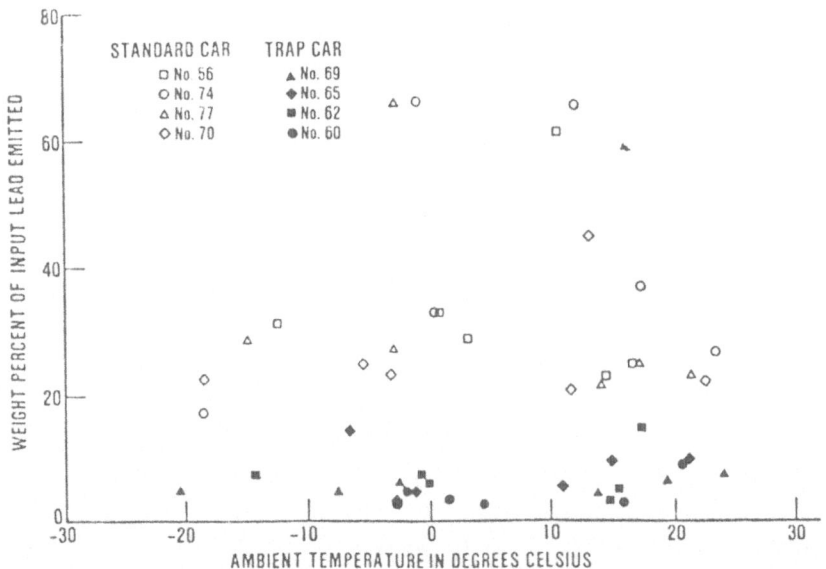

Figure 8. Lead emissions from standard and trap cars as a
          function of ambient temperature. Trapping efficiency
          is not affected by winter or summer driving conditions.

TABLE 2.   Effectiveness of lead traps in urban and
           highway operation.

	Percent of Input Lead Emitted	
	Urban	Highway
Standard Car No. 56	13.9	120.7
Trap Car No. 60	4.5	9.7
Percent Reduction Due to Traps	67.6	92.0

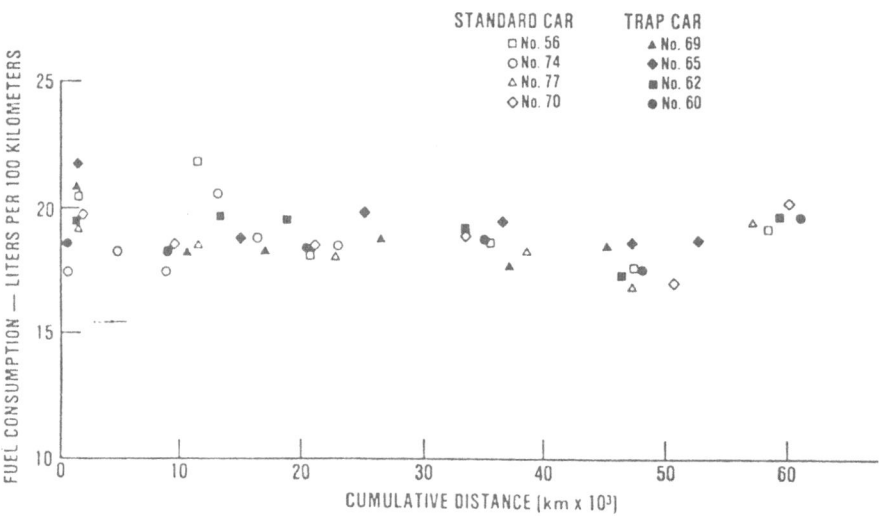

Figure 9.   Fuel consumption of test cars as a function of
            distance traveled.

low during the urban trial; however, during the highway trial on the standard car, purging of previously retained lead salts increased lead emissions almost eight-fold for this untrapped vehicle.

For the lead trap cars, lead emissions were only one-third those of the standard cars over the urban cycle and one-twelfth those of the standard cars over the highway cycle. This test demonstrates that purging of previously collected exhaust particulate matter does not take place with properly designed traps when switching from sustained mild driving conditions to more severe conditions of continuous operation at highway speeds. Other tests have shown higher lead emissions from cars with standard exhaust systems than the 33% shown in Figure 7. This low value is probably due to varying lead salt purging levels of standard vehicles due to different test cycles.

Car Performance

Muffling, Drivability and Maintenance. During the 27 months of test operation, there were no complaints about excessive exhaust muffling or drivability of either the standard or trap cars. From this, it is considered that the trap systems sounded essentially the same as the standard mufflers and that there was no apparent difference in drivability. The traps required no maintenance over the test program.

CONCLUSIONS

1. Automotive exhaust lead traps can replace standard mufflers and portions of the exhaust system and can reduce automotive lead emissions by 80%.

2. There is no discernible deterioration of trapping efficiency with mileage accumulation.

3. Automotive exhaust lead traps are equally effective over the wide range of temperatures encountered during the extremes of winter and summer driving conditions in Canada.

4. Purging of previously collected lead salts occurs with standard exhaust systems, but does not occur with well-designed lead traps, such as the type tested in this program.

5. The use of lead traps will not affect fuel consumption, vehicle drivability and exhaust muffling. Automotive exhaust lead traps are relatively

simple, have no moving parts, require no
maintenance and can be designed to last
the life of the vehicle.

6.   If the use of leaded gasoline is increased
     in Canada to obtain improved refinery
     efficiency and lower automotive fuel con-
     sumption, automotive lead traps offer a
     viable means to control lead emissions into
     the environment, if such control is considered
     necessary.

REFERENCES

1.   Transport Canada, "Fuel Consumption Guide, 1980, Edition 2",
     Ottawa, Canada, January 1980.

2.   International Lead Zinc Research Organization, Inc.,
     "Automotive Exhaust Filters", New York, July 1979.

3.   Environmental Conservation Executive, "Energy Conservation
     and Exhaust Gas Filter Developments", Parts I and II,
     Australian Institute of Petroleum Ltd., Melbourne,
     Australia, April 1977.

4.   Casella et al., "Evaluation of Lead Trap Performance in
     Exhaust Engine Control", Accordo di Ricerca (FEEMAS),
     January 1978.

ACKNOWLEDGEMENTS

     The cooperation of the Department of National Defence Canada,
in particular, Base Transport, Ottawa, in supplying the vehicles
for the detailed testing when required throughout the test program,
is gratefully acknowledged.

     The efforts of Mr. G.C. Anderson and Mr. R.E. Dureau, Solid
Fuels Analysis Laboratory, CANMET, in carrying out the many
sample analyses required throughout the experimental program are
most appreciated.

CARCINOGENIC IMPACT FROM AUTOMOBILE EXHAUST CONDENSATE AND THE DEPENDENCE OF THE PAH-PROFILE ON VARIOUS PARAMETERS

Prof.Dr.Jürgen Jacob

Biochemisches Institut für Umweltcarcinogene
Sieker Landstraße 19
D-2070 Ahrensburg
F.R.G.

The analysis of balance of the carcinogenic impact from automobile exhaust condensate using a carcinogen-specific detector (mouse skin painting) clearly shows that predominantly (85%) polycyclic aromatic hydrocarbons are responsible for the carcinogenic potency of this matrix. Only some of the numerous PAH contribute, however, to this effect; 50% of the total carcinogenicity could be provoked by a synthetic mixture of 15 selected individuals.
   The PAH-content of automobile exhaust condensate correlates to (a) the gasoline/air ratio, and (b) to the percentage of aromatic compounds (benzene, toluene, xylene) in gasoline. The results are compared with those obtained from other environmental matrices, such as used engine oil and coal heating exhaust.

Nowadays automobile exhaust is claimed to be one of the most important evironmental PAH sources and thus a particular relevant pollutant for the carcinogenic risk of man. The absolute annual benzo(a)pyrene masses emitted from vehicles in the FRG have been calculated to be about 1.85 tons (1), but presently no serious estimation of the benzo(a)pyrene ratio from vehicle exhaust to the total benzo(a)pyrene concentration originating also from other sources is available. In the USA, however, preferentially other sources, such as domestic coal heating (about 33%) and refuse burning (45%) have been shown to contribute to the PAH-pollution in the 60ies of this century, whereas vehicle ex-

101

D. Rondia et al. (eds.), Mobile Source Emissions Including Polycyclic Organic Species, 101–114.
© 1983 by D. Reidel Publishing Company.

haust counts only for 2% (as reviewed in (2)). Despite
of this yet to be confirmed estimation which will be
dealed with later on in this paper, the questions have
to be answered : (a) What portion of the total carcinogenicity of automobile exhaust originates from PAH at
all, and (b) what portion of the total carcinogenicity
may be associated with benzo(a)pyrene?

    The carcinogenic activity of automobile exhaust
gas condensate has been tested repeatedly using the
long-term mouse skin painting test (3-7) which is considered to be a carcinogen-specific test model. These
results have been confirmed in the meantime with other
test systems, such as implantation in the lung. The
precondition for the determination of the biological
effect resulting from various fractions of the total
material and hence the establishment of a balance of
the carcinogenic effect is a clear dose-response relation, which actually has been obtained for automobile
exhaust condensate (fig. 1).

Amount (mg/year)	Animals with Carcinomas
53	1.3%
106	15.0%
158	29.7%
316	60.0%
438	71.8%

Figure 1. Correlation of the amount of automobile ex-
          haust condensate (285 μg BaP/g) and the fre-
          quency of carcinoma.
          (Dropping onto the skin of mice (80 mice
          per dose)).

    In order to associate the biological activity
with certain chemical classes, the condensate then is
fractionated by liquid/liquid-distribution and subsequent chromatography on Sephadex LH 20 using various
solvent systems, and this procedure results in PAH-free,
2-3 ring PAH, and 4-7 ring PAH fractions (fig. 2).

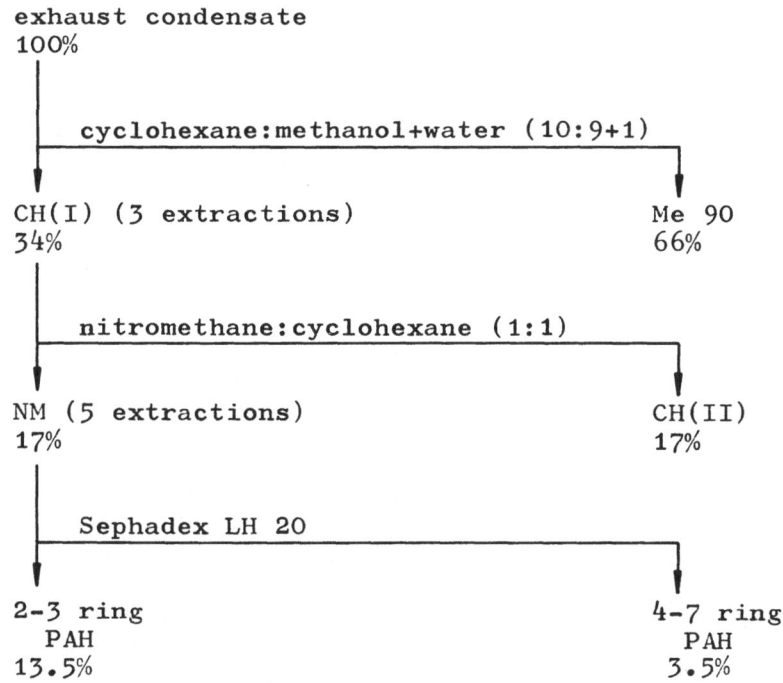

Figure 2. Fractionation of automobile exhaust.

The carcinogenic effects of these fractions are presented in figure 3 which clearly shows that almost the full
activity remains in the small fraction containing 4-7
ring PAH, which is only 3.5% by weight of the total
condensate. Moreover, this fraction follows the same
dose-response relation as the crude condensate (fig. 4).

    A first attempt to an estimation of the contribut-
ion of benzo(a)pyrene of the total biological activity
of condensates is recording their PAH-profiles. Typical
profiles from OTTO- and DIESEL-engines after Europa-
test cycles are given in figures 5 and 6.

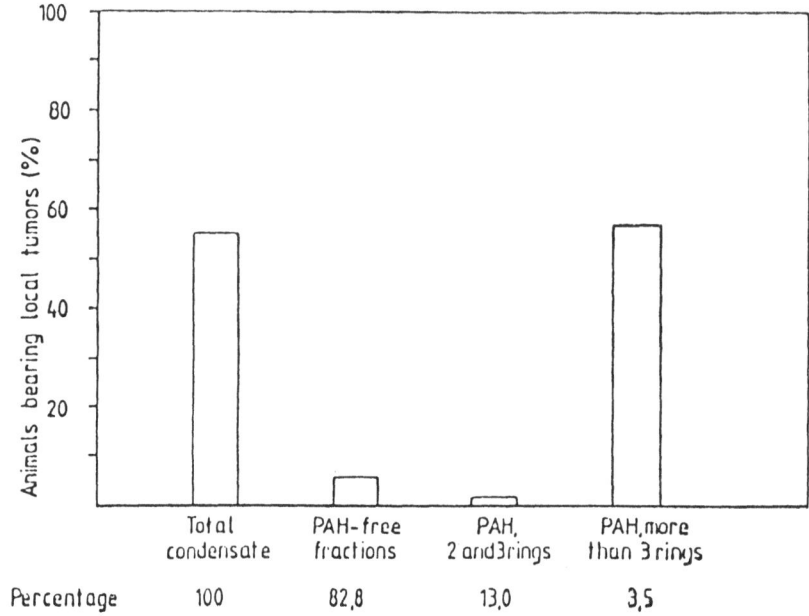

Figure 3. Biological activities of automobile exhaust
fractions tested by mouse skin painting.

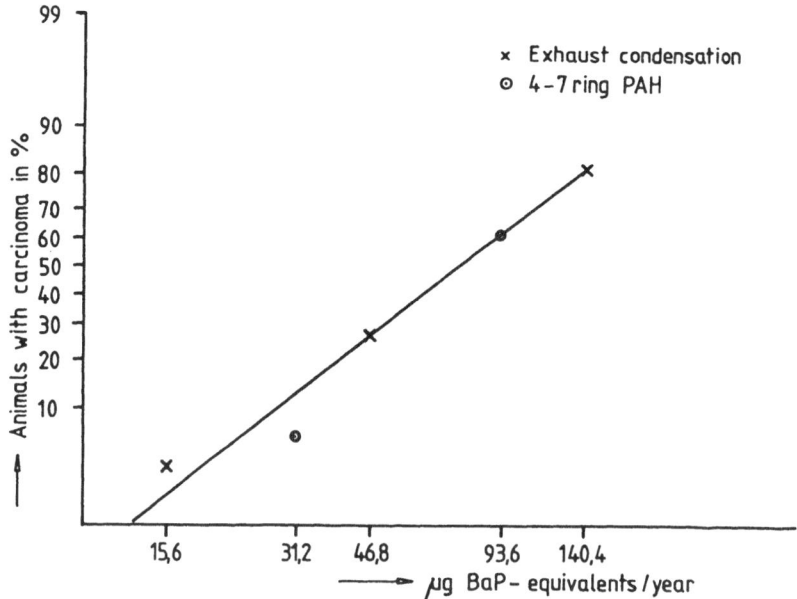

Figure 4. Dose-response relation of the 4-7 ring
fraction of automobile exhaust.

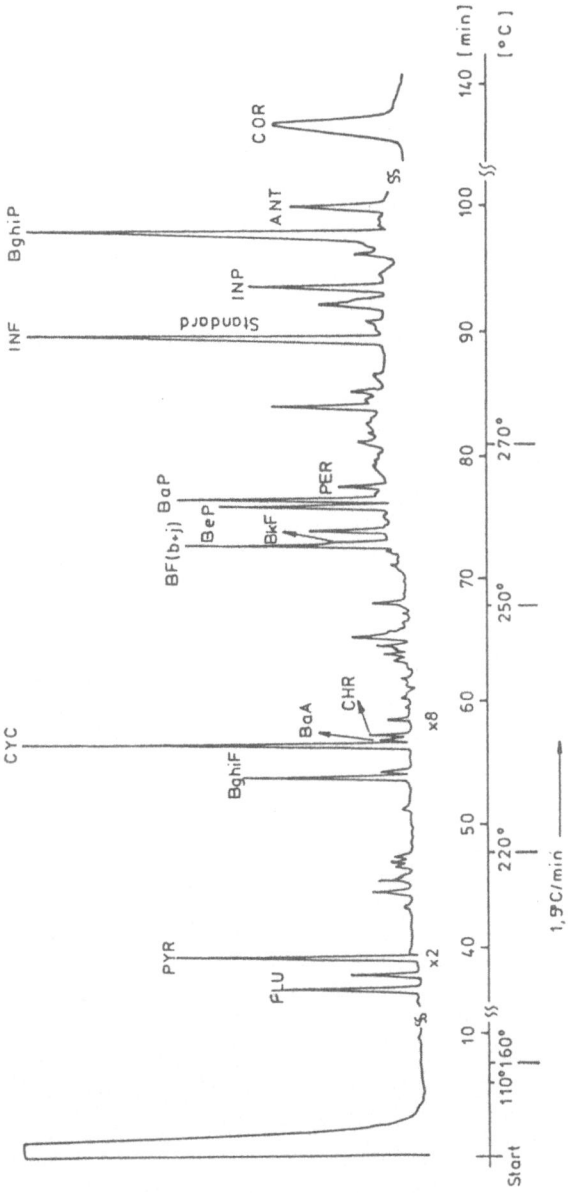

Figure 5. PAH-profile of automobile exhaust from OTTO-engine car
(25 m glass-capillary column CPsil5)

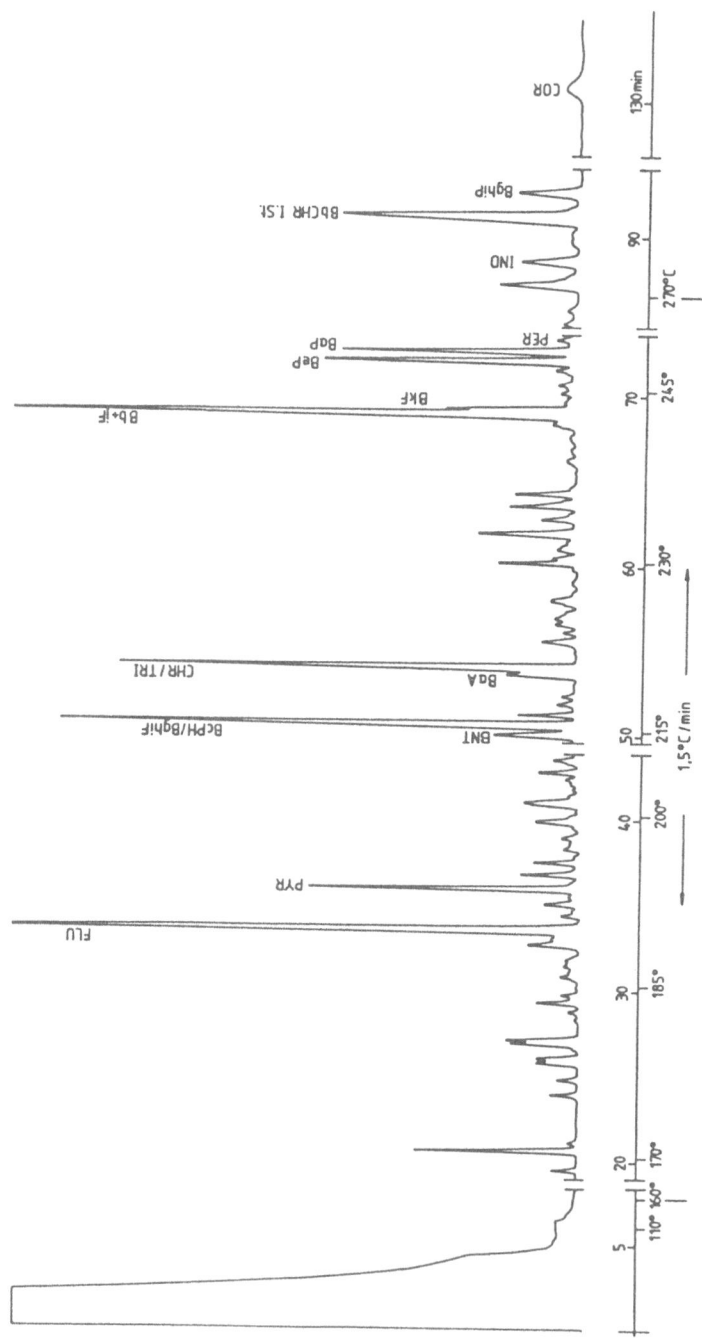

Figure 6. PAH-profile of automobile exhaust from DIESEL-engine car
(25 m glass-capillary column CPsil5)

They demonstrate (a) that, apart from benzo(a)pyrene,
several potent carcinogens such as cyclopenta(cd)pyrene,
benzo(b)- and (j)fluoranthene, and anthanthrene are detectable, and (b) that there are significant differences between both profiles inasmuch as e.g. cyclopenta-
(cd)pyrene is low or absent in DIESEL, but most intense
in OTTO exhaust, or benzo(b)naphtho(2,1-d)thiophene
(BNT) occurs in DIESEL only. Anthanthrene, however, is
high in OTTO but low in DIESEL and the reversed ratio
is observed for chrysene. 10-15% of the carcinogenic
effect of the original condensate actually can be simulated by a mixture of 6 different PAH (benz(a)anthracene,
chrysene, benzo(b)fluoranthene, benzo(j)fluoranthene,
dibenz(a,h)anthracene, and indeno(1,2,3-cd)pyrene) (8),
whereas a 50% activity could be attributed to a mixture
of 15 different PAH (benzo(c)phenanthrene, cyclopenta-
(cd)pyrene, benz(a)anthracene, chrysene, benzo(b)fluor-
anthene, benzo(j)fluoranthene, benzo(k)fluoranthene,
benzo(a)pyrene, 11H-cyclopenta(qrs)benzo(e)pyrene, 10H-
cyclopenta(mno)benzo(a)pyrene, dibenz(a,h)- and dibenz-
(a,j)anthracene, indeno(1,2,3-cd)pyrene, PAH-M300 A and
B) (9). Although it has been claimed that there is an
additivity of the carcinogenic activities for the
various PAH of automobile exhaust and that the remaining
50% originates from other individuals not present in
the afore-mentioned mixture, this has not yet been de-
finitely proved and synergistic effects via e.g. enzyme
induction might also be involved in the total biological effects. Benzo(a)pyrene, however, counts for only
10% of this effect. Other matrices such as coal exhaust,
used lubricating oil, or cigarette smoke condensate exhibit similar trends showing that the PAH are the most
relevant compounds for tumor production activity
(fig. 7). However, the contribution of benzo(a)pyrene
to the total activities of these condensates varies
considerably (fig. 8).

A variety of parameters can influence, both the
mass concentration and the profile of PAH.
(A) Ratio of gasoline-air mixture : An excess of gasoline in the gasoline-air mixture above the stoichio-
metric ratio (λ=1) significantly increases the rate of
PAH emitted. When starting, the engine working temperature has not yet been reached and a high fuel concentration in the above mixture is inevitable. The PAH
emission per combusted amount of gasoline thus is increased many times.

		Carcinogenicity explained by PAH-portion
Automobile exhaust condensate		91%
Domestic hard coal	about	90% +
Used lubricating oil		86%
Cigarette smoke condensate	about	50%

+preliminary, not yet finished

Figure 7. Percentage of tumor-producing activity
          of the 4-ring PAH fraction to the total
          activity from various matrices.

		Carcinogenicity of the BaP-portion
Automobile exhaust condensate (gasoline engine)		9.6%
Automobile exhaust condensate (DIESEL engine)		16.7%
Domestic hard coal	about	6.0% +
Domestic brown coal (briquets)	about	9.0%
Lubricating oil from cars (used)		10.5%
Sewage sludge (extract)		22.9%
Cigarette smoke condensate	about	1.0%

+preliminary, not yet finished

Figure 8. Percentage of carcinogenic activity
          of benzo(a)pyrene related to the to-
          tal activity from various matrices.

(B) The percentage of aromatic compounds in gasoline :
Variation of the percentage of aromatic parts in gasoline significantly influence the amount of PAH emitted
(fig. 9).

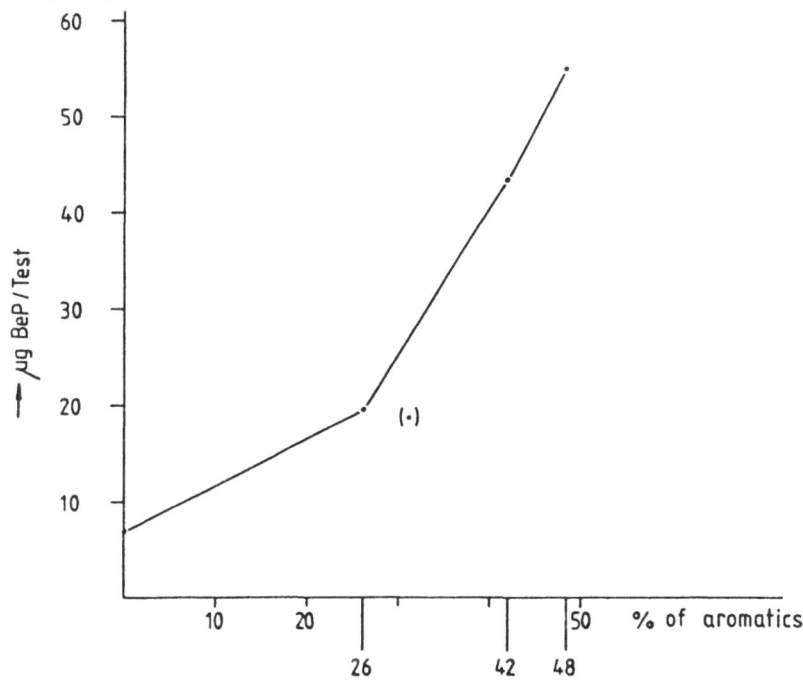

Figure 9. Influence of the percentage of aromatic com-
          pounds on the PAH-emission.

Although up to a ten-fold increase of PAH output may
be observed if aromatic-free gasoline is compared with
e.g. gasoline composed of 50% aromatics, the increase
in praxi may be estimated to be about two- or three-fold
only, since gasoline normally used already contains
aromatic compounds. It is, however, remarkable that the
increase is different for various PAH (benzo(ghi)pery-
lene and coronene 3fold; benzo(e)- and (a)pyrene, perylene, and the benzofluoranthenes 5-7fold; fluoranthene,
pyrene, chrysene, and cyclopenta(cd)pyrene 10fold) (10).

(C) Content of PAH present already in the gasoline :
Several times repeated tests with PAH-free and analo-
gous but PAH-containing gasoline did not give any evidence for a relevant influence on the PAH mass concentration or the PAH profile of the exhaust. MEYER et al.
(11) did not find differences in the mass concentrations

of the ten most abundant PAH before and after removal
of PAH from the gasoline by distillation. The opposite,
however, has been claimed by various authors (for reference see : Chapter : A. Candeli 'PAH content of exhaust gases from fuels with different aromatic fractions').

(D) Age of the engine : The result from a total of 140
Europa tests (12) with two different new cars showed
no significant increases of the PAH output after distances of 5 000, 10 000, 20 000 km, respectively.
There was a slight PAH increase (10%) after 10 000 km
city driving which was compensated already after
1 000 km high speed driving.

(E) PAH concentration in lubricating oil : In lubricating oil a considerable accumulation of PAH occurs with
increasing time of operation. Under the described operation conditions a PAH-profile is formed in used oil
which is similar to the PAH-profile of gasoline (13).

(F) PAH emission during operation with fresh and used
oil : The period of use and the corresponding accumulation of PAH in lubricating oil do not significantly
affect the PAH emission.

     What can be presently said about the contribution
of automobile exhaust to the general PAH pollution in
our environment? This question certainly can not be
answered as it has been tried in the introduction of
this paper, calculating the benzo(a)pyrene concentration from automobile exhaust profiles and fuel consumption, since temporary and local variations have to be
taken into account. In 1967 a concentration of 160 ng
benzo(a)pyrene/m^3 was measured in the Essen area of
Northrhine-Westphalia (FRG) (14), whereas nowadays the
mean concentration there varies between 1.2-27 ng/m^3
(15); and this holds true for other PAH as well. Hence,
it may be stated that the PAH concentration has decreased to about one tenth within the last 25 years,
possibly due to the changes in heating habits (e.g.
substitution of hard coal by heating oil). The benzo(a)-
pyrene concentration, on the other hand, varies within
a wide range up to more than 300 fold of the concentration of days with low pollution, if the data obtained
from a series of subsequent weeks are compared (fig. 10).
Moreover, the benzo(a)pyrene concentration and even
the PAH-profiles vary considerably from one area to another, which is demonstrated in figure 11, showing the

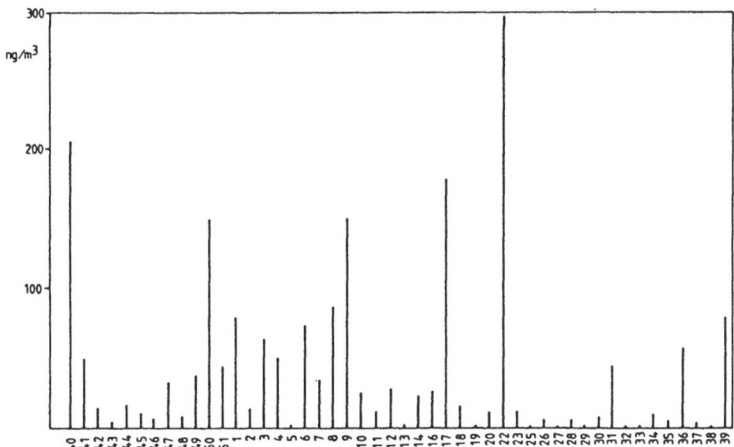

Figure 10. Benzo(a)pyrene concentration of air sampled
from a coal heated area over a period of 51
weeks.

profiles of (I) a residential area with preferential coal heating,
(II) a second one with preferential oil heating, (III) a tunnel
with automobile traffic and (IV) a coke plant area.

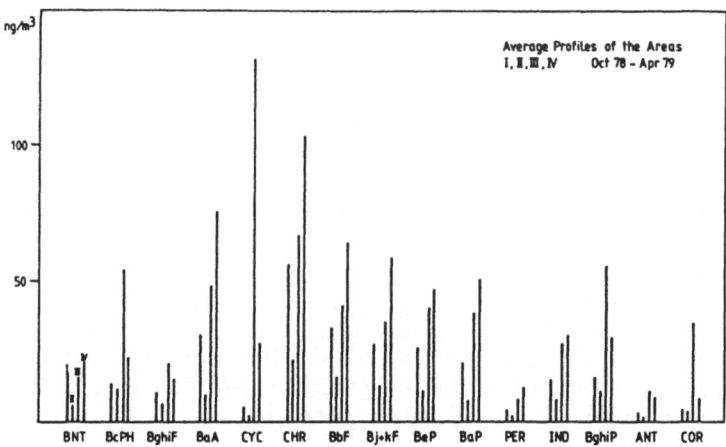

Figure 11. PAH-profiles of air samples in various areas
of Essen (Northrhine-Westphalia)

In some cases the emission source can be readily recognized from the PAH-profile which may be seen from the
'tunnel-profile' which resembles the automobile exhaust
PAH-profile (fig. 12).

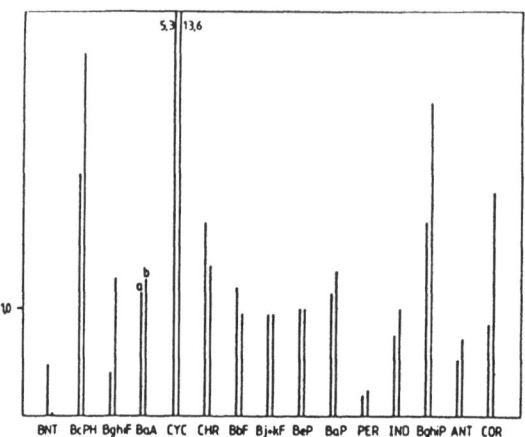

Figure 12. Comparison of the PAH-profiles of (a) air
           in a automobile tunnel and (b) emissions
           of a passenger car on a chassis dynamometer,
           simulating city-traffic (ECE-reglement 15).

It seems promising to extend these investigations to
other PAH emissions such as hard coal, heating oil, and
brown coal combustion to find specific markers for the
various emission sources.

To summarize :
(I) The carcinogenic activity of automobile exhaust gas
almost exclusively originates from the fraction containing 4-7 ring PAH.

(II) Preferentially two parameters influence the mass
concentration and the profiles of PAH in the exhaust :
(a) ratio of gasoline-air mixture, and (b) percentage
of aromatic compounds in the gasoline.

(III) A predominant contribution of automobile exhaust
to the air pollution in certain areas may be recognized from the similarities of the PAH-profiles.

(IV) DIESEL and OTTO exhausts differ significantly in their PAH-profiles. Hence, the balance of effect may look different in both cases. An investigation on the balance of effect of DIESEL exhaust fractions using carcinogen-specific test models seem to be highly desirable and has still to be performed.

References

1. Grimmer, G. and Hildebrandt, A.: 1975, Zbl.Bakt.Hyg., I.Abt.Orig.B161, pp. 104-124.
2. Berichte 1/79 'Luftqualitätskriterien für ausgewählte polyzyklische aromatische Kohlenwasserstoffe' Umweltbundesamt (Ed.), E. Schmidt Verlag, Berlin 1979.
3. Kotin, P., Falk, H.L., and Thomas, M.: 1954, Arch. Ind. Hyg. 9, pp. 164-177.
4. Wynder, E.L. and Hoffmann, D.: 1962, Cancer 15, pp. 103-108.
5. Brune, H.: 1977, IARC Sci. Publ. 16, pp. 41-47, Lyon.
6. Brune, H., Habs, M., and Schmähl, D.: 1978, J. Environ. Path. Toxicol. 1, pp. 737-746.
7. Grimmer, G. and Böhnke, H.: 1978, J. Environ. Path. Toxicol. 1, pp. 661-667.
8. Grimmer, G.: 1977, IARC Sci. Publ. 16, pp. 29-39, Lyon.
9. Grimmer, G.: 'Wirkungsbilanzanalyse' in: Berichte 1/79 'Luftqualitätskriterien für ausgewählte polyzyklische aromatische Kohlenwasserstoffe' Umweltbundesamt (Ed.), E. Schmidt Verlag, Berlin 1979, pp. 241-253.
10. Grimmer, G. and Voigtsberger, P.: 1980, Erdöl und Kohle-Erdgas-Petrochem. 33, p. 126.
11. Meyer, J.P. and Grimmer, G.: 1974, Einflüsse PAH-haltiger und PAH-freier Kraftstoffe auf die Emission von polycyclischen aromatischen Kohlenwasserstoffen eines Fahrzeugs mit Ottomotor im Europa-Test, DGMK-Forschungsbericht 4547, Hamburg
12. Meyer, J.P., Grimmer, G., Misfeld, J., Müller, K., Heidemeyer, P., and Janssen, O.: 1977, Einflüsse der Betriebsart und der Betriebszeit des Motors sowie des Motorenschmieröls auf die Emission von polycyclischen aromatischen Kohlenwasserstoffen (PAH) aus Kraftfahrzeugen mit Ottomotoren; Teil I. BMI-DGMK-Projekt 110, 39 p.
13. Behn, U., Meyer, J.P., and Grimmer, G.: 1980, Erdöl und Kohle-Erdgas-Petrochem. 33, p. 135.

14. Hettche, H.O. and Grimmer, G.: 1968, Schriftenreihe der Landesanstalt für Immissions- und Bodennutzungsschutz des Landes Nordrhein-Westphalen in Essen, Heft 12, pp. 92-108.

15. Grimmer, G., Buck, M., and Ixfeld, H.: 1982, Immissionsmessungen von polycyclischen aromatischen Kohlenwasserstoffen (PAH); Minister für Arbeit, Gesundheit und Soziales des Landes NW (Ed.), Schottedruck, Krefeld, 104 p.

RECENT ADVANCES IN EPA'S MONITORING AND METHODS DEVELOPMENT RESEARCH

Robert H. Jungers

Data Management and Analysis Division
Environmental Monitoring Systems Laboratory
U. S. Environmental Protection Agency (EPA)
Research Triangle Park, NC 27711   U.S.A.

ABSTRACT

Several areas of advanced research related to sampling, analysis, and human exposure assessment of exhaust emissions in ambient air have been developed. These include studies of new methods for volatile organic compounds (VOCs), and the development and application of personal exposure monitors (PEMs) in screening for polynuclear aromatics (PNAs) and carbon monoxide (CO). These new methods for screening PNAs are fast, economical, and accurate. The more expensive and time consuming traditional methods of analysis may be judiciously applied to those samples which the screening methods indicate are high in PNAs. Carbon monoxide, an emission product directly related to automotive emissions, is being monitored using personal exposure monitors in urban scale studies to obtain data on population exposures on a real time basis. Such data may ultimately be used in assessing more accurately human exposure to mobile source and other emissions.

INTRODUCTION

Currently exposure assessment is made utilizing fixed site monitoring and the results are extrapolated to the general population. Personal exposure monitoring is essential in order to better assess exposure of the population to various pollutants. At present a personal exposure monitor for carbon monoxide is available. In order to advance personal monitoring for other pollutants such as volatile organic compounds and polycyclic organic compounds, research is being conducted on solid sorbents

115

D. Rondia et al. (eds.), Mobile Source Emissions Including Polycyclic Organic Species, 115–125.

for improved sampling of VOCs and analytical methodology for
rapid screening of PNAs.

ADVANCES IN SAMPLING: SOLID SORBENTS FOR VOCs

   Although TENAX-GC has been considered the polymer of choice
for air sampling, it has poor adsorptive capability for certain
low molecular weight polar compounds (1), e.g., aldehydes,
nitriles and amines.

   The initial new solid sorbent development was performed at
North Carolina State University (NCSU) under an EPA grant (2).
The sorbent copolyamide, COP III, was developed,. It has a
primary amine and high unsaturation which increases its affinity
for polar volatile organics. These properties also resulted in
COP III adsorbing excessive amounts of water. A second sorbent,
Thermid 600, was reacted with bisphenol A resulting in a polyimide ether with these characteristics: 1) packs well, 2)
thermally stable, 3) low water absorption, and 4) high breakthrough volume. This polar porous polymer is considered useful
for trapping both polar and non-polar compounds.

   The second phase of the solid sorbent development was
conducted at the Research Triangle Institute (RTI) (3) where
many aromatic polyimides were synthesized with the objective of
finding a solid sorbent or sorbents that collect VOCs in ambient
air. Four polyimides met the study objectives and will be
further tested.
       1.   2,6-dichloro-p-phenyline-diamine
       2.   3,3',5,5'-tetra-methyl-benzidene
       3.   bis-4-amino-phenyleneone
       4.   bis-4,4'-diamino-diphenyl-methane

   These four polyimides and the one polyimide ether will be
evaluated by Battelle Columbus Laboratories (4).

ADVANCES IN ANALYTICAL METHODS: SCREENING METHODS FOR PNAs

   Environmental assessment studies require characterization
of PNAs in large numbers of samples over extended periods. Gas
chromatography-mass spectrometry (GC/MS) and high performance
liquid chromatography (HPLC) have demonstrated their capability
to provide specific information for samples containing complex
mixtures of pollutants. The GC/MS and HPLC methods, however,
require sophisticated and expensive instrumentation and elaborate
experimental procedures. Two techniques, synchronous luminescence (SL) and room temperature phosphoresence (RTP), developed
at Oak Ridge National Laboratory have been applied to the work

place atmosphere, but are considered adaptable for use in routine screening of atmospheric samples and personal exposure monitors (PEMs) (5).

The SL method is a new approach in luminescence spectroscopy. Whereas conventional luminescence measurements are conducted using either a fixed excitation wavelength or a fixed emission wavelength, an SL measurement is based on scanning both emission and excitation wavelengths while maintaining a constant wavelength interval. As a result, the emission bands from individual components in a complex mixture become better resolved in an SL spectrum than in conventional luminescence spectra.

The RTP technique is based upon the detection of the phosphorescence emission from organic compounds adsorbed on a filter paper disc impreganated with a heavy-atom chemical mixture. Unlike conventional phosphorimetric procedures, the RTP technique does not require low-temperatures, refrigerants and cryogenic equipment. The main advantages of the RTP technique include simplicity, rapidity, and cost-effectiveness.

The merits of using SL and RTP as screening techniques for analysis of field samples were examined. These two methods were evaluated in conjunction with traditionally accepted methods: thin layer chromatography (TLC), HPLC, and GC/MS (6).

The two field tests used to evaluate SL and RTP were:

1. An exposure assessment in a wood-burning community, Petersville, Alabama, conducted by the Tennessee Valley Authority. This study involved outdoor monitoring with the collection of daily total suspended particulates (TSP) and dichotomous samples at two sites.

2. An exposure assessment in a wood-burning community, Waterbury, Vermont, conducted by Harvard University. This study involves indoor and outdoor monitoring with the collection of respirable particulate samples, TSP and dichotomous samples.

The collected samples from both of the above studies were extracted with dichloromethane (DCM) and sent for PNA analysis to: 1) Oak Ridge National Laboratory for SL and RTP analysis, 2) PEDCO Environmental for GC/MS analysis, and 3) Rockwell International for TLC and HPLC analysis.

Tables 1 and 2 show the relative ranking of total PNA content by SL and RTP. Both SL and RTP, as a rule, were consistent in their ranking of PNAs and generally agreed with

independent conventional PNA analytical methods. These results
suggest that both SL and RTP will provide good screening techniques which may be applicable to future use in PEMs.

TABLE 1.   ORNL RANKING USING SL WITH $\Delta\lambda=3$

PEDCO SAMPLE	358-362	405-406	436	471-473
CG654	23	71	183	42
CG543	7	25	35	12
CG626	5.5	22.5	31.5	23
CG547	5	22	16	6
CG634	4	13.5	20	9
CG573	8	6	6	7
CG656	4	4	4.5	4.5

Relative Weighted Peak Heights in Indicated
Wavelength Range Using A Dilution of 1:10.

PNA CONTENT RANKING

654>543>626>547>634>573>656

TABLE 2.   ORNL RANKING USING RTP EMISSION AND
           EXICATION AT $\lambda=340$ and $\lambda=390$

SAMPLE	$\lambda=340$	$\lambda=390$
CG654	227.5	112
CG543	255	93
CG547	203	71
CG634	117	40
CG626	108	33
CG573	62	20
CG656	32.5	11

Heavy Metal Used — Tl/Pb Mixture (Saturated)
Above Numbers Indicate Average Peak Heights
of Two Determinations.

PNA CONTENT RANKING

654>543>547>634>626>573>656

ADVANCES IN EXPOSURE ASSESSMENT: PERSONAL EXPOSURE MONITORING

Advances in personal monitoring for exposure assessment studies are currently dependent upon the development of reliable analytical techniques and instrumentation applicable to PEMs. Currently PEMs for carbon monoxide are the most highly developed and evaluated. Carbon monoxide is an emission product which can be directly related to lead emissions from automotive combustion. Personal exposure monitors for CO are being used in urban studies to obtain population exposure data on a real time basis. This information can be used in risk assessment.

A study to evaluate the many available carbon monoxide (CO) personal exposure monitors (PEMs) was conducted in Los Angeles utilizing nine participants (7). The results showed that it was possible to survey CO in the ambient atmosphere and that the concentration did not correspond to data reported by fixed sites in that geographical area. A second exposure study was designed to encompass four cities of different geographical and topographical features (8). The four cities were Los Angeles, Phoenix, Denver, and Stamford. This study was conducted in each city during the winter season when CO would be at its highest concentration. General Electric Model CO-I was the PEM selected for this study because of ruggedness and reliability displayed in the Los Angeles study. Lack of agreement was found between the concentration values at the fixed site monitors and the PEMs. It was found that certain activities could be identified by personal exposure concentration. Three basic activities identified were residential, indoor, and commuting. The results of the fixed sites vs. PEMs by hourly average concentration ranges in each of these activities are shown in Figures 1, 2, and 3. In all cases the PEM indicated more exposure to CO in excess of the hourly or daily standard than did the fixed site data.

Based on this evidence an urban scale population exposure program was designed as follows:

1.   To develop a methodology for measuring CO exposures of a representative sample of the population of an urban area, with the intention that the data produced is to be useful for exposure assessment.

2.   To test, evaluate, and validate this methodology by using it in the execution of urban field studies in Denver, Colorado (9, 10) and in Washington, D. C. (9), two urban areas with different ambient pollution levels.

DENVER
GROUP–RESIDENTIAL

FREQUENCY BLOCK CHART

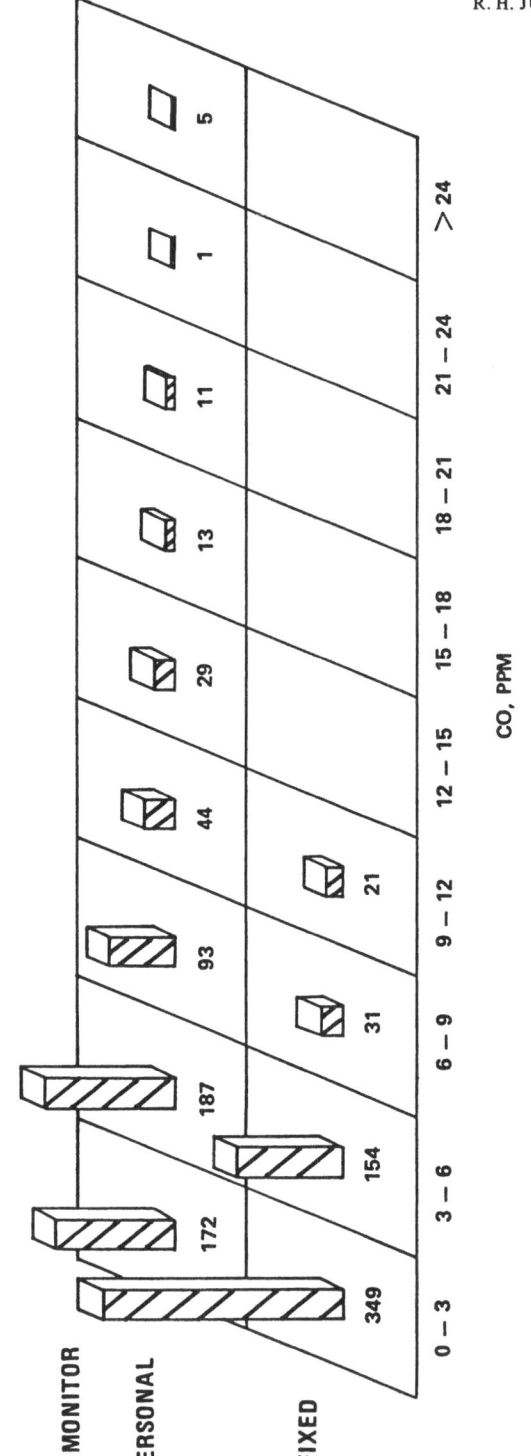

FIGURE 1. COMPARISON OF FIXED SITE AND PERSONAL MONITOR DATA FOR DENVER, COLORADO – AMBIENT DATA

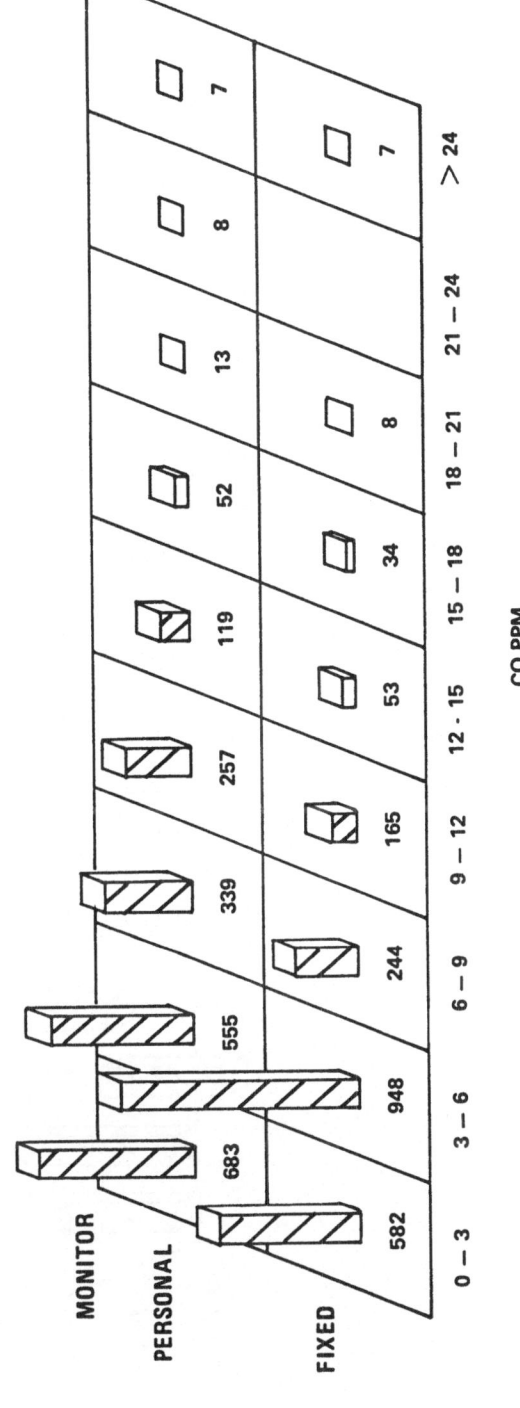

FIGURE 2. COMPARISON OF FIXED SITE AND PERSONAL MONITOR DATA FOR DENVER, COLORADO — INDOOR DATA

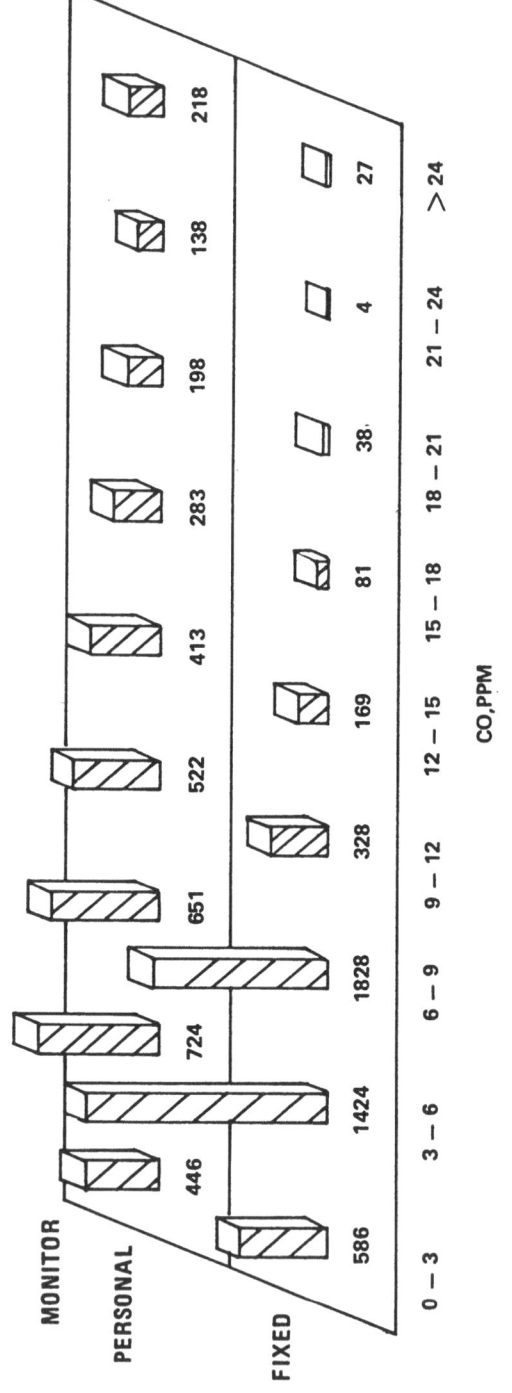

FIGURE 3. COMPARISON OF FIXED SITE AND PERSONAL MONITOR DATA FOR DENVER, COLORADO — COMMUTING DATA

3. To collect data suitable for use in EPA computer
    models of CO exposure and for developing a data base
    for determining pollution-related activities.

4. To collect additional fixed-site data for purposes of
    comparison with personal monitoring as well as with
    the designated fixed site air monitoring data.

The objectives and proposed approach of each urban field
study is listed below:

1. Develop and apply methodology to estimate population
    exposure to CO for an urban area.

2. Obtain exposure estimates and compare to fixed site
    measurements.

3. Add fixed sites.

4. Develop methodology for selecting relevant micro-
    environment.

5. Select and obtain measurements in microenvironments.

6. Develop computational techniques (models).

7. Develop methodology for using PEMs for analysis of
    CO   in breath.

8. Sample during expected highest exposures (winter 82-83).

9. Proposed Approach:

Site	Screen	Select	Sample
Washington, D. C.	10,000	1,000	1 day
Denver, Colorado	5,000	500	2 day

CONCLUSIONS

Recent advances in EPA's monitoring and methods development
research has focused on improving human exposure assessment
through personal exposure monitoring. The availability of
personal carbon monoxide monitors has facilitated EPA's first
large scale urban personal exposure monitoring studies. These
studies will provide a data base for evaluating the utility of
personal monitoring in comparison to fixed site exposure
monitoring. Future studies to evaluate personal exposure to

organic compounds (e.g., VOCs and PNAs) will depend upon the development of sampling and analytical methodologies applicable to the development of PEMs for organics. The PNA screening method employing filter paper impregnated with heavy metals and analysis by RTP or SL may provide such a personal PNA monitor for future assessment of human exposure. Further advances in sampling are required to address the exposure assessment to other organic compounds (e.g., VOCs) found in the ambient air.

REFERENCES

1. Krost, K.J., Pellizzari, E.D., Walburn, S.G., and Hubbard, S.A.: 1980, Environmental Science and Technology, in press.

2. U.S. Environmental Protection Agency, Grant No. CR-807922-01-1: 1980-1981, "Porous Polymer-Adsorption of Volatile Organic Toxicants in Ambient Air. The Synthesis and Evaluation of Aromatic Copolyamides Containing N-Propargyl Groups."

3. U.S. Environmental Protection Agency Contract No. 68-02-3440: 1982, "Preparation and Evaluation of New Sorbents for Environmental Monitoring."

4. U.S. Environmental Protection Agency Contract No. 68-02-3847, Work Assignment 15: 1982, "Evaluation of Solid Sorbents for the Collection of Volatile Organics from Ambient Air."

5. Vo-Dinh, T.: 1982, Second Annual National Symposium on Recent Advances in Measurement of Pollutants in Ambient Air and Stationary Sources, "A Personal or Area Dosimeter for Polynuclear Aromatic Vapors," in press.

6. U.S. Environmental Protection Agency Contract No. 68-02-3496, Work Assignment 21: 1982, "Analysis of Air Samples and Field Test of the Synchronous Luminesence and Room Temperature Phosphoresence Methods."

7. U.S. Environmental Protection Agency Contract No. 68-02-3412: 1981, "Pilot Field Study - Carbon Monoxide Exposure Monitoring in the General Population."

8. U.S. Environmental Protction Agency Contract No. 68-02-3173, 68-02-3168, 68-02-3629, 68-02-3412 (report is compilation of four separate contractual efforts): 1982, "Carbon Monoxide Concentrations in Four U.S. Cities During the Winter of 1981," in press.

9. U.S. Environmental Protection Agency Contract No. 68-02-
    3679: 1982, "Study of Exposure to Carbon Monoxide of
    Residents of Washington, D.C. and Denver, Colorado."

10. U.S. Environmental Protection Agency Contract No. 68-02-
    3755: 1982, "Carbon Monoxide Exposure in Denver Colorado."

STANDARDISATION ASPECTS IN PAH/POM ANALYSIS

W. Karcher

Commission of the EC, JRC Petten Establishment,
Post Box 2, 1755 ZG Petten, The Netherlands.

ABSTRACT

For reproducible analyses and reliable control of polycyclic
aromatic hydrocarbons (PAH) and related heterocyclic compounds,
the availability of standardised analytical procedures is
essential. When national or international regulations are issued
or developed, standardised analytical methods are usually
mandatory.
In the light of such standardisation requirements, the following
aspects are reviewed and discussed:
- sampling and sample treatment
- analytical techniques (including screening methods)
- priority list for PAH/POM analysis.
The capabilities of various analytical methods are reviewed, such
as capillary gas chromatography, high pressure liquid
chromatography, thin layer chromatography and low temperature
luminescence spectrophotometry. A candidate list of high priority
PAH/POM compounds, including some representative heterocyclic
compounds, which is based on occurrence in environmental
emissions, health effects and analytical considerations, is
discussed.
Finally, some activities carried out in support of harmonisation
and standardisation efforts in PAH/POM analysis are described:
- preparation and characterisation of high purity compounds which
  is oriented on priority list for PAH/POM analysis;
- systematic determination and collection of molecular spectra
  under standard conditions of a comprehensive range of PAH/POM
  compounds of environmental significance.

127

D. Rondia et al. (eds.), Mobile Source Emissions Including Polycyclic Organic Species, 127–149.
© *1983 by D. Reidel Publishing Company.*

1.   INTRODUCTION

     In order to permit the consistent assessment of the health
effects of environmental pollutants with wide ranging
distributions, and to ensure reliable analytical control,the
harmonisation or standardisation of sampling and analytical
procedures is desirable. Where national or international
regulations are in force or being developed, the availability of
accepted standard methods is usually mandatory.

     Guidelines and recommendations for this purpose are
described in an ISO Standard (35). The analytical procedures
chosen are usually tested in a round-robin exercise to determine
repeatability and reproducibility criteria before final approval
and adoption as a standard method.

     Traditionally, standardised methods are based on techniques
which are widely available and easy to handle for routine
applications, which usually excludes advanced techniques
requiring expensive equipment or special experience. In
addition, standardised or recommended methods of analysis are
often supported by the development of reference materials and
related data for calibration and intercomparison purposes.

     Polycyclic aromatic hydrocarbons (PAH) and their
heterocyclic homologues (the term polycyclic organic matter - POM
- was recently introduced to classify both compound types) have
attracted much environmental interest in the last decade due to
their ubiquity as by-products of almost all combustion processes
of organic material and in view of their mutagenic and/or
carcinogenic activity. An international standard, which is based
on the determination of the sum of six PAH compounds by Thin
Layer Chromatography (TLC) (also the subject of an EEC directive
(12)), was adopted in 1971 for the quality control of drinking
water on the initiative of the World Health Organisation (70).

     In PAH/POM analysis, harmonisation and standardisation
efforts are complicated by the large variety of emissions and
matrices containing a broad and variable spectrum of PAH and
heterocyclic compounds. This frequently requires the development
and application of specific sampling or extraction techniques for
a particular emission source.

     In this context, a general consensus on a concise list of
PAH and POM compounds to be reported in all emissions might be
seen as a first step towards a more general approach. In turn,
this would allow the restricted number of compounds chosen to be
made available in certified purity for calibration and reference
purposes.

It is the aim of this paper to review harmonisation and
standardisation aspects in the various stages of PAH/POM
analysis, and to describe briefly some activities carried out in
the frame of R and D programmes of the Commission of the EC in
support of harmonisation and standardisation efforts.

2.   ANALYTICAL PROCEDURES

To ensure full comparability of results, all the various
steps which are involved in the analytical procedures (sampling,
sample extraction, enrichment, clean-up, analytical method and
data presentation) should be considered in harmonisation and
standardisation discussions.

2.1   Sampling

Due to the low PAH/POM levels which are encountered in most
environmental matrices, sampling occupies a critical place   since
errors and losses incurred in this initial step of PAH analysis
will influence all later analytical results.

An overview of sampling methods for PAH analysis from mobile
and stationary sources was given recently by Stenberg (62). A
review of recommended sampling techniques for various sources is
given in Table 1.

As can be seen, in the USA, a recommended sampling method
was proposed for PAH-analysis from car exhausts in 1971 by Gross
et al. (66). However, it was soon recognised that losses of the
low molecular compounds can easily occur in sampling at low flow
rates over prolonged periods (56). In Europe, sampling methods
for automobile exhausts were described by Grimmer (21) and Shabad
(59). The earlier methods relied largely on the total collection
of exhaust after a cooling step, on glass-fibre, silica gel or
cellulose filters. To avoid artifacts and to simulate realistic
conditions in sampling procedures, dilution techniques were
introduced for PAH-collection from automobile exhaust gases. For
sampling PAH from air particulate matter, high-volume filtration
techniques were developed and standardised methods were
recommended by WHO (69) and EPA (15) which were updated
repeatedly to improve retention of gaseous PAH-species. However,
most of the updated sampling arrangements have not been tested
and validated for aspects of effiency, repeatability and
reproducibility.

Optimisation and standardisation of sampling procedures is
complicated by the number of experimental parameters which must
be considered; such as filter type and temperatures, dimensions,
filter weighing conditions (temperatures and relative

TABLE 1 Review of Standard or Recommended Procedures in PAH Analysis

SOURCE	Car Exhaust	Lubricating oil/fuels	Air particulates/ stack emissions	Water	Soils/sewage sludge	Food
SAMPLING	EPA (66) Gross Europa tests (21) CEC method for diesel emissions (under development)		EPA Method 5 (15) ASTM (2) BS (4)	EPA Method 610 625 1625		AOAC (1)
ANALYSIS	Method 2 IARC (37)	Method 3 IARC (37)	Methods 6 and 8 IARC (37)	EPA Method 610 (13) 625 (14) and 1625 (9) Method 1 IARC (37)	Method 4 IARC (37)	Methods 4/5 IARC (37) AOAC (1)
			(Method 7 (IARC))	(LU – all emissions))		
TECHNIQUE	GC^2	GC^2	TLC and LU	GC/HPLC and GC/MS (isotope dilution) (EPA) TLC (IARC)	GC^2	TLC (AOAC/IARC) GC^2 (IARC Method 4)
DATA REPORTING				WHO (56) 6 PAH EPA 16 PAH		
REFERENCE MATERIALS			NBS (34)	NBS (34) NBS 16 EPA prior. poll.		

EEC–BCR (21 available/19 in preparation – see table 8)

humidity), particle size fractions to be collected, flow rates, collecting time and efficiency. Other factors which can have a pronounced effect on the repeatability and reproducibility of sampling are the chemical reactivity and stability of PAH during collection and the distribution of PAH between particulates and gas phase.

Recently, two criteria were suggested for assessing the repeatability and reproducibility of sampling arrangements (32):
a) Identical flow rates should result in identical mass
   fractions and POM profiles,
b) Under constant flow rates, POM profiles collected should
   be independent of collecting period.
At present, a standardised sampling method for diesel emissions is being discussed in a working group of CEC. Preferably, candidate sampling methods, including suitable extraction, enrichment and clean-up procedures, should be extensively tested and validated for collection and recovery efficiency and checked for repeatability and reproducibility in collaborative tests before acceptance, where necessary using ^{14}C-labelled PAH-tracers for recovery and stability trials.

2.2   Extraction

A wide range of techniques and solvents has been used for PAH extraction (see tables 2 and 3). In most cases, acetone, cyclohexane and benzene with some reservations (see table 3) are reported to show high extraction efficiencies (61).

For some specific purposes other solvents are preferred, as for example methanol for extracting O-containing POM from particulates and tetrahydrofuran (THF) or n-pentane for POM extraction from Tenax filters or polyurethane plugs (table 3). In view of toxicological considerations, cyclohexane was recommended as extractant both by WHO (69) and ISC (Interscience Committee) (57).   Apart from Soxhlet extraction, ultrasonic vibration is also being increasingly used as it seems to afford higher recoveries (64). This technique is recommended for the determination of coal tar volatiles by NIOSH (54). A short list of recommended solvents could conceivably be established on the basis of satisfactory and reproducible extraction efficiencies of the entire PAH profile rather than single components, and used in combination with suitable enrichment or clean-up procedures.

2.3   Analytical Methods

A number of recommended methods for PAH analysis are described in the IARC/WHO manual (37), occasionally including specifications for sampling and sample pretreatments (see Table 1). These comprise: TLC methods for water and food analysis and

TABLE 2   PAH Extraction Procedures for Various Matrices

MATRIX	SOLVENT	TECHNIQUE
Lubricating, mineral, vegetable oils, fats	Cyclohexane	Solvent partition
Water	Cyclohexane	Liquid- liquid partition
Foods (meat, fish, etc)	Cyclohexane	Saponification/extraction
Air particulates, car exhaust and sediments, soils	Acetone, xylene	Soxhlet extraction, ultrasonic vibration, thermal methods
Carbonaceous materials	Boiling xylene, toluene	

TABLE 3   Recoveries in Solvent Extraction of PAH from Air Particulates and filter stripping.

PAH	MATRIX	SOLVENT	RECOVERY (%)	REFERENCE
Pyrene	Air particulates	Benzene	96	(7)
	Air sampling (Polyureth. plug)	THF	97.9	(46)
	Air sampling (Tenax GC)	n-Pentane	91-104	(63)
Benzo(a) pyrene	Air particulates	Cyclo-hexane	98.2	(19)
Benzo(a) pyrene	Air particulates	Benzene	75	(20)
Benzo(a) pyrene	Fly ash	Benzene	25	(20)
Benzo(ghi) perylene	Air particulates	Benzene	39	(7)
Benzo(ghi) perylene	Air samples	n-Pentane	101-106	(63)

for PAH analysis in air particulates; GC methods for car exhaust
condensates, for lubricating and cutting oils and fuels, for
foods, plants, soils and sewage sludge; and a low temperature
luminescence method for the determination of benzo(a)pyrene,
which can be applied in combination with suitable extraction and
enrichment procedures to practically all matrices containing
traces of PAH. Precision, repeatability and reproducibility
criteria in the analysis of PAH have been tested in several
analytical campaigns, mainly for car emissions and food analysis
(see Table 4). In most cases, GC was chosen as the analytical
method. In trace analysis, the following criteria are of
importance in selecting and validating recommended methods of
analysis:

- Recovery (>80% at level >100ppb)
  (>60% at level <100ppb)
- Detection limit
- Coefficient of variation
- Rate of outliers (5-15%)
- Number of laboratories for collaborative studies (5)
- General availability of methodology and equipment.

Originally, when the need for harmonisation and
standardisation of analytical methods in response to
international regulations arose, TLC was the preferred choice
since GC^2 and HPLC were neither fully developed nor universally
available.

As GC^2 techniques became more refined and were generally
accepted, due to the superior separation and resolution potential
of modern capillary columns, they took preference, especially in
the analysis of PAH in car exhaust emissions and air
particulates. In combination with an FID-detector, which, in
contrast to the UV-detectors used frequently in HPLC analysis,
has a nearly uniform response factor for hydrocarbons, or coupled
to mass spectrometry, this technique must now be considered the
method of first choice for a reliable and reproducible
determination of PAH traces in a wide range of matrices. The
introduction of more sensitive (photo-ionisation detector - PID)
and specific detectors, such as the nitrogen-phosphorus (NPD)
(48) and sulphur-phosphorus (FPD) detectors, has extended the
application range of this technique further to the determination
of heterocyclic derivatives (55).

For some matrices, e.g.those in water and food analysis,
HPLC methods are gaining gradually at the expense of TLC
in recommended procedures. Thus, EPA recommends an alternative
HPLC method for waste water analysis of priority PAH pollutants
(13) and HPLC methods have been developed recently by FDA for
various foods (33, 39). This development reflects the adaptation

TABLE 4   Round-robin Analyses of PAH and Heterocyclic Compounds in Various Matrices

MATRIX	ANALYTICAL METHOD	NUMBER OF PAH DETERMINED	NUMBER OF DETERMINATIONS	ORGANISATION	COEFFICIENT VARIANCE (%)	REFERENCE
Car Exhaust	GC^2	10	120	DGMK	7–24	(22)
Lubricating Oil	GC^2	8	30	DGMK	11.7–36	(24)
Used Motor Oil	GC^2	11	30	DGMK	3–7.5	(24)
Heating Oil	GC^2	8	30	DGMK	11–36	(38)
Food (minced meat)	UV/Fluorescence	4	20	IUPAC	5–14	(34)
Food (smoked ham)	UV/Fluorescence	4	36	IUPAC/AOAC	7–13	(34)
Food (minced meat)	GC^2	8	12	IUPAC	7–27	(23)
Food (sunflower oil)	GC^2	7/8	13	IUPAC	9–25	(23)
Food (minced ham)	GC^2	4 (N-heterocycles)	17	IUPAC (in progess)		

of existing methods to technical progress, an aspect which should
not be overlooked in standardisation efforts. In some cases, an
alternative approach has been taken by accepting a TLC method for
screening purposes in combination with quantitative analysis by
GC^2.

If a restricted number of components is to be determined,
the separation potential of currently available HPLC columns
is normally sufficient and the technique affords greater speed
and automation facilities in comparison with TLC. For the
analysis of PAH and heterocyclics with molecular masses above
350, HPLC is a first choice as decomposition or column retention
may be significant in GC for these heavier compounds. Thus, HPLC
methods play an important role in the characterisation of
residues and effluents from coal conversion technologies.

In addition, a number of advanced analytical techniques,
such as low temperature luminescence spectroscopy, tandem mass
spectrometry (MS/MS), Fourier-Transform IR-spectroscopy and
nuclear magnetic resonance spectrometry have been successfully
applied to PAH analysis. For instance, low temperature
luminescence spectrometry, sometimes in combination with laser
excitation, was used for the analysis of PAH in various matrices
without prior separation, which is attractive especially for
screening or finger-printing purposes (10, 73). However, a wider
application for routine analysis is at present inhibited by the
limited availability of the required equipment. The same remark
applies to tandem mass spectrometry, FT-IR spectroscopy and NMR.
All three techniques, however, are increasingly used for the
detection and identification of novel PAH species and derivatives
and efforts are continuing towards coupling IR and NMR as
detectors to GC and HPLC (74) respectively.
Thus, a broad range of sometimes complementary analytical
techniques is available at present for the characterisation of
the various PAH/POM emissions. For standardisation purposes,
candidate methods must be tested extensively in a collaborative
exercise to determine and evaluate repeatability, reproducibility
and recovery criteria before final definition and approval.
(Recently, the method detection limit, defined as the
concentration which can be detected at a specific confidence
level, was proposed as one criterion for assessing the
performance of an analytical method (18)).

3.   HARMONISATION OF DATA REPORTING

In reviewing the range of individual PAH and
heterocyclic compounds which are reported for identical
emissions, considerable variations are apparent. As mentioned
above, a short list of six individual PAH-compounds was

introduced for the quality control of drinking water (70) and in
1976 16 PAH were defined as primary pollutants in waste water by
EPA (13). More recently, IARC has included 37 PAH, 6
N-heterocyclic compounds and 6 Nitro-derivatives in an updated
list for the evaluation of the carcinogenic risk of chemicals to
man (72).

In general, most researchers rely on reporting between 15
and 30 PAH-compounds for characterising specific emissions. In
order to facilitate intercomparison of results between the large
number of laboratories which are engaged in the analysis of PAH
and POM emissions and to permit a comprehensive assessment and
evaluation of analytical data in terms of environmental,
occupational and health impact, a concensus on a minimum number
of representative compounds to be reported in all PAH emissions
is highly desirable.

In fact, a restricted number of PAH tends to occur in
practically all matrices which are analysed for PAH. To evaluate
their environmental and health significance, rankings can be
allocated for range and concentration levels of occurrence, for
carcinogenic and mutagenic effects, for inclusion in existing
regulations and specific analytical considerations (internal
standards or tracers/markers) (see Table 6).

Therefore, it is proposed to identify between 15-20
PAH-compounds (see Table 5) which, on account of their widespread
occurrence (including concentration aspects) (27-31, 41, 47, 50,
58, 60) and their mutagenic or carcinogenic potential (11, 17)
are of primary importance for assessing the significance of PAH
emissions.

For the various emissions which are regularly controlled for
PAH, this general list could then be complemented by a few
compounds which are specific for individual sources.
Thus, 1-nitropyrene and 1,8-dinitropyrene may be included as
examples for car exhaust emissions; 5-methylchrysene,
1-nitropyrene and 10-azabenzo(a)pyrene for air particulates;
benz(c)acridine, dibenzo(c,g)carbazole, dibenzo(a,h)pyrene and
some representative aromatic amines for coal conversion
processes.

4.   REFERENCE MATERIALS AND SPECTRAL DATA

There are many applications of certified reference materials
and analytical reference data for PAH (see Table 7). In order to
facilitate the analysis and control of PAH and related
heterocyclics in environmental emissions and occupational
hazards, and in support of harmonisation procedures, the

Commission of the EC initiated two complementary R & D projects
with the following aims:
- to develop and certify high purity compounds for the
  calibration of analytical methods and apparatus,
- to determine and collect systematically various molecular
  spectra of PAH in the form of a "Spectral Atlas".

TABLE 5

Tentative List for General Reporting of PAH/POM Emissions

		Mass
1.	Fluoranthene	178
2.	Pyrene	202
3.	Cyclopenta(cd)pyrene	226
4.	Benz(a)anthracene	228
5.	Chrysene/Triphenylene	228
6.	Benzo(c)phenanthrene	228
7.	Benzo(b)naphtho(2,1-d)thiophene	234
8.	Benzo(b)fluoranthene/Benzo(j)fluoranthene	252
9.	Benzo(k)fluoranthene	252
10.	Benzo(a)pyrene	252
11.	Benzo(e)pyrene	252
12.	Anthanthrene	276
13.	Indeno(1,2,3-cd)pyrene	276
14.	Benzo(ghi)perylene	276
15.	Dibenz(a,h)anthracene	278
16.	Coronene	300

TABLE 6   Tentative Scoring System for Selecting High Priority PAH/POM Compounds

Compound (Scale)	Occurrence (0-10)	Biological Activity (0-10)	Subject of Regulations (0-5)	Analytical Standard or Marker (0-5)	Score
Phenanthrene	8	0	3	0	11
Cyclopenta(cd)pyrene	6	7	0	4	17
Chrysene	10	5	3	1	19
1-Nitropyrene	4	5	0	0	9
Benzo(a)pyrene	10	10	5	0	25
Dibenz(a,h)anthracene	8	9	3	0	20

TABLE 7   Field of Application for PAH Certified Reference Materials

SUBJECT	APPLICATION	REGULATIONS
Car industry	Health assessment of PAH emissions in car exhaust	Under consideration in USA
Ambient air	Control of PAH levels in air	Recommended level for BaP in FRG (65)
Coal combustion	Monitoring of PAH emissions to ambient air	(6)
Surface waters	Quality control of drinking water	EEC Directive 75/440 (12) WHO-Standard (70) EPA Quality Criteria for water (13)
Occupational exposure	Control of exposure limits	US-limit for coke oven emissions (67) TLV for coal conversion (6)
Foods	Quality control	Directive on chewing gum in FRG (68) Directive on smoked meat in FRG (16)
Sediments, marine environments	Control of PAH levels in food chains	
Biological activity	Testing for mutagenicity, carcinogenicity and synergistic effects	

4.1   High Purity PAH Compounds

The analysis of the many PAH and heterocyclic compounds
which are found in air, water, soils and food and in specific
emissions is often handicapped by the unavailability of the
species to be identified and quantified. Therefore, it was
decided to make the substances of primary importance available in
sufficient purity (better than 99%). The selection of compounds
to be prepared and certified was oriented mainly by the following
criteria:
- occurrence
- mutagenic/carcinogenic activity
- availability
- national and international regulations.

When a consensus can be obtained on an international level
on a restricted number of PAH and heterocyclic compounds for
general analysis, as was proposed in the previous section, future
work on PAH reference materials should consider these substances
as a first priority. (For PAH reference materials available or in
preparation, see Table 8.)

In order to provide sufficient material for a wide
distribution over an extended period, a batch size of between 5g
and 50g is used. Materials are prepared mostly by multi-stage
synthesis sometimes using novel synthetic routes (64). In a few
cases, purification of commercial materials was preferred for
reasons of economy. Quantitative analysis, including
determination of impurities, is carried out in the frame of a
collaborative exercise, involving a number of laboratories in the
EC.   Purity certification is based on the statistical evaluation
of at least 60 individual analytical results (44, 45). Although
the objective of the analytical campaign is the determination of
impurities in a PAH-matrix, which differs markedly from the trace
analysis of PAH in foreign matrices, GC² and HPLC again proved to
yield the most reliable and reproducible results. (Differential
scanning calorimetry is used for homogeneity and long term
stability control.)

Of the 16 PAH included in the list of EPA priority
pollutants, solutions in acetonitrile at certified concentrations
have become available through the National Bureau of Standards
(34). Also available is an urban dust and a shale oil (51) both
containing certified levels of 5 PAH (see also (52)).

4.2   Molecular Spectra

Although the availability of calibration materials of known
purity is highly desirable for the analysis of PAH in complex
environmental matrices, such a comprehensive enterprise is not

TABLE 8   Certified Reference Materials for Analysis of Polycyclic
           Aromatic Hydrocarbons and Heterocyclic Compounds

Compound Available	Certified Purity
Anthanthrene	0.995 ± 0.003
10-Azabenzo(a)pyrene	0.995 ± 0.004
Benzo(b)chrysene	0.995 ± 0.003
Benzo(a)fluoranthene	0.995 ± 0.003
Benzo(b)fluoranthene	0.995 ± 0.003
Benzo(j)fluoranthene	0.995 ± 0.003
Benzo(k)fluoranthene	0.995 ± 0.003
Benzo(ghi)perylene	0.990 ± 0.004
Benzo(a)pyrene	0.993 ± 0.004
Benzo(e)pyrene	0.990 ± 0.005
Dibenz(a,c)anthracene	0.995 ± 0.003
Dibenz(a,j)anthracene	0.997 ± 0.003
Dibenzo(a,l)pyrene	0.996 ± 0.003
Indeno(1,2,3-cd)pyrene	0.990 ± 0.005
1-Methylchrysene	0.990 ± 0.003
2-Methylchrysene	0.992 ± 0.003
3-Methylchrysene	0.992 ± 0.003
4-Methylchrysene	0.992 ± 0.003
5-Methylchrysene	0.995 ± 0.003
6-Methylchrysene	0.998 ± 0.002
1-Methylbenz(a)anthracene	0.993 ± 0.004

In Preparation

Benz(a)acridine	Dibenz(a,c)acridine
Benz(c)acridine	Dibenz(a,h)acridine
Benzo(c)chrysene	Dibenz(a,i)acridine
Benzo(ghi)fluoranthene	Dibenz(a,j)acridine
Benzo(b)naphtho(2,1-d)thiophene	Dibenz(c,h)acridine
Benzo(b)naphtho(2,3-d)thiophene	Dibenz(a,h)anthracene
Benzo(b)naphtho(1,2-d)thiophene	Dibenzo(a,e)pyrene
Benzo(c)phenanthrene	Dibenzo(a,h)pyrene
Cyclopenta(cd)pyrene	Fluoranthene
	1-Nitropyrene

Orders for reference materials and requests for information
should be addressed to: Community Bureau of Reference - BCR,
Directorate General XII, Commission of the European Communities,
200 rue de la Loi, B-1049 Brussels.

feasible for all of the large number of compounds which occur in environmental emissions and occupational exposure. A more realistic and economic approach may consist of a comprehensive collection of the molecular spectra of these compounds. This information would be of considerable value to the analyst in the identification and quantitative determination of PAH and heterocyclic compounds. For instance, the UV and fluorescence spectra are important for HPLC analysis. Also a wider application of low temperature luminescence techniques depends to some extent on the availability of good reference spectra. Some PAH compounds do not exhibit quasi-linear luminescence spectra at low temperature and thus the spectrum of a major impurity may be mistaken for the main constituent. The utility of mass spectra in PAH analysis, especially for the identification of newly detected species, is evident and a broader availability of IR spectra may further the application of FT-IR analysis.

Finally, NMR spectra often offer the only means of unambiguous identification and distinction of closely related isomers as illustrated in Fig. 1 which shows the differentiation of two closely related isomers of S-containing homologues of benzo(a)pyrene (42).

For the determination of the spectra, standardised conditions are applied wherever possible. Thus, UV and fluorescence spectra are recorded in standard solutions (same solvent, approximately constant concentration levels) to avoid the considerable variation of experimental conditions and equipment which has been observed in the corresponding literature. The following molecular spectra will be included in the projected "Spectral Atlas" (43):

- mass spectra (as determined by both quadrupolar and magnetic instruments)
- UV-, IR- and fluorescence spectra (including some low temperature luminescence spectra)
- NMR-spectra (both ^{1}H- and ^{13}C).

Spectra are accompanied by tables, giving molar extinction values and wavelength of absorption peaks, relative fluorescence intensities and wavelengths, relative mass fragment abundances, chemical shifts and coupling constants etc. together with essential details of experimental conditions and purity or melting point of the particular substance (for an example, see Fig.2).

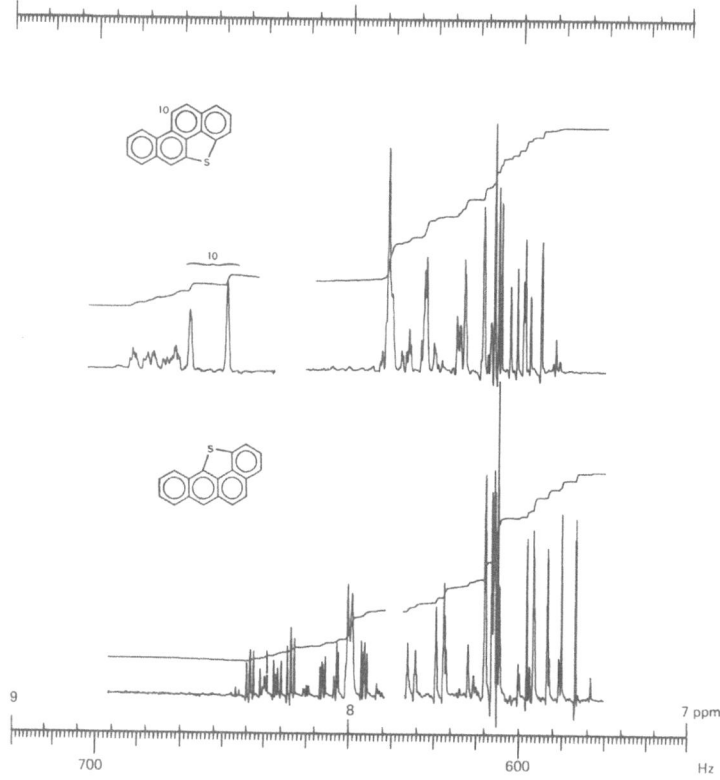

Fig. 1   H^1-NMR spectra of chryseno(4,5-bcd)thiophene
and benzo(2,3)phenanthro(4,5-bcd)thiophene in
CDCl$_3$

5. CONCLUSIONS

The need to harmonise and standardise analytical procedures
in order to allow the reliable assessment of health effects and
to permit intercomparison of PAH and POM determinations at an
international level is well recognised. However, standardisation
efforts in PAH analysis are complicated by the significant
variety of matrices which regularly have to be analysed and
monitored. Also, in the past, attention was sometimes
concentrated only on particular analytical stages, neglecting
some other steps which are essential elements of the procedures
(e.g. sampling, extraction, enrichment and clean-up techniques).
In addition, a general agreement on the number and type of PAH
and related compounds which should be reported in all emissions,
in order to permit a comprehensive evaluation of their health
impact, seems to be lacking. In this context, a consensus on a

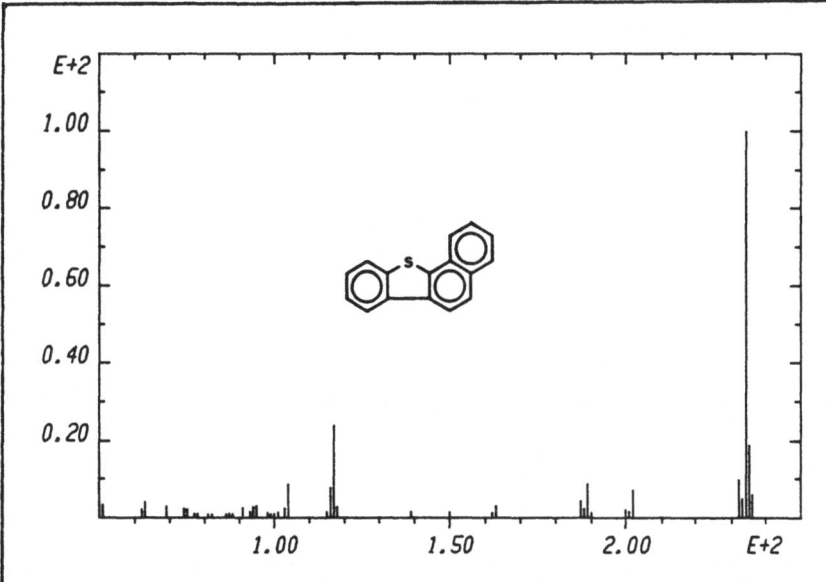

m/z	relative abundance	m/z	relative abundance
51	3.5	118	2.9
62	2.4	163	3.3
63	4.3	187	4.6
69	3.1	188	2.5
74	2.5	189	8.7
75	2.4	200	2.2
91	2.8	202	7.3
94	2.9	232	9.8
95	3.3	233	5
103	2.6	234	100
104	8.7	235	18.9
116	7.9	236	6.2
117	24		

Mass spectrum   Formula : $C_{16}H_{10}S$
Instrument : Ribermag R10-10C   Mol. wt.: 234.3
Inlet system : GC/MS
Source temperature : 200°C
Electron voltage : 70eV

JRC
Petten

Benzo(b)naphto(2,1-d)-
thiophene

short list of PAH and heterocyclic compounds which should be the
basis for analysing and characterising PAH/POM emissions is
highly recommended, to be complemented by a few representative
species for specific emissions. A more general approach to the
harmonisation of PAH analysis may be seen in the elaboration,
testing and validation of source-specific sampling methods which,
in combination with suitably tested and evaluated extraction,
enrichment or clean-up methods, may be coupled to standardised GC
or HPLC methods. Analytical procedures and the priority list of
PAH, including heterocyclic and related derivatives, should be
reviewed at regular intervals in order to take into account new
discoveries and developments. In turn, this regular updating
would facilitate the development and preparation of the
analytical support (reference materials and spectral data) which
are often an essential element in harmonisation and
standardisation procedures.

Finally, as the ultimate objective of environmental and
occupational PAH analysis and monitoring is the interpretation
and assessment of their health impact, it might be desirable to
extend harmonisation and standardisation efforts into the field
of biochemical characterisation of PAH/POM emissions.

REFERENCES

1. Association of Official Analytical Chemists, Official
    Methods of Analysis of the AOAC, 12th edition, (1975).

2. ASTM Annual Book of Standards: Standard Methods of Test for
    Sampling Stacks for Particulate Matter, D2928-71, Part 23,
    Am. Soc. Testing Mat., (1973).

3. ASTM Annual Book of Standards, Part 31, Water, (1975).

4. British Standard Simplified Methods for Measurement of Grit
    and Dust Emissions (Metric Units), B.S. 3405: 1971.

5. P. Brookes, Mutation Research, 39, pp 257-284, (1977).

6. N.E. Bolton, C.L. Hunt, T.A. Lincoln and W.E. Porter,
    Occ. Health and Safety, March/April 1977, p 30.

7. W. Cautreels and K. Van Cauwenberghe, Water Air Soil
    Pollution, 6, p 103, (1976).

8. W. Cautreels and K. Van Cauwenberghe, Atmos. Environ., 12,
    p 1133, (1978).

9. B.N. Colby, R.G. Beimer, D.R. Rushwick and W.A. Telland,
    Anal. Chem., in press.

10. A.L. Colmsjö and C.E. Östman, Anal. Chem., 52, pp
    2093-2095, (1980).

11. A. Dipple, Polynuclear Aromatic Carcinogens.
    In:"Chemical Carcinogens" (C.E. Searle, ed.) ACS Monograph
    173. Am. Chem. Soc. Washington DC, p 245, (1976).

12. EEC Council Directive 75/440/EEC, O.J. L 194, 25 July, 1975.

13. EPA: Quality criteria for water, US Fed. Reg. 3 Dec. 1979,
    pp 69464-69 for Survey of Industrial Effluents for Priority
    Pollutants Sampling and Analysis Procedures (rev. April
    1978)

14. EPA, EMSL, Method 625: Base/Neutrals, Acids and Pesticides,
    Cincinnati, Ohio, 1979

15. Fed. Reg. 36, No. 247, 24878 (1977)

16. Fleischverordnung, 4/7/1978, Bundesgesetzblatt I. S.1003.

17. H.V. Gelboin and P.O.P Ts'o in "Polycyclic Hydrocarbons and
    Cancer", Acad. Press, New York, p 110, (1978).

18. J.A. Glaser, D.L. Foerst, G.D. McKee, S.A. Quave and W.L.
    Budde, Environmental Scientific Technology, 15, pp
    1426-1435, (1981).

19. C. Golden and E. Sawicki, International J. Environ. Anal.
    Chem., 4, pp 9-23, (1975).

20. W.H. Griest, L.B. Yeatts Jr. and J.E. Caton, Anal. Chem.
    52, pp 199-201, (1980).

21. G. Grimmer, A. Hildebrandt, and M. Böhnke, Erdöl, Kohle,
    Erdgas, Petrochem., 25, pp 442-531, (1972).

22. G. Grimmer, A. Hildebrandt and H. Böhnke, Zentralbl.
    Bakterrol. Hyg. Abt. I (Orig. B), 158 , pp 35-49, (1973).

23. G. Grimmer and H. Böhnke, J. Assoc. Off. Anal. Chem.,
    58, pp 725-733, (1975).

24. G. Grimmer and H Böhnke, Chromatographia, 9, pp
    30-40, (1976).

25. G. Grimmer, H. Böhnke and A. Glaser, Zbl. Bakt. Hyg.,
     I. Abt. Orig. B 164, p 218, (1977).

26. G. Grimmer, H. Böhnke and A. Glaser, Erdöl, Kohle, Erdgas
     und Petrochem., 30, pp 411-417, (1977).

27. G. Grimmer and P. Voigtsberger, Erdöl und Kohle, Erdgas
     Petrochem. ver. mit Brennstoffchem. 33, p 226, (1980).

28. G. Grimmer, J. Jacob, K.W. Naujack, and G. Dettbarn,
     Fres. Z. Anal. Chem., 309, pp 13-19, (1981).

29. G. Grimmer, G. Dettbarn and D. Schneider, Z. Wasser
     Abwasser Forsch 14, Nr. 3, pp 100-106, (1981).

30. G. Grimmer, J. Jacob, and K.W. Naujack, Fres. Z.
     Anal. Chem., 306, pp 347-355, (1981).

31. G. Grimmer, K.W. Naujack and D. Schneider, Fres. Z.
     Anal. Chem., 311, pp 475-484, (1982).

32. G. Grimmer, OECD workshop on PAH in air, Paris, October
     1981.

33. P. Hanus, H. Guerrero, R. Biehl and T. Kenner, J. Assoc.
     Off. Anal. Chem., 62, No. 1, pp 29-35, (1979).

34. J. Howard, J. Assoc. Off. Anal. Chem., 61, pp 122-129,
     (1968).

35. ISO Draft International Standard DIS 78/II (1974) Geneva.

36. IARC/WHO: Environmental Carcinogens - "Selected Methods of
     Analysis", Vol. 3, IARC Publications, No. 29, Lyon, (1979).

37. IARC: Monograph of carcinogenic risk of the chemical to man,
     Vol. 3 "Certain polycyclic aromatic hydrocarbons and
     heterocyclic compounds", International Agency for
     Research on Cancer, Lyon, (1973).

38. O. Janssen, Erdöl, Kohle, Erdgas, 28, pp 624-638, (1975).

39. L. Joe, Jr., L. Roseboro and T. Fazio, J. Assoc. Off. Anal.
     Chem., 64, pp 641-646, (1981).

40. J. Josephson, Environ. Sci. Technology, 15, pp 1408-
     1412, (1981).

41. G.A. Junk and C.S. Ford, Chemosphere 9, pp 187-230, (1980).

42. W. Karcher, A. Nelen, R. Depaus, J. van Eijk, P. Glaude, and
     J. Jacob in "Polycyclic Aromatic Hydrocarbons Chemical
     Analysis and Biological Fate" (M. Cooke and A.J. Dennis,
     eds.) Battelle-Press, Columbus, Ohio, p 317, (1981).

43. W. Karcher, R.J. Fordham, A. Nelen, R. Depaus, J. Dubois and
     P. Glaude in "Polynuclear Aromatic Hydrocarbons, Chemical
     Analysis and Biological Fate" (M. Cooke and A.J. Dennis,
     eds.) Battelle-Press. Columbus, Ohio, (1982).

44. W. Karcher, J. Jacob, and L. Haemers, EUR-report
     6967, (1980).

45. W. Karcher, J. Jacob, and R. Fordham, EUR-report
     7175 (1981), and EUR-report 7812 (1982).

46. A.M. Krstulovic, D.M. Rosic and P.R. Brown, Anal. Chem., 48,
     p 1348, (1976).

47. R.C. Lao, R.S. Thomas, H. Oja, and L. Dubois, Anal. Chem.,
     45, p 908, (1973).

48. C.J. Least, G.F. Johnson and H. M Solomon, Clin. Chem., 22,
     p 765, (1976).

49. M.L. Lee, M. Novotny and K.D. Bartle, Anal. Chem., 48, p
     1566, (1976).

50. M.L. Lee, M. Novotny, and K.D. Bartle, "Analytical
     Chemistry of Polycyclic Aromatic Compounds" Academic Press,
     New York, p.448, (1981).

51. W.E. May, J. Brown-Thomas, L.R. Hilpert and S.E. Wide in
     "Polynuclear Aromatic Hydrocarbons, Chemical Analysis and
     Biological Fate" (M. Cooke and A.J. Dennis eds.)
     Battelle Press, Columbus, Ohio, pp 1-16, (1981).

52. W.E. May, J.M. Brown, S.N. Chester, F. Guenther, L.R.
     Hilpert, H.S. Hertz and S.A. Wise in 'Polynuclear Aromatic
     Hydrocarbons, Chemistry and Biology, Carcinogenesis and
     Mutagenesis (P.W. Jones and P. Leber eds.) Ann Arbor
     Science, pp 411-418, (1979).

53. F.W. McLafferty, Science, 214, pp 280-287, (1981).

54. NIOSH Manual of Analytical Methods, HEW Publications, NIOSH,
     Washington DC, pp 77-157, (1977).

55. T. Ramdahl, K. Kreseth and G. Becher, J. High Resolution
     Chromatogr. and Chrom. Comm., 5, pp 19-26, (1982).

56. D. Rondia, Int. J. Air Water Pollution, 9, pp 113-121, (1965).

57. E. Sawicki, R.C. Carey, A.E. Dooley, J.B. Giscland, J.L. Monkman, R.E. Neligan and L.A. Ripperton, Health Laboratory Science, 7, p 60, (1970).

58. I. Schmeltz and D. Hoffmann, Chem. Review, 77, p 295, (1977).

59. L.M. Shabad, A. Khesina, G.A. Smirnow, N.E. Styopina, W. Prietsch, N. Yaskulla and M. Narmann, Gig. Sanit., 10, pp 50-53, (1976).

60. M.E. Snook, R.F. Stevenson, H.C. Higman, R.I. Arrendale and O.T. Chortyk, Beitr. Tabakforsch., 8, p 250, (1976).

61. T.W. Stanley, J.E. Meeker and M.J. Morgan, Environ. Sci. Technol., 1, p 927, (1967).

62. U. Stenberg, OECD Workshop on PAH, Paris, October 1981.

63. P.E. Strup, R.D. Giammar, T.R. Stanford and P.W. Jones in "Polynuclear Aromatic Hydrocarbons: Chemistry, Metabolism and Carcinogenesis" (R.I. Freudenthal and P.W. Jones, eds.), Raven Press, New York, p 241, (1976).

64. K. Tintel, J. Lugtenburg and J. Cornelisse, J. Chem. Soc., Chem. Communications, pp 185-186, (1982).

65. Umweltbundesamt (ed.): Berichte 1/79. Luftqualitätskriterien für ausgewählte polyzyklische aromatische Kohlenwasserstoffe. Erich Schmidt Verlag, Berlin.

66. US-clearing House Fed. Science Technol. Inf. PB rep. No. 200266 (G.P. Gross) (1971).

67. US Fed. Register, Dept. of Labor OSHA, 22 October 1976.

68. Verordnung über Zulassung von Zusatzstoffen für die Herstellung von Kaugummi, 20/9/1972, Bundesgesetzblatt I, S.1825.

69. WHO: Working group on Air Standarisation of Sampling and Analytical Procedure for estimation of PNA in the Environment, WHO, Geneva, (1969).

70. WHO International Standard for drinking water, 3rd edition, WHO, Geneva, (1971).

71. WHO: Selected Methods of Measuring Air Pollutants, WHO
     Offset Publication No. 24 (1970).

72. N. Wilburn, IARC Lyon, Private Communication.

73. Y. Yang, A.P. D'Silva and V.A. Fassel, Anal. Chem., 53,
     pp 894-899, (1981).

74. E. Bayer, K. Albert, M. Nieder, E. Grom, G. Wolff and M.
     Rindlisbacher, Anal. Chem., 54, pp 1747-1750, (1982).

PUBLIC HEALTH ASPECTS OF POLYAROMATIC HYDROCARBONS (PAH) IN
BELGIUM AND REFLECTIONS ON THE PROBLEM RELATED TO THE PAH IN
EXHAUST GASES

Prof. Dr. A. Lafontaine

Institute for Hygiene and Epidemiology,
Brussels, Belgium.

I will divide my short communication into two parts. The
first part will be devoted to some Belgian data on the atmospheric PAH-concentrations and their evolution. The second part will
summarize the reactions of a public health official, who is
responsible for the evaluation of the risks related to the PAH
in the exhaust gases, and this in the context of the air pollutants
in general and of the various risks related to PAH in the modern
style of life.

x

x   x

The general air pollution surveillance related to exhaust
gases in Belgium has put the emphasis on lead, carbon monoxide
and NO_x. The PAH pollution data are relatively restricted and
the most interesting are those from the Laboratory for Industrial Toxicology and Ecotoxicology of the Environment, under
the direction of our host Professor Rondia.

In the period 1958 - 1962, benzo(α)pyrene has been measured
in different sectors of Liege. The results show levels of 100 ng/
m^3 during winter, both in residential districts and industrial
areas; in the rural regions near Liege the levels were eight
times lower. The seasonal variation was evident : low in summer
(ten times less than in winter) and medium in autumn and in
spring (a third of the concentrations measured in winter).
(Table 1).

D. Rondia et al. (eds.), Mobile Source Emissions Including Polycyclic Organic Species, 151–158.
© 1983 by D. Reidel Publishing Company.

Table I
BaP in Liege, ng/m^3, years 1958-1962
(Prof. Rondia - personal communication)

Period	Residential	Industrial	Rural
Winter	105	113	17
Spring	49	31	8
Summer	16	17	3
Autumn	49	46	8
m	54.7	51.5	9

In 1973, the levels were evaluated in the same places and
show a reduction of $\pm$ 50 % of the BaP in the residential areas
(56.1 ng/m^3 for 105 ng/m^3 in the decade before). The mean values
for all seasons also show a reduction in the industrial area
(39.6 ng/m^3 for 51.5 ten years before). A parallel seasonal variation was observed (Table II). The explanation put forward was the
important reduction of coal used as a fuel for domestic heating.
During the same period, a survey has been done in the five most
important urban centers (Antwerp, Ghent, Brussels, Liege and
Charleroi) where an automatic network for the systematic measurement of the pollution has been developed (Table III). This confirmed the values of Liege for the residential area with an
exception for Charleroi.
In the rural areas used as reference, much lower results were
observed, although at higher population density a higher pollution appeared. It seems that one of the reasons of the higher
levels in Charleroi was the importance of coke ovens.
A new approach done in 1980 shows in the Liege area in December
a maximum around 50 ng/m^3 BaP : during this winter the general
level of pollution measured by SO_2, fumes and NO_x was more than
two times higher than the other years. But in June the concentration was lower than 1 ng/m^3 and in September it was 3.5 ng/m^3.
One of the proposed reasons of such decrease is the more important
use of natural gas as a domestic and industrial fuel.

Table II
BaP in Liege, ng/m^3, year 1973
(Prof. Rondia - personal communication)

Period	Residential	Industrial
Winter	56.1	94.6
Spring	14.8	21.2
Summer	< 2	< 2
Autumn	27.1	42.4
m	26.0	39.6

Table III
BaP in Belgium, ng/m^3, year 1973
(Prof. Rondia - personal communication)

Urban	Residential	Industrial	Rural	
Antwerp	25,1	12,9	Ploegsteert	16.5
Ghent	20,7	12,4	Mol	16,3
Brussels	35.4	-	Houffalize	5,9
Liege	26.7	38.3	Zeebrugge	‹ 9,4
Charleroi	63,8	-	Dourbes	9,2
m	34,3	21.2		11,4

It is interesting to look at the number of vehicles in Belgium from 1970 to 1981 (Table IV) and the variation from 1973 to
1981 in the consumption of benzine and diesel for road transport
(Table V) : the increase of vehicles and the increase of diesel
oil comsumption had no parallel influence on the content of BaP
in the atmosphere.
It may also be important to notice that during the same period NO_2
increased and that the concentration of Pb in the air was reduced
from ± 1 µg/m^3 in 1974 to below 0.4 µg/m^3 in 1982 (especially after
the enforcement of the new regulation in 1977).
The Belgian surveillance of PAH in the atmosphere was not particularly directed on the exhaust gases but if we consider the part
of domestic heating and of the fuels used for industrial energy in
the production of PAH, we must admit that the exhaust gases must
be considered as a minor source (between 2 to 3 %) of direct
atmospheric exposure to PAH, except perhaps at some specific areas
(high density traffic, urban tunnels, bus stations) and in the
case of professional exposure (including the use of some instruments such as chain saws-professional or homeowners).

x
x   x

Table IV

Number of vehicles in Belgium from 1970 till 1981 (National Institute of Statistics, Brussels)

year	cars	buses	tractors	trucks	cars (benzine)	cars (diesel oil)	trucks (benzine)	trucks (diesel oil)
1970	2059616	16169	39249	212156	1883246	41291	112321	94190
1972	2273163	17717	38306	219642	2135082	48118	118295	75641
1973	2389544	18549	37906	224029	2256131	57078	119717	22364
1974	2502356	19346	37753	229872	2366182	68732	121544	89183
1975	2613835	19553	37007	235360	2461091	83268	121665	95239
1976	2737989	19854	36397	237325	2574131	97166	120005	100474
1977	2871332	19733	35138	242263	2688211	117732	120485	106564
1978	2973418	19745	31657	247454	2765275	142325	120665	113158
1979	3076570	19753	31514	258553	2806846	180139	123410	121708
1980	3158737	19560	31415	287669	2823066	228837	123935	130141
1981	3206472	18948	27842	255765	2784635	280955	113331	130272

Table V
Consumption of benzine and diesel for road transport
(in thousands of tons)
(Belgian Petroleum Federation)

	1973	1977	1978	1979	1980	1981
Benzine	2573	2978	3108	3132	2948	2719
Diesel oil	1251	1400	1410	1723	1795	1838

x

x   x

In the second part of my intervention, I will try to express
some personal opinions on the topics of the meeting in the convic-
tion that some progresses will be realized during this week.

As chairman of the Scientific Committee for Toxicology and
Ecotoxicology of the EEC, as member of the Scientific Committee
of IARC or as consultant of WHO, and being confronted for several
years with the problems of pollution and human exposure to altero-
genic chemical, physical and biological agents, I think it is my
duty to express some reflections on the subject which may be of
some importance during the discussions.
If we exclude two other well defined PAH origins of main importance, consumption of contaminated food (e.g. smoked fish or meat)
and some habits as smoking, the sources of PAH in the air may be
divided into two categories : the stationary sources (residential
heating and cooking, industry as coke production or petroleum
catalytic cracking, power and heat generators, incinerators and
open fires, including forest fires and agriculture burning) and
the mobile sources (gasoline and diesel engines, rubber tirewear,
air and sea traffic).
Perhaps that besides these categories of sources, we will identify
other activities such as sidewalk tarring or coal tar pitch
working areas ... It demonstrates that it is urgent to know with
precision all the emissions of PAH in the atmosphere in a certain
place. And if we consider only the exhaust gases, we will be obliged to recognize their dependence of a large variety of factors :
type and adjustment of engine and fuel, running of the engine, way
of driving and so on.

If we consider now the emission values, at the point where man is exposed, we notice that they are not only function of the emissions of PAH but also of a lot of other parameters : meteorological and topographic conditions, sampling methods and analytical techniques (1), choice of the reference substance(s) submitted to the analyses.

Furthermore, the emission may depend on the chemical characteristics of the different PAHs present in the atmosphere and the effects may be influenced by external factors as the presence of particulates and in this case soot is of a direct significance. The very recent publication of Li and its colleagues in Toxicology confirm the importance of the last on the cytotoxicology (2).

Finally, we should remember that the atmosphere may contain other components, toxic or not, chemical or not, which can interfere with the action of PAH.

Coming back to the chemical characteristics, we should also keep in mind that PAHs influence chronic effects, including enzyme reduction, and mutagenic, carcinogenic and teratogenic effects. This is true not only for the different PAH (including the heteroatomic which probably play a non negligible role), but also for the other POM (Polycyclic Organic Matter).

---

(1) Discarding the basic and acid fraction of the extract, the aromatic fraction can be directly evaluated by column - or thin layer chromatography of the sample.
Other methods of PAH analyses are :

a) two dimensional thin layer chromatography and fluorometric or U.V. spectrophotometric detection;
b) gas chromatography (GC) with electron-capture detection;
c) gas chromatography - mass spectrometry (GC-MS);
d) high performance liquid chromatography (HPLC).

The last one seems the upcoming technique, especially suited for heavy compounds and derivatives of high polarity but it shows the difficulties which still exist to evaluate what we are really exposed to.

(2) A.P. Li, R.E. Royer, A.L. Brooks and R.O. Mc Lellan - Cytotoxicity of Diesel Exhaust Particle Extract - A comparison amoung five diesel passenger cars of different manufacturers - Toxicology 24 (1982) 1 - 8.

The exposure to PAH air pollution in general surely has to be considered in a different way in the urban and in the non urban atmospheres : in this case, a significant difference exists for the outdoor pollution. But if we consider the indoor pollution, we are surprised that a room polluted by tobacco smoke has a level of pollution as high as a polluted urban area and that in some huts in the mountains of Kenya we reach levels 3 to 5 times higher than the concentrations measured in bus garages.
Finally, man may also be indirectly exposed to PAH, the air being an agent of transfer to soil, water and plants from which diffe-rent food chain contaminations are possible, including contamina-tions of meat, milk or egg. The same pathways must also be consi-dered for the polluted waters and sediments. We must consider the global exposure to PAH from different types of exposure.

If we admit that experimental approaches at the cellular or at the animal level have opened the way to some knowledge, then we also have to recognize that there are still a lot of gaps in the evaluation of the dangers of the different chemical structures of PAH and still more in the evaluation of the morbidity and the mortality in man. If we are able to identify the sources and the ways of exposure, we must admit that the epidemiology of the diseases and incomforts related to PAH remain difficult for two reasons : the problems already evocated and the interferences with other factors of agression.

The best data available for the effects on man are coming from studies performed in different occupational environments (coke and gas workers, iron and steel workers, aluminium plants, carbon black industry and asphalt workers, etc ...) : a good review has been proposed by G. Lindstedt and J. Soelenberg in Search J. Environmental Health 8 (1982) 1 - 19. Although some criticism can be expressed on the referred substance and the analytical technique they have chosen.

We must accept that a better epidemiological approach has to be developed on risks related to PAH and, for our meeting, on risks related to exhaust gases, taking into account not only the lung cancers but also the other carcinogenic (bladder) and possibly genetic and teratogenic consequences.

x

x   x

I will stop here. My remarks don't have the purpose of limi-
ting the importance of research on the effects of PAH in exhaust
gases but to reach in the recommendations realistic proposals.
Although we still need better chemical research on the sources
and sinks of PAH, more toxicological experimentation and more
epidemiological data, we already have to recognize that exhaust
gases present a global risk related to PAH and to other components.
We also have to recognize that exhaust gases as a PAH source are
counting for only 2 % or maximum 3 % of the total PAH atmospheric
emissions.
However, this low percentage is not a reason to forget that our
major purpose must be to understand the production of these
substances and to control them, to identify the most dangerous
among them and to reduce the direct and the indirect exposure.

     The research and the actions relative to exhaust gases are
an inevitable step before envisaging the possibilities of evaluating
the risks of the PAH for the population and for the professionally
exposed people.
Not only the PAH in the air or transported by air will be
considered but all the exposures from air, water and food to
PAH have to be studied including all origins, the direct or
indirect biosynthesis by microorganisms and plants and some PAH
generated in the past centuries.
But we cannot wait for the results of all the experiences to
consider realistic measures to reduce the production and to
limit the exposures taking into account the advantages and dis-
advantages of the techniques responsible for the production of
PAH. And finally, we will perhaps need to consider other approaches
of the problem such as better use of space, new conception in
housing and better sanitary education.
If some limits of concentration and standards are proposed, they
must take into account the purposes : protection of the professionally exposed subjects or of groups of the population.
They must be based on well defined standardized analytical techniques (and when necessary on the determination of specific components). They must be coordinated with the standards and limits
proposed for other ways of exposure as for instance, the European
Standards of EEC for drinking water : "It is recommended that six
compounds (fluoranthène; 3,4 benzfluoranthène; 11,12 benzfluoranthène; 11,12 benzperylène; 3,4 benzopyrène, indeno (1,2,3-cd)
pyrène) will be analysed and their concentrations should not
exceed 0,2 µg/l".

THE PAH-EMISSION OF SPARK IGNITION ENGINES

Gerhard Lepperhoff

FEV - Forschungsgesellschaft für Energietechnik und
Verbrennungsmotoren m.b.H. Aachen, Germany.

The influence of motor-specific parameters on the mass of PAH in
the exhaust gas was investigated for three different spark ignition
engines in steady state tests.
No generally valid correlation exists between PAH and HC-
emissions. For the investigated engines however, the share of
PAH of the HC-emissions increases with increasing HC-emission
levels, when parameters such as air-fuel ratio, ignition timing
or coolant are varied.
The air-fuel ratio has a significant influence on PAH emissions.
With leaner mixtures the mass of PAH in the exhaust gas drops
noticably. The PAH profile changes to the extent that the share
of carcenogenic components of total PAH are lower, whereas
especially the share of Pyren increases. The absolute levels of
the PAH emissions have no effect on the PAH profile for equal
air-fuel ratios.
Through the following measures it is possible to reduce the PAH
emissions per exhaust gas mass as well as per unit of energy:
- operation predominantly with lean air-fuel ratios
- smaller quench distances in the combustion chamber
- increasing cylinder wall temperatures.

These measures lead to a reduction of the emission levels as well
as a shift in the PAH profile, whereby the emission of carcinogenic components are reduced by a greater margin than the non-
carcinogenic components.

D. Rondia et al. (eds.), Mobile Source Emissions Including Polycyclic Organic Species, 159–163.
© *1983 by D. Reidel Publishing Company.*

1. INTRODUCTION

The carcinogenic effects of selective polycyclic aromatic hydrocarbons (PAH) found in the exhaust gas of internal combustion
engines has only indirectly been determined through animal tests.
Studies with internal combustion engine have focussed essentially on the level and composition of PAH emissions as well as the
influence of fuel for instationary engine operation. In this
presentation the influence of engine parameters on PAH emissions
of spark ignition engines and possibilities to reduce these
emissions will be discussed.

2. TEST PROGRAM

The investigations were carried out on three different spark
ignition engines (1-, 4-, 6-cylinders) at various speeds (n) and
throttle positions (α_D) under steady-state operation. The PAH
were collected on filters from cooled raw exhaust and subsequently separated into 13 substances with the aid of a gas chromatograph. The sum of the 13 PAH and their total mass will be
referred to as ΣPAH and Σm_{PAH} respectively. Total hydrocarbons
as measured with a flame ionisation detector (FID) are referred
to as HC.

3. RESULTS

3.1 Relationship Between PAH and HC-Emissions

Fig. 1 illustrates the relationship of ΣPAH and HC-emissions
as a function of HC-emissions (g/kg exhaust gas). As HC-emissions
are also influenced by the equivalent air-fuel ratio (λ), the
equivalent air-fuel ratio is also specified for some of the
results. The results reveal not only an influence of the equivalent air-fuel ratio on the ΣPAH portion of the HC-emissions, but
furthermore an influence of the throttle position (α_D) and the
design of engine. Thus, a generally valid correlation between
ΣPAH and HC-emissions does not exist. However, a reduction in
HC-emissions results in a qualitative reduction of the ΣPAH
portion of the HC-emissions.

3.2 Influence of the Equivalent Air/Fuel Ratio

The relative change of the PAH emissions as a function of the
equivalent air-fuel-ratio (λ) is presented in Fig. 2. ΣPAH
values are related to the corresponding values at $\lambda = 1$. Regard-
less of the ΣPAH emissions level, as influenced by the design

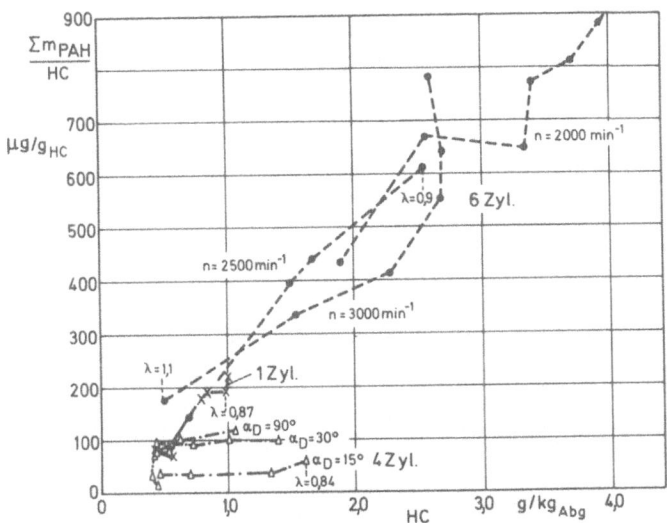

Figure 1. Portion of ΣPAH emissions of HC-emissions as
a function of the HC-emissions

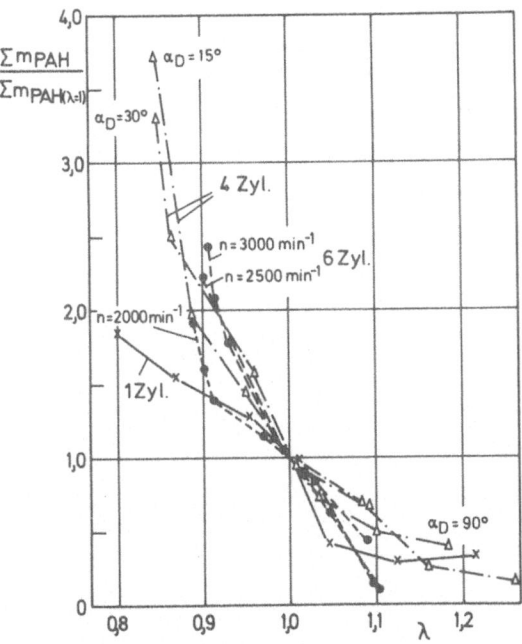

Figure 2. Relative change in ΣPAH emissions (based on
Σm_{PAH} at $\lambda = 1$) as a function of the air-
fuel ratio λ

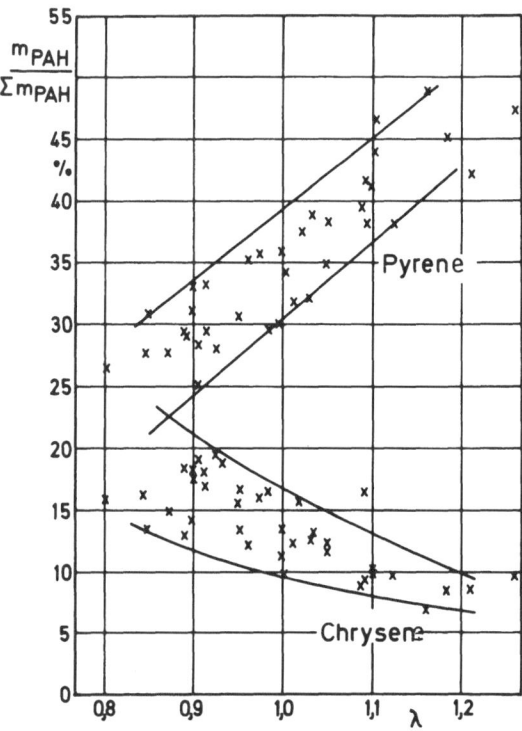

Figure 3. Pyrene and "Chrysene"-portion of the ΣPAH
as a function of λ

of engine, e.g. combustion chamber layout, ΣPAH emissions can in
general be reduced by approx. 80 % for comparable engines by increasing the equivalent air-fuel ratio from λ = 0,9 to λ = 1,1.

The equivalent air-fuel-ratio also influences the PAH profile. The profile describes the mass relationship of the different PAH components. The portion of carcinogenic components of
ΣPAH drops with increasing λ. In Fig. 3 the reduction of "Chrysene"
(consisting of Chrysene + Benzo(a)anthracene + Cyclopenta(cd)pyrene
+ Triphenylene) is illustrated for mixture leaning. Benzo(a)pyrene
also drops from 3 % at λ = 0,9 to approx. 2 % at λ = 1,1. The
non-carcinogenic component Pyrene is mainly the component which
proportionally increases with increasing λ.

In summary it can be stated that:

a) Mixture leaning leads to a considerable reduction of the PAH
   mass in the exhaust gas

b) No quantitative correlation exists between ΣPAH and the HC-
   emissions, however, a reduction of HC-emissions through mixture leaning leads to a reduction of the ΣPAH portion of the
   HC-emissions

c) The air-fuel-ratio influences the PAH profile considerably

d) The portion of the carcinogenic components in ΣPAH drops with
   leaner mixtures

c) The level of emitted PAH mass has no influence on the PAH
   profile

4. CONCLUSIONS

The total as well as the specific PAH emissions of modern engines
can be reduced through the following measures:

. Operation primarily with lean mixtures
. small squish volumes in the combustion chamber
. increased cylinder-wall temperatures.

In the effort to reduce emissions as well as fuel consumption,
the development of modern production engines has followed these
guidelines. This has resulted not only in a considerable reduction in PAH pollution through spark ignition engines but also
has shifted the PAH profile towards the non-carcinogenic components.

EVALUATION OF MOTOR VEHICLE AND OTHER COMBUSTION EMISSIONS USING
SHORT-TERM GENETIC BIOASSAYS

Joellen Lewtas

Genetic Bioassay Branch, Genetic Toxicology Division,
Health Effects Research Laboratory, U.S. Environmental
Protection Agency, Research Triangle Park, NC   27711
U.S.A.

ABSTRACT

   Short-term genetic bioassays have been useful in evaluating
unregulated organic combustion emissions from motor vehicles.
Identification of mutagens and carcinogens in complex exhaust
emissions has been greatly facilitated by the use of bioassay-
directed chemical fractionation and characterization methods.   It
has also been possible to evaluate the effect of fuels, engine
types, and control technologies on the rates of mutagenic
emissions from motor vehicles. Greater differences in the rate
of mutagenic emissions have been observed between different
engines (e.g., diesel vs. gasoline) and control technologies
(e.g., with and without catalyst) than between different fuels.
A comparative evaluation of various combustion sources indicates
that motor-vehicle emissions make a major contribution to the
mutagenicity observed in ambient air.

INTRODUCTION

   Combustion emissions from both motor vehicles and stationary
sources contain a complex mixture of organic compounds. Chemical
characterization of these organics shows that they contain
carcinogenic polycyclic aromatic hydrocarbons (PAH), such as
benzo(a)pyrene. Recently, chemical characterization studies of
motor-vehicle emissions have identified the presence of
methylated PAHs (e.g., methylphenanthrenes) (1), nitrated PAHs
(e.g., nitropyrene) (2), oxidized PAHs (e.g., 4 oxapyrene-5-one)

165

D. Rondia et al. (eds.), Mobile Source Emissions Including Polycyclic Organic Species, 165–180.

(3), and a variety of other polycyclic organic compounds not yet evaluated in animal cancer bioassays.

The development of short-term genetic bioassays has provided relatively simple, sensitive, and rapid bioassays for mutagenic and potential carcinogenic activity. Short-term genetic bioassays have been particularly useful in evaluating combustion emissions. This paper summarizes the results of studies where short-term genetic bioassays have been used in the following areas:

(1) identification of mutagens and carcinogens in complex exhaust emissions,

(2) evaluation of the effect of fuels, engine types, and control technologies on the mutagenic activity of the emissions, and

(3) comparative assessment of mutagenicity and carcinogenicity of various combustion sources and their contributions to the mutagenic activity of ambient air.

IDENTIFICATION OF MUTAGENS AND CARCINOGENS IN COMPLEX EXHAUST EMISSIONS

Bioassay-directed chemical fractionation closely coupled to chemical characterization has been shown to be the most efficient and effective approach to identifying the specific chemical compounds in a complex mixture that exhibit a particular biological activity (4). This approach has been used to identify tumor initiators and tumor promoters in cigarette-smoke condensates (5), automotive exhaust emissions (6), and urban-air particles (7). More recently, this approach has been coupled with short-term genetic bioassays, including both microbial and mammalian-cell mutation assays, to identify mutagens and potential carcinogens in complex mixtures (8). This method was employed to identify the chemical compound classes and specific components associated with diesel particulate emissions that were mutagenic in the Ames Salmonella typhimurium mutagenesis assay (9).

Diesel particles collected by the dilution-tunnel method (10) were Soxhlet-extracted with dichloromethane and solvent-partitioned into organic acids, bases, and neutral components. The neutral components were further fractionated into paraffins (hexane), aromatics and transitional compounds (1% ether/hexane), and oxygenated compounds (50% acetone/methanol). The mutagenic activity of each fraction was determined using the Ames Salmonella typhimurium/microsome assay with strains TA1535, TA1537, TA1538, TA98, and TA100 (9). The distributions of the

mass of each fraction and of its mutagenic activity using TA98
are shown in Figure 1 and Table 1 for a heavy-duty four-stroke
V-8 Caterpillar 3208 engine used in urban service vehicles. The
moderately and highly polar neutral compounds in the transitional
(TRN) and oxygenated (OXY) fractions account for 89-94% of the
mutagenic activity of the extractable organics and only 32% of
the mass. Conventional gas chromatography/mass spectroscopy
identified many nonmutagenic methylated fluorenones as major
constituents of these fractions. None of these, nor other
identified constituents (e.g., PAHs), account for the direct-
acting frameshift mutagenic activity observed. The mutagenic
activity of the total extract from diesel particles was found to
be significantly less in nitroreductase-deficient strains of
Salmonella typhimurium (e.g., TA98FR1), suggesting that nitrated
polycyclic compounds are present (11). Nitrated polycyclic

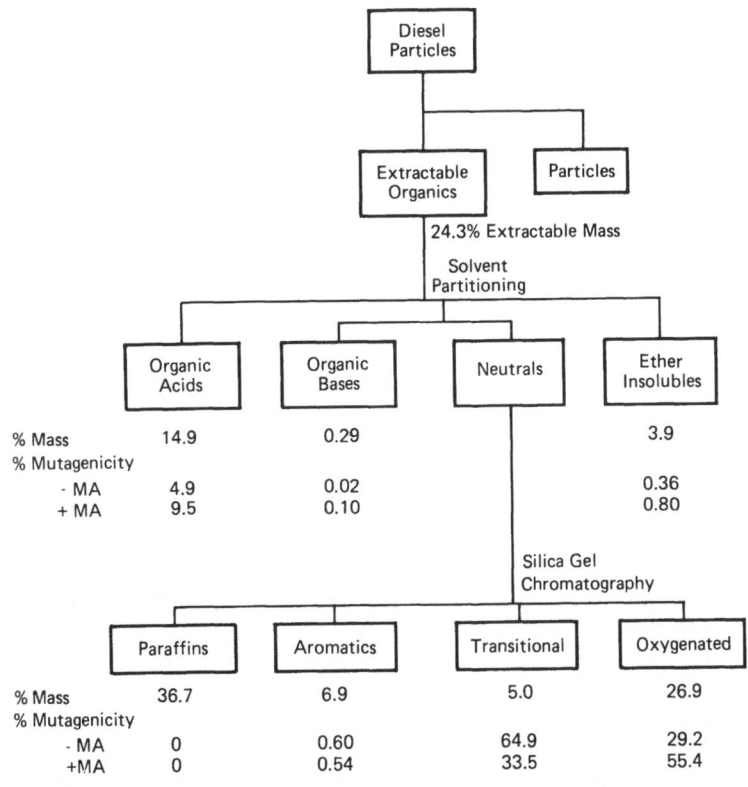

Figure 1. Distribution from diesel particles of
mass and mutagenic activity in Salmonella
typhimurium TA98.

Table 1. Distribution of the mass and mutagenic activity of fractionated diesel particle organics[a]

Fraction	Mass (%)	Mutagenic Activity[b] (rev/mg)		Weighted Mutagenic Activity[c] (rev/mg)		Distribution of Mutagenic Activity (%)	
		-MA	+MA	-MA	+MA	-MA	+MA
Organic acids	14.9	193	248	28.8	37.0	4.9	9.5
Organic bases	0.3	43.8	132	0.13	0.40	0.02	0.10
Ether insolubles	3.9	53.9	80.9	2.1	3.2	0.36	0.80
Paraffins	36.7	Neg.	Neg.	0.0	0.0	0.0	0.0
Aromatics	6.9	49.5	30.1	3.42	2.1	0.60	0.54
Transitionals	5.0	7520	2620	376	131	64.9	33.5
Oxygenates	26.9	629	798	169	215	29.2	55.4

a-MA = without metabolic activation; +MA = with metabolic activation.
b Slope determined from linear regression analysis of the initial portion of the dose-reponse curve.
c Determined by multiplying the mutagenic activity by the percent mass.

aromatic hydrocarbons (NO_2-PAHs) are potent direct-acting frameshift mutagens detected in xerographic toners (12). Identification and quantification of a series of NO_2-PAHs in diesel extracts has made it possible to estimate their contribution to the mutagenic activity of diesel particulate emissions (13). Particulate emissions from catalyst-equipped gasoline-engine vehicles using unleaded fuel contain significantly less of these NO_2-PAHs (13). The mutagenic activity of both leaded- and unleaded-gasoline emissions is substantially increased with the addition of an exogenous metabolic activation (MA) system, suggesting that the classical PAHs may play a more important role than do NO_2-PAHs in the mutagenicity and carcinogenicity of gasoline emissions (14).

EVALUATION OF THE EFFECTS OF VARIOUS FUELS, ENGINES, AND CONTROL TECHNOLOGIES

Short-term bioassays have proven useful in evaluating the effects of various engines, fuels, and control technologies on the mutagenicity of emissions. To draw meaningful conclusions from such comparisons, however, Claxton and Kohan (15) studied the normal day-to-day variations in the emissions (e.g., particle emission rate and percent organic extractables) and in bioassay results (mutagenicity slope) for one engine under standard conditions. For this study, an Oldsmobile 350 diesel vehicle was run on repeated Highway Fuel Economy Test (HWFET) cycles during one day and on separate days. The coefficients of variation (CV) for these parameters, as shown in Table 2, ranged from 0.07 to 0.11. A computerized statistical method recently developed by Stead et al. (16) for analysis of dose-response data from the Ames Salmonella typhimurium bioassay greatly facilitated comparisons between different vehicles and fuels. An example of this

Table 2. Coefficients of variation of assay parameters for standard operation of one diesel vehicle[a]

Parameters	All Days (CV)	Range of Separate Days (CV)
Particle emission rate	0.07	0.02 – 0.05
% Organic extractables	0.09	0.07 – 0.08
Mutagenicity slope	0.11	0.01 – 0.11

[a]Oldsmobile 350 diesel vehicle operating on repeated HWFET cycles (from Claxton and Kohan [15]).

analysis is shown in Figures 2a and 2b for the dichloromethane-extractable organics from a gasoline vehicle (Ford Van) operated with leaded fuel. The average non-linear-model slope, shown with 95% confidence limits, is then used in the comparisons.

To compare the mutagenicity of particle emissions from different sources or fuels, the percent of organic material extractable from the particles and the vehicle's particle emission rate must also be included in the analysis. The final rate of mutagenic emissions is determined from all three parameters (Table 3). The significantly lower rate of mutagenic emissions from the catalyst-equipped, gasoline-engine vehicle is due primarily to its much lower particle emission rate (0.0033 g/km), compared with the leaded gasoline and diesel emissions. The significantly lower rate of mutagenic emissions from the GM bus (Table 3) than from the diesel truck and car, however, is due to the substantially lower mutagenic activity of the organics. As the bus and truck were operated on the same fuel, this difference appears to be due to the combustion characteristics of the two engines. The GM bus has a two-stroke-cycle engine (1977 Detroit diesel, 8V-71), whereas the Mack truck has a four-stroke-cycle engine. The lower mutagenicity of the emissions from the two-stroke cycle engine (GM bus) reported here are consistent with the results of an earlier study (9).

The mutagenic and carcinogenic activities of the extractable organics from a series of diesel and gasoline particle emissions have been compared in a battery of short-term bioassays (14). The bioassays that provided the best quantitative and reproducible dose-response data were (1) the Ames Salmonella typhimurium mutagenesis assay, (2) the mouse lymphoma mutagenesis assay, and (3) the Chinese hamster ovary-cell sister-chromatid-exchange assay. The results of these short-term genetic bioassays were compared with those of a skin-tumorigenesis assay in SENCAR mice (17). Within the diesel and gasoline vehicle emission samples examined, a very high correlation was observed among the results of these three genetic bioassays and the mouse skin tumor initiation assay (14).

The influence of fuel on the mutagenicity of emissions was initially examined for five fuels, including four No. 2 diesel fuels and one No. 1 jet fuel, in two light-duty diesel vehicles (a Volkswagen Diesel Rabbit and a Mercedes 240D) (9). In the VW Rabbit, the emissions from the minimum-quality fuel, with a higher aromatic and nitrogen content, were significantly (approximately 5 times) more mutagenic per milligram particulate emission than were the emissions from the other fuels. The emissions from these fuels did not differ significantly in mutagenicity, however, when they were burned in the Mercedes (9).

Table 3. Comparison of the rates of mutagenic emissions from motor vehicles

Source	Mutagenicity of Organics (rev/µg)[a]	Extractable Organics (%)	Particle Emission Rate (g/km)	Mutagenic Emission Rate (rev/km)
Diesel fuel				
Car (Mercedes)[b]	12.0	8	0.24	240,000
Truck (Mack)[c]	2.3	11	1.3	320,000
Bus (GM)[c]	0.1	17	2.1	37,000
Gasoline fuel				
Non-catalyst (Ford Van)[d]	32.0	19	0.03	180,000
Catalyst (Mustang II)[e]	3.5	43	0.0033	5,000

[a] Salmonella typhimurium TA98 with metabolic activation; non-linear-model slope analyzed by the method of Stead et al. (16).

[b] Mercedes 300D, 1977 model, operated on the HWFET cycle using No. 2 diesel fuel obtained from Union 76.

[c] Mack ENDT 676 diesel engine in a dual-drive tandem-axle truck and GM bus with a Detroit diesel DD8V-71 engine were operated with the same average No. 2 diesel fuel (EM-239-F) on the 1983 transient heavy-duty cycle.

[d] Ford Van, 1970, in-line 6-cylinder engine, operated with leaded gasoline (Premium A) on the HWFET.

[e] Ford Mustang II-302, 1977, operated with unleaded gasoline on the HWFET.

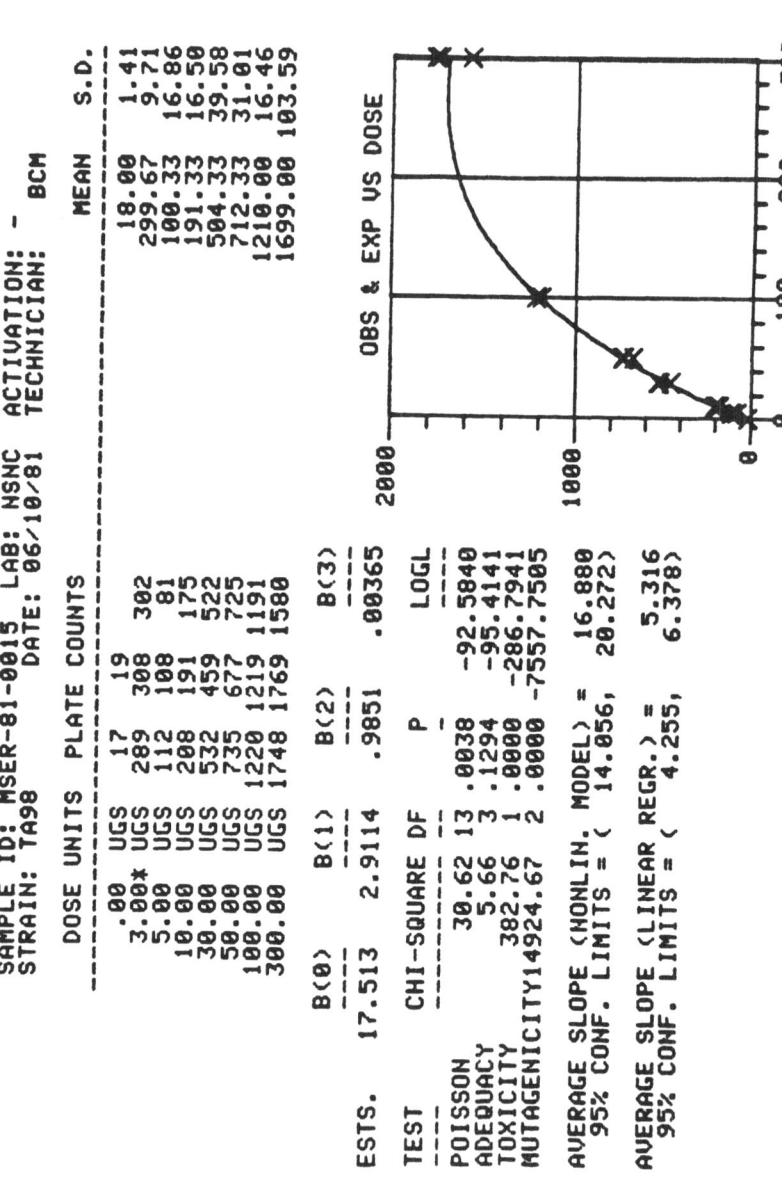

SAMPLE ID: MSER-81-0015 LAB: NSNC ACTIVATION: -
STRAIN: TA98 DATE: 06/10/81 TECHNICIAN: BCM

DOSE	UNITS	PLATE COUNTS			MEAN	S.D.
.00	UGS	17	19		18.00	1.41
3.00*	UGS	289	308	302	299.67	9.71
5.00	UGS	112	108	81	100.33	16.86
10.00	UGS	208	191	175	191.33	16.50
30.00	UGS	532	459	522	504.33	39.58
50.00	UGS	735	677	725	712.33	31.01
100.00	UGS	1220	1219	1191	1210.00	16.46
300.00	UGS	1748	1769	1580	1699.00	103.59

	B(0)	B(1)	B(2)	B(3)
ESTS.	17.513	2.9114	.9851	.00365

TEST	CHI-SQUARE	DF	P	LOGL
POISSON	30.62	13	.0038	-92.5840
ADEQUACY	5.66	3	.1294	-95.4141
TOXICITY	382.76	1	.0000	-286.7941
MUTAGENICITY	14924.67	2	.0000	-7557.7505

AVERAGE SLOPE (NONLIN. MODEL) = 16.880
95% CONF. LIMITS = (14.056, 20.272)

AVERAGE SLOPE (LINEAR REGR.) = 5.316
95% CONF. LIMITS = (4.255, 6.378)

OBS & EXP VS DOSE

Figure 2a. Example of a computerized statistical method for the analysis of dose-response data from Ames Salmonella typhimurium without metabolic activation.

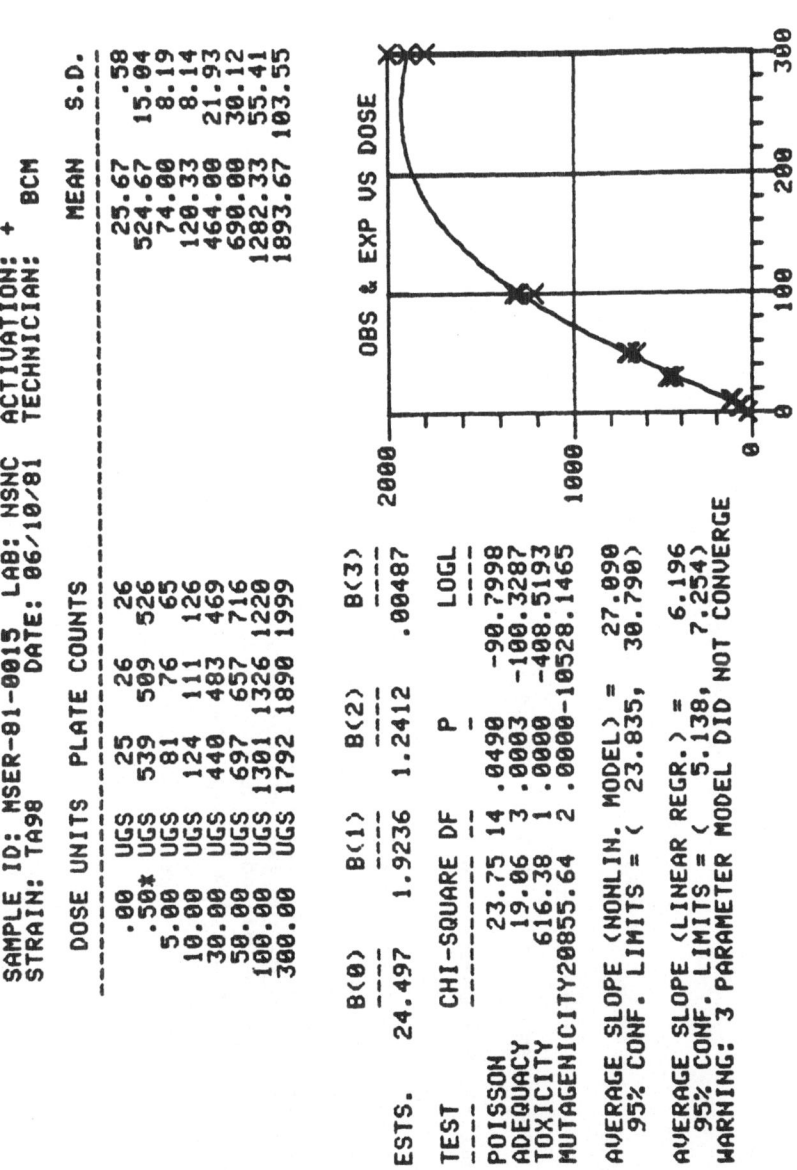

Figure 2b. Example of a computerized statistical method for the analysis of dose-response data from Ames Salmonella typhimurium with metabolic activation.

Table 4. Effect of fuel quality on rate of mutagenic emissions

Fuel	Fuel Characteristics			Mutagenic Emission Rate[a] (rev/km x 10^5)[b]		
	Aromatics (% by FIA)	Nitrogen (wt. %)	Cetane No.	Ford/Cat[c]	Mack Truck[c]	GM Bus[c]
Jet, JP-7, EM-401-F	2.7	0.025	56.0	-	-	0.54
Diesel #1, EM-400-F	10.5	0.006	49.0	-	-	0.88
Diesel #2						
Premium, EM-242-F	16.9	0.046	52.1	7.3	4.2	0.22
Average, EM-329-F	21.3	0.040	48.0	6.8	5.6	0.79
Minimum, EM-241-F	35.8	0.61	42.0	14.0	3.7	1.1

[a] Salmonella typhimurium TA98 without metabolic activation; non-linear-model slope determined by the method of Stead et al. (16).
[b] Determined from the model slope (rev/μg organics), extractable organics (%), and particle emission rate (g/km) for each test condition.
[c] Heavy-duty vehicles operated on the 1983 transient cycle.

Similar studies were recently conducted (Table 4) for five fuels in the three heavy-duty vehicles:

(1) GM city bus with a Detroit DD8V-71 engine;

(2) Mack truck with a Mack ENDT 676 engine; and

(3) Ford Bob Tail Van (LN 7000) with a Caterpillar 3208 engine.

The five fuels ranged in aromatic content from 2.7% to 35.8% and in nitrogen content from 0.006% to 0.61%. The minimum-quality fuel produced a higher rate of mutagenic emissions in the Ford/Cat Van and the GM bus than did the higher-quality fuels. This difference was not observed for the Mack Truck (Table 4). The fuels with the lowest aromatic and nitrogen content, jet and diesel No. 1, did not produce less mutagenic emissions in the GM bus than did the diesel No. 2 premium fuel. However, the mutagenic activity (revertants per microgram) for the extractable organics from the bus particle emissions was much lower than that for the emissions from the other two vehicles.

These studies suggest that although poorer-quality fuel, with relatively high aromatic and nitrogen content, can increase the rate of mutagenic emissions in certain vehicles, such differences are not observed in all vehicles. In general, the observed changes in mutagenic activity of emissions as a function of fuel quality were much smaller than the differences between different types of engines (e.g., diesel vs. gasoline) and control technologies (catalyst-equipped vs. non-catalyst-equipped vehicles).

COMPARATIVE ASSESSMENT OF VARIOUS COMBUSTION SOURCES

The extractable organics from diluted and cooled combustion particle emissions from both stationary and mobile sources have been evaluated in a battery of mutagenesis and carcinogenesis bioassays (Table 5). The mutagenic or carcinogenic activity per microgram of extractable organics within each assay was generally within two orders of magnitude (10^2). The emission rates for the particle-bound organics, however, differed in some cases by as much as five orders of magnitude (10^5). To compare the mutagenicity of emissions from various sources, it is critical to compare the rates of mutagenic emissions on the same basis (e.g., fuel consumption, distance driven, energy consumption, or yield). To compare rates of mutagenic emissions of mobile sources, we have compared the mutagenic activity per kilometer, as shown previously in Table 3. The rates of mutagenic emissions from stationary sources are compared on the basis of fuel or energy consumption (Table 6).

Table 5. Comparative mutagenicity and carcinogenicity of extractable organics

Source	Mouse Skin Tumor Initiation[a] (papillomas/mouse/mg)	Mutation in L5178Y Mouse Lymphoma Cells (TK mutants/10^6 surviving cells/µg/ml)		SCE in CHO Cells[b] (SCE/cell/µg/ml)		Ames Salmonella TA98[c] (rev/µg)	
		-MA	+MA	-MA	+MA	-MA	+MA
Diesel							
Mercedes	0.37	0.03	1.5	0.09	0.16	10.0	12.0
Nissan	0.59	4.2	2.9	0.30	0.071	11.0	13.0
Volkswagen Rabbit	0.24	0.98	0.72	0.075	0.030	3.8	3.0
Oldsmobile	0.31	1.2	1.3	Neg.	0.017	2.2	1.5
Caterpillar	Neg.	0.25	0.063	0.011	Neg.	0.38	0.31
Gasoline catalyst							
Mustang II	0.17	0.38	1.1	0.076	–	1.6	3.5
Gasoline non-catalyst							
Chevrolet 366	–	1.2	4.9	0.72	0.22	2.9	6.2
Ford Van	–	2.1	5.7	0.62	0.47	17.0	32.0

(continued)

[a] Based on papilloma multiplicity data in SENCAR mice (17).
[b] -MA was 21.5 h exposure and +MA was a 2 h exposure.
[c] Determined from simple linear regression analysis.
[d] IP = In progress.

Table 5, continued

Source	Mouse Skin Tumor Initiation[a] (papillomas/ mouse/mg)	Mutation in L5178Y Mouse Lymphoma Cells (TK mutants/ 10^6 surviving cells/μg/ml)		SCE in CHO Cells[b] (SCE/cell/μg/ml)		Ames Salmonella TA98[c] (rev/μg)	
		-MA	+MA	-MA	+MA	-MA	+MA
Residential heaters							
Oil	0.12	1.2	2.6	0.06	0.04	1.3	2.1
Wood	-	-	-	-	-	0.15	0.93
Coal	-	-	-	-	-	IP[d]	IP
Utility power plants							
Coal, conventional	-	IP[d]	IP	IP	IP	3.1	-
Coal, FBC	-	IP	IP	IP	IP	9.4	5.2
Oil	-	-	-	-	-	IP	IP

aBased on papilloma multiplicity data in SENCAR mice (17).
b-MA was 21.5 h exposure and +MA was a 2 h exposure.
cDetermined from simple linear regression analysis.
dIP = In progress.

Table 6. Rate of mutagenic emissions from stationary sources

Source	Fuel	Mutagenicity[a] of Organics (rev/μg)	Organic Emission Rate (mg/kg fuel)	Organic Emission Rate (ng/J)	Mutagenic Emission Rate (rev/kg fuel)	Mutagenic Emission Rate (rev × 10⁻³/J)
Residential heaters						
Woodstove	Pine	1.3	8940	508.0	12,000,000	660.0
	Oak	0.9	3096	187.0	2,800,000	168.0
Residential oil furnace: #1	No. 2 fuel oil	2.0	21	0.5	40,000	1.0
#2	No. 2 fuel oil	5.1	70	1.5	360,000	7.6
Utility power plants (coal)		3.1	–	0.01	–	0.031

[a]Salmonella typhimurium TA98 with metabolic activation.

The relative contributions of these various combustion sources to the mutagenicity observed in ambient-air particles can be estimated using emission inventories for a particular geographic region or source-apportionment data from ambient-air samples. Such studies, now in progress (18), suggest that combustion emissions from motor vehicles contribute nearly half of the mutagenic activity observed from respirable air particles (< 2.5 μm) collected in urban and suburban street-level locations. The relative contributions from diesel, leaded-gasoline, and unleaded-gasoline vehicles would depend upon their distribution at the particular location.

ACKNOWLEDGMENTS

The author acknowledges the editorial assistance of Dan Tisch and Susan Dakin, of Northrop Services, Inc., and the technical assistance of Katherine Williams and Ann Austin, of the Environmental Protection Agency.

REFERENCES

1. Yu, M.-L., and Hites, R.A.: 1981, Anal. Chem. 53, pp. 591-954.

2. Schuetzle, D., Lee, F.S.-C., Prater, T.J., and Tejada, S.B.: 1981, Int. J. Environ. Anal. Chem. 9, pp. 93-144.

3. Pitts, J.N., Jr., Lokensgard, D.M., Harger, W., Fisher, T.S., Mejia, V., Schuler, J., Scorziell, G.M., and Katzenstein, Y.A.: Mutation Res. Lett. (in press).

4. Claxton, L.D.: 1982, "Genotoxic Effects of Airborne Agents," R.R. Tice, D.L. Costa, and K.M. Schaich, eds., Plenum Press, New York, pp. 19-34.

5. Hoffman, D.L., and Wynder, E.L.: 1976, "Chemical Carcinogens," C.E. Searles, ed., ACS Monograph 173, American Chemical Society, Washington, DC, pp. 324-365.

6. Grimmer, G., Naujack, K.-W., Dettborn, G., Brune, H., Deutsch-Wenzel, R., and Misfield, J.: 1982, "Polynuclear Aromatic Hydrocarbons," Battelle-Columbus Press, Columbus, OH, pp. 335-345.

7. Hueper, W.C., Kotin, P., Tabor, E.C. Payne, W.W., Falk, H., Sawicki, E.: 1962, Arch. Pathol. 74, pp. 89-116.

8. Epler, J.L., Clark, B.R., Ho, C.-h., Guerin, M.R., and Rao,
    T.K.:   1979, "Application of Short-Term Bioassays in the
    Fractionation and Analysis of Complex Environmental
    Mixtures," M.D. Waters, S. Nesnow, J.L. Huisingh, S.S.
    Sandhu, and L. Claxton, eds., Plenum Press, New York,
    pp. 269-290.

9. Huisingh, J., Bradow, R., Jungers, R., Claxton, L.,
    Zweidinger, R., Tejada, S., Bumgarner, J., Duffield, F.,
    Waters, M., Simmon, V.F., Hare, C., Rodriguez, C., and Snow,
    L.:   1979, "Application of Short-Term Bioassays in the
    Fractionation and Analysis of Complex Environmental
    Mixtures," M.D. Waters, S. Nesnow, J.L. Huisingh, S.S.
    Sandhu, and L. Claxton, eds., Plenum Press, New York,
    pp. 381-418.

10. Bradow, R.L.:   1982, "Toxicological Effects of Emissions
    from Diesel Engines," J. Lewtas, ed., Elsevier Science
    Publishing Co., New York, pp. 33-47.

11. Claxton, L.D. and Huisingh, J.L.:   1980, "Pulmonary
    Toxicology of Respirable Particles," Department of Energy,
    U.S. Government Printing Office, CONF-791002, pp. 453-465.

12. Rosenkranz, H.S., McCoy, E.C., Sanders, D.R., Butler, M.,
    Kiriazides, D.K., and Mermelstein, R.:   1980, Science 209,
    pp. 1039-1043.

13. Nishioka, M.G., Petersen, B.A., and Lewtas, J.:   1982,
    "Polynuclear Aromatic Hydrocarbons: Physical and Biological
    Chemistry," M. Cooke, A.J. Dennis, and G.L. Fisher, eds.,
    Battelle-Columbus Press, Columbus, OH, pp. 603-613.

14. Lewtas, J.:   1982, "Toxicological Effects of Emissions from
    Diesel Engines," J. Lewtas, ed., Elsevier Science Publishing
    Co., New York, pp. 243-264.

15. Claxton, L. and Kohan, M.:   1981.   "Short-Term Bioassays in
    the Analysis of Complex Environmental Mixtures II," M.D.
    Waters, S.S. Sandhu, J. Lewtas Huisingh, L. Claxton, and S.
    Nesnow, eds., Plenum Press, New York, pp. 299-317.

16. Stead, A.G., Hasselblad, V., Creason, J.P., and Claxton, L.:
    1981, Mutation Res. 85, pp. 13-27.

17. Nesnow, S., Triplett, L.L., and Slaga, T.J.:   1982, J. Natl.
    Cancer Inst. 68, pp. 829-834.

18. Lewtas, J., manuscript in preparation.

PASSENGER CAR PAH EMISSIONS AS A FUNCTION OF ENGINE DISPLACEMENT,
FUEL, AND DRIVING CYCLE

K.-H. Lies, A. Hartung, J. Kraft, J. Schulze

Volkswagenwerk AG (Wolfsburg/Germany)

ABSTRACT

Individual polycyclic aromatic hydrocarbons (PAH) were
measured for 18 light-duty vehicles representing 1978 through 1981
model-year production cars for the European and US market. In detail, the test fleet included the following classes: 1) gasoline
non-catalyst vehicles, 2) gasoline catalyst vehicles, 3) diesel
vehicles, 4) neat methanol vehicles with and without a catalyst,
and 5) methanol-gasoline blend vehicles.
Eleven PAH were quantified by means of glass capillary gas
chromatography and two-dimensional thin-layer chromatography
coupled with in situ fluorescence spectrometry.
All exhaust emission tests were carried out on a chassis
dynamometer using different driving cycles.
Sampling conditions with regard of the different fuels are
discussed.

INTRODUCTION

In automobile exhaust gases polycyclic aromatic hydrocarbons
(PAH) are identified only at low concentrations. But for the purpose of comparison there is an interest in investigating these
substances in the exhaust of automobiles in dependency of different parameters (1-3). A representative group of important PAH has
been selected as a basis for evaluating the composition of an exhaust sample (4, 5) (Table 1).
Presently, the sampling of PAH in the exhaust gas of internal
combustion engines used world wide is performed by two methods.
The first one is based on the described technique of the total exhaust sampling by condensing and filtering the raw exhaust gas flow
(5, 6, 7). The other procedure makes use of collection of particu-

181

D. Rondia et al. (eds.), Mobile Source Emissions Including Polycyclic Organic Species, 181–192.
© 1983 by D. Reidel Publishing Company.

Table 1 - Polycyclic aromatic hydrocarbons, 4-6 rings

Name PAH	Formula	M.w.[*]	B.p.[**] °C
Fluoranthene	$C_{16}H_{10}$	202	384
Pyrene	$C_{16}H_{10}$	202	394
Benzo(ghi)fluoranthene	$C_{18}H_{10}$	226	-
Benz(a)anthracene	$C_{18}H_{12}$	228	438
Chrysene	$C_{18}H_{12}$	228	448
Benzo(k)fluoranthene	$C_{20}H_{12}$	252	481
Benzo(b)fluoranthene	$C_{20}H_{12}$	252	481
Benzo(a)pyrene	$C_{20}H_{12}$	252	496
Benzo(e)pyrene	$C_{20}H_{12}$	252	493
Perylene	$C_{20}H_{12}$	252	497
Benzo(ghi)perylene	$C_{22}H_{12}$	276	500
Indeno (1,2,3-cd)pyrene	$C_{22}H_{12}$	276	534
Anthanthrene	$C_{22}H_{12}$	276	547
Coronene	$C_{24}H_{12}$	300	602

[*] M.w. Molecular weight
[**] B.p. Boiling-point

late samples via proportional streams from the diluted exhaust (8, 9). The goal of this investigation was to study the PAH exhaust emissions as a function of engine displacement, fuel, and driving cycle.

EXPERIMENTAL

    There are two advanced sampling methods for the quantitative collection of the PAH compounds in the exhaust gas of passenger cars.

Total Exhaust Collection

    The first one is based on the described technique of the
total exhaust sampling by condensing and filtering the raw exhaust
gas flow (5, 7). The three PAH samples obtained in this way (condensed material, acetone-soluble phase resulting from the cleaning up process, particulate filter extract) are concentrated by
liquid-liquid extractions and purified by silica gel into aromatic fractions. The PAH components isolated are quantified by thin-
layer chromatography coupled with in situ fluorescence spectrometry and glass capillary gas chromatography. Figure 1 shows the
corresponding flow diagram for the determination of PAH compounds
from undiluted exhaust.

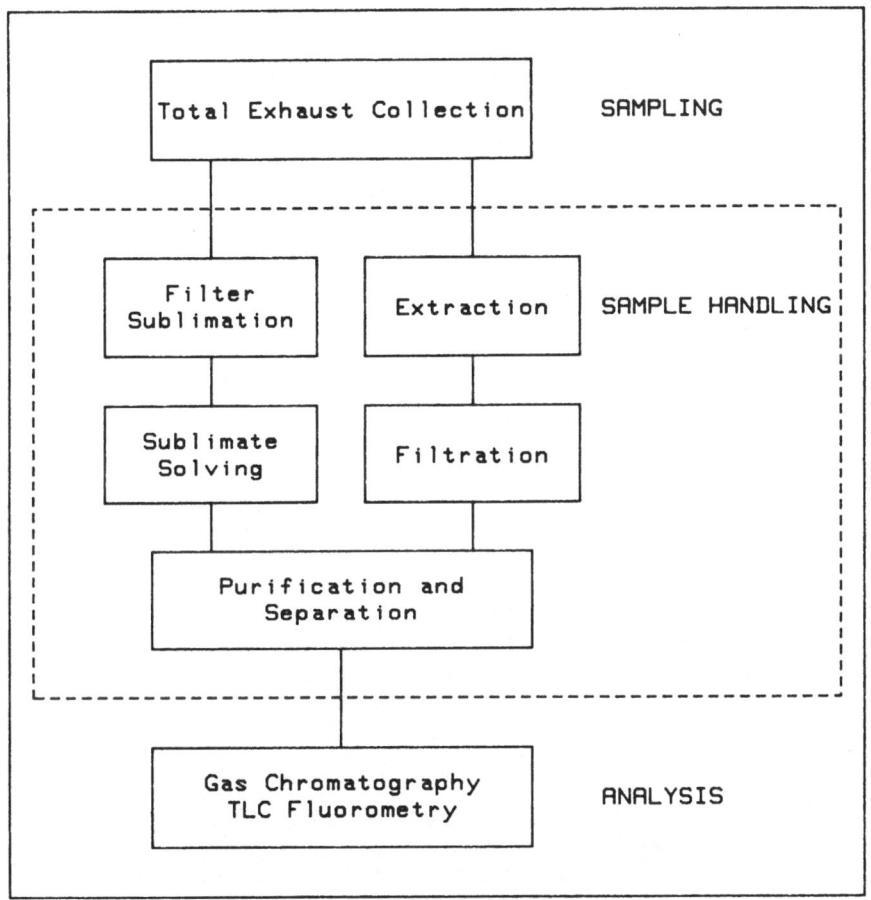

Fig. 1 - Flow diagram for PAH determination from undiluted ex-
        haust

Diluted Exhaust Collection

    The other procedure makes use of collection of particulate
samples via proportional streams from the diluted exhaust, known
as dilution tunnel technique (8, 9). Figure 2 displays a flow
diagram for this procedure.

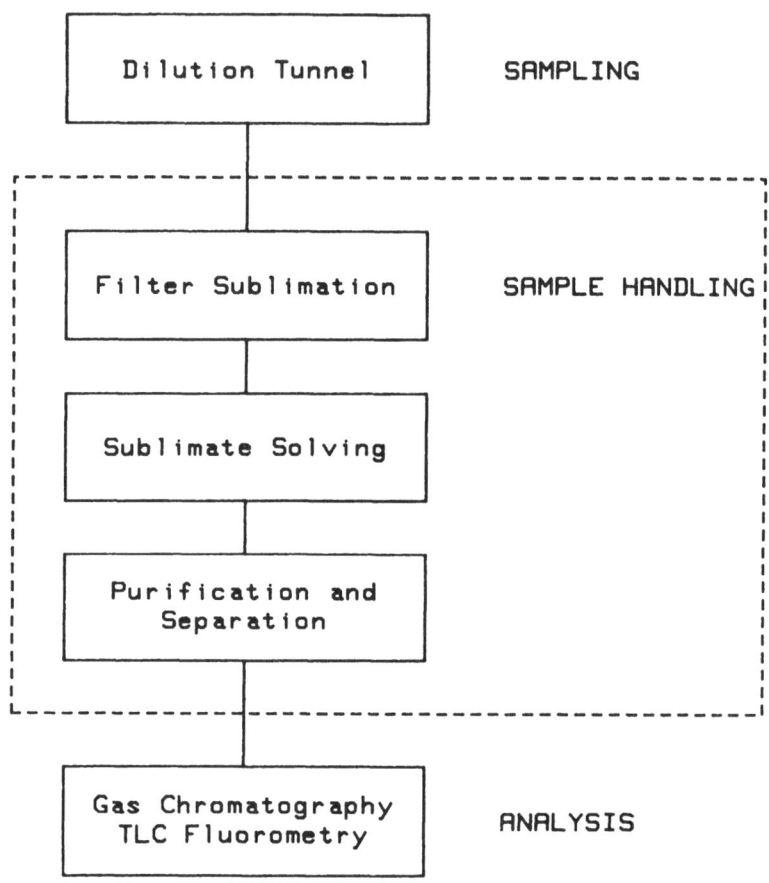

Fig. 2 - Flow diagram for PAH determination from diluted exhaust

    The dilution tunnel for the PAH collection has been described
(9, 10). This system consists of a stainless steel dilution tunnel,
a CVS device, and the unit for drawing off proportional streams
from the total gas flow. The length of the tunnel is 6 m and its
diameter is 0.45 m. At the point where the exhaust gas enters the
tunnel there is an orifice creating turbulence for mixing the exhaust more thoroughly with the filtered air. The velocity of flow

in the tunnel is 1.3 m/s and the capacity of the CVS unit positive displacement pump is 13 m³/min. Representative particulate samples are taken at the end of the tunnel through Teflon filters (diameter 47 and 70 mm) with vacuum pumps. For this purpose, a multihead dust measuring apparatus is used, in which the flow rate is regulated independently of the pressure.

Sample Handling

The methods for the sample preparations have also been published (5, 10).

Sample Analysis

Thin-Layer Chromatography Coupled with Fluorometry. Aliquots of the prepared samples stored in the pear-shaped flasks are spotted on the thin-layer plates. The development of these plates for separation of the individual compounds can be done in one or two directions which takes place in an equilibrated chamber first with a mixture of toluene/n-hexane/n-pentane (5:90:5) to a distance of 15-16 cm. After drying, the plates are developed in the second direction with a mixture of methanol/diethyl ether/water (6:4:1). A typical two dimensional TLC of a PAH standard is shown in Figure 3.

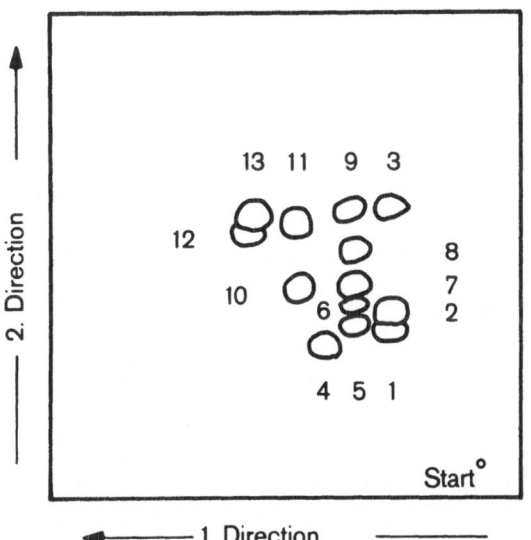

1. Anthanthrene
2. Indeno (1,2,3,-cd) pyrene
3. Benzo (ghi) perylene
4. Benzo (a) pyrene
5. Benzo (b) fluoranthene
6. Benzo (j) fluoranthene
7. Benzo (k) fluoranthene
8. Perylene
9. Benzo (e) pyrene
10. Chrysene
11. Benzo (a) anthracene
12. Fluoranthene
13. Pyrene

Fig. 3 - Two dimensional TLC of a PAH reference blend

The fluorometric determination of PAH is based on the advanced technique of thin-layer chromatography coupled with in situ fluorescence spectrometry using optical scanners. A detailed description of this procedure has already been published (11).

The repeatability of the thin-layer chromatographic determination amounts to 1 through 5 % (relative standard deviation) depending on the PAH component for the range of the concentration between 0.5 and 200 ng per sample.

Glass Capillary Gas Chromatography. For the quantitative analysis of the PAH a Varian 3700 gas chromatograph equipped with a flame-ionization detector and coupled with a data system of Spectra Physics was used. The glass capillary columns statically coated with different stationary phases (OV 101, "SE 30", SE 52, SE 54, Cp^{tm} Sil 5 (SE 30), Dexsil 300, 400, 410) were conditioned at 320 °C for 16 hours. The operating conditions have been described (5, 10). As an example, Figure 4 shows the high resolution of a SE 54 non-polar glass capillary column (25 m/ 0.3 mm/0.3 μm) of the PAH emission profile from a diesel engine driven in the cycle of the 1975 Federal Test Procedure (FTP). The relative standard deviation ranges between 1 and 6 % (5, 7, 10).

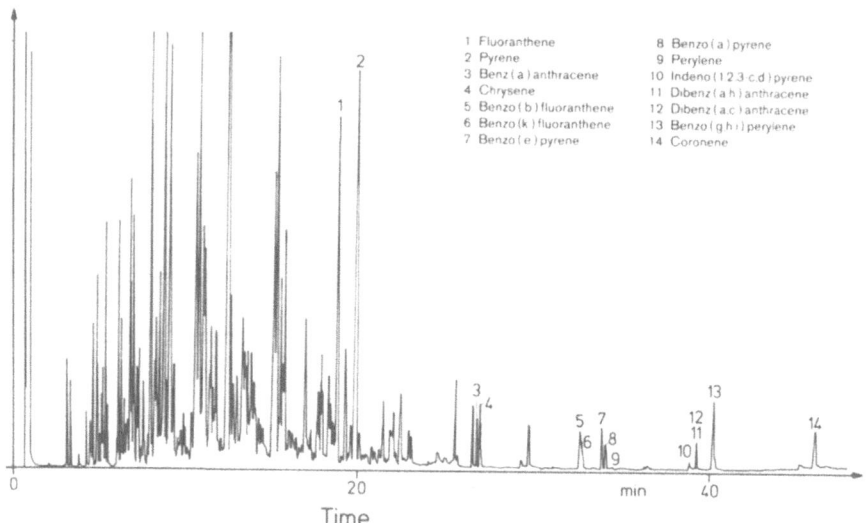

Fig. 4 - Gas chromatogram of PAH from diesel engine exhaust

Comparison of the Analytical Methods

In order to compare the analytical methods used in this study the same exhaust gas sample was analysed employing the thin-layer and the glass capillary gas chromatographic procedure taking the TLC method as the reference procedure. The relative standard deviation resulting from 11 double determinations gives a value of about 9 %, which is a satisfactory result considering the very low concentrations of the individual PAH compounds (5, 7, 12, 13).

Test Cars

A fleet of 18 vehicles was tested in this program. The description of the individual types is given in Table 2.

Table 2 - Vehicle description

Vehicle No.	Model Year	Displacement litre	Inertia Weight lbs.	Type
1a	1979	1.1	2.000	Gasoline
1b	1981	1.1	2.000	"
2	1979	1.3	2.500	"
3	1980	1.5	2.250	"
4a	1981	1.6	2.500	"
4b	1981	1.6	2.500	"
5a	1981	1.6	2.500	Gasoline 3-Way-Cat.
5b	1981	1.6	2.500	"
6a	1978	1.5	2.250	Diesel
6b	1979	1.5	2.250	"
7a	1980	1.6	2.500	"
7b	1979	1.6	2.500	"
8a	1979	2.0	3.250	"
8b	1980	2.0	3.250	"
8c	1980	2.0	3.250	"
9	1981	1.6	2.500	M 15
10	1981	1.6	2.500	M 100
11	1981	1.6	2.500	M 100 3-Way-Cat.

The preconditioning was realized by driving the cars at about 80 % of their maximum speed over a distance of about 120 km on the motor road. After each test the cars were conditioned in the same way.

Each car was tested over three driving cycles including the driving cycle of the 1975 Federal Test Procedure (FTP), of the Highway Fuel Economy Test (FET) and of the European Test Procedure (ECE).

RESULTS AND DISCUSSION

The results of the PAH emission measurements are plotted in Figures 5 - 7.

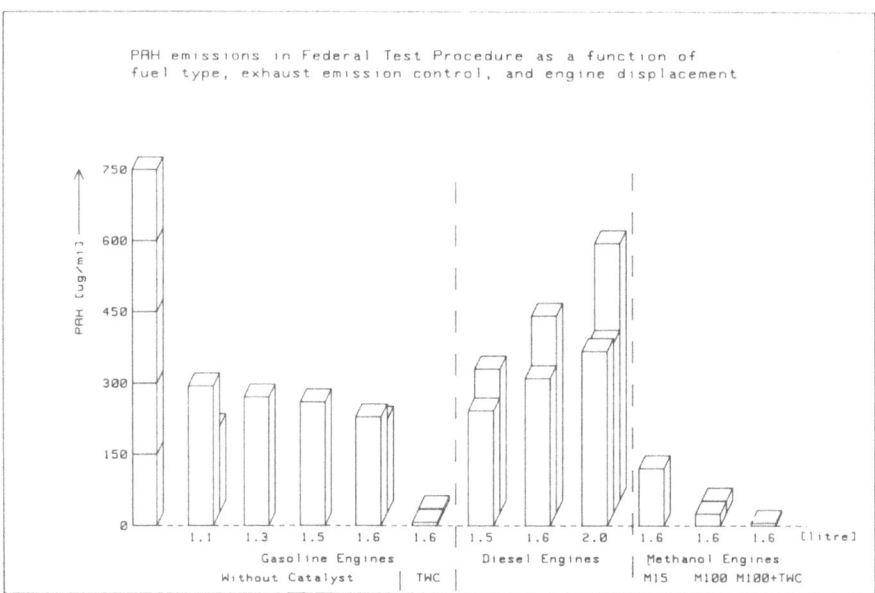

Fig. 5 - Exhaust gas PAH emissions of the VW/Audi test fleet in the Federal Test Procedure (FTP). On the abscissa the cars are depicted with regard to the parameters; engine displacement, fuel, and exhaust control device. Columns in the second horizontal direction represent measuring values for the PAH of different vehicles of the same type. Each column represents the mean value of three re- petetive measurements of the same vehicle. For each car group two or three different cars were measured in most cases. TWC = Three Way Catalyst; M 15 = methanol-gasoline blend fuel (15 % methanol); M 100 = pure methanol fuel

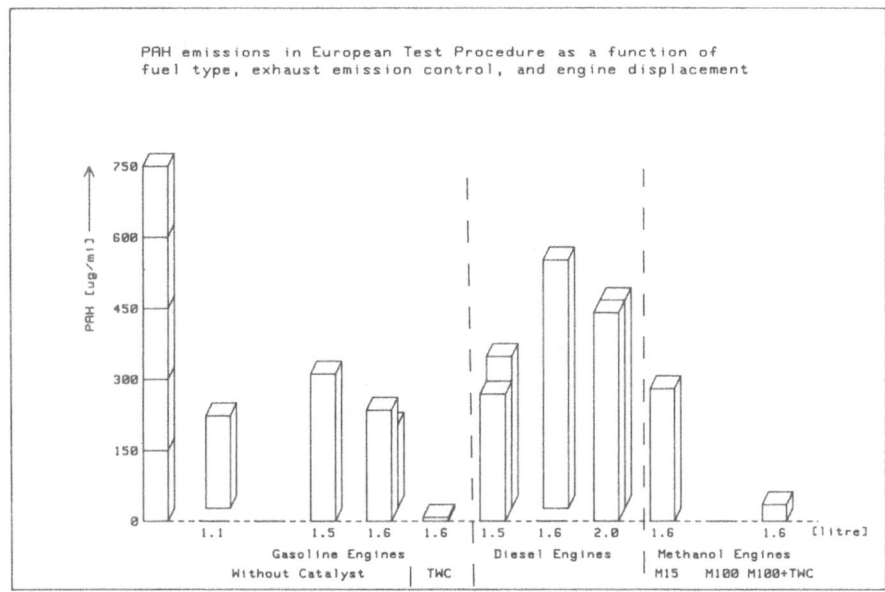

Fig. 6 – Exhaust gas PAH emissions of the VW/Audi test fleet in
the European Test Procedure (ECE)

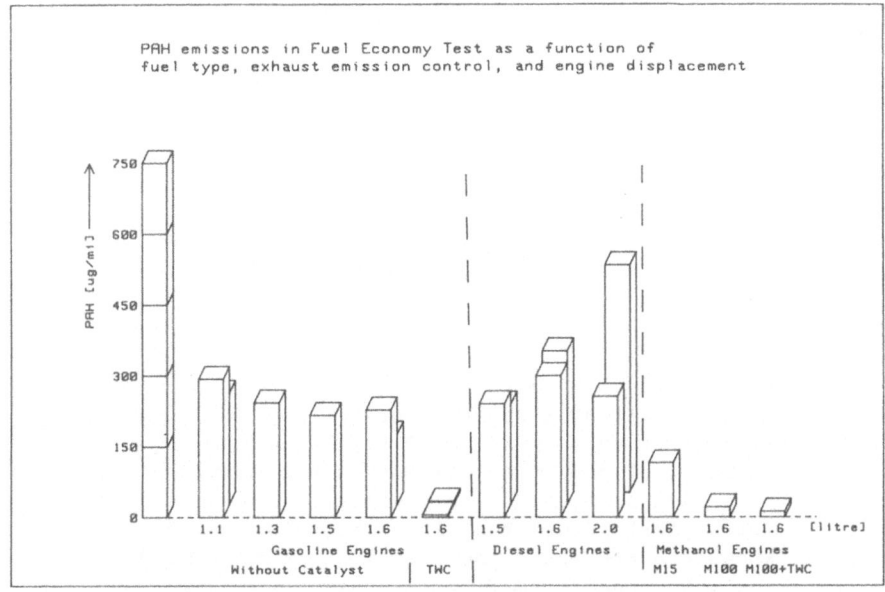

Fig. 7 – Exhaust gas PAH emissions of the VW/Audi test fleet in
the Fuel Economy Test Procedure (FET)

In some cases not all three test procedures have been performed. Especially the ECE test procedure was not carried out in
all cases. With regard to the test cars not all types of VW/Audi
production have been measured. Especially in the case of methanol
blend and pure methanol only a few cars were used.

In the figures the sum of the emissions for the following
eleven PAH are plotted: fluoranthene, pyrene, chrysene, benzo(b)-
fluoranthene, benzo(k)fluoranthene, benzo(a)pyrene, benzo(e)-
pyrene, perylene, indenopyrene, benzo(ghi)perylene, coronene. In
such cases where the individual tests for vehicles have not yet
been carried out, the corresponding block is missed in the figures.

The emissions of the limited exhaust gas components are well
below the applicable standards and are not presented here.

For a comprehensive overview all values of the Figures 5
through 7 are combined to one diagram shown as Figure 8.

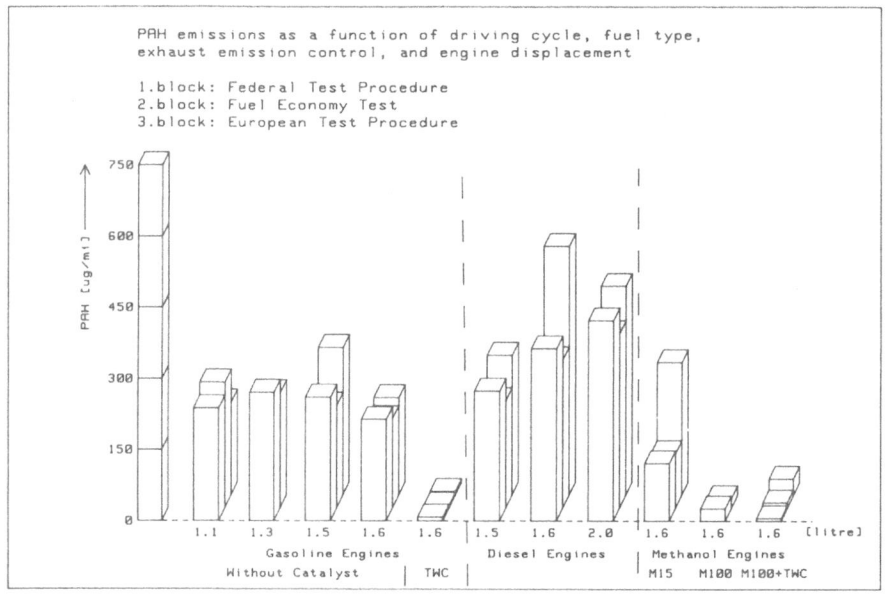

Fig. 8 – Exhaust gas PAH emissions of the VW/Audi test fleet in
three different test procedures. Each block represents
the resulting mean value of the PAH emissions from all
vehicles of one type. Columns in the second horizontal
direction show in this Figure the dependency on the different test procedures. The first block of the group
shows the FTP test, the second block displays the FET
test, and the third block represents the ECE test. ECE
test measurements for the gasoline engine 1.3 l and the
methanol (M 100) engine 1.6 l have not yet been performed

The quantitative aspects of the results so far obtained serve to give some significant indications on the one hand and some trends only on the other hand. The main points of both are the following:
- There is an indication of a tendency that diesel cars emit more PAH than gasoline cars without a control device. But the results range in the same order of magnitude. Most of the values were found between 300 - 450 µg/mi.
- Significantly lower results are obtained when gasoline cars are equipped with an exhaust control device (three way catalyst). In these cases there are PAH values of about 10 µg/mi. This is true for all three cycles run in this study.
- The cars fueled with a gasoline/methanol blend (M 15) show lower values only in the FTP and the FET.
- Pure methanol (M 100) fueled cars emit in both cases, without and with a catalyst, very low amounts of PAH. Significant influences from the driving cycle cannot be observed.

ACKNOWLEDGEMENT

The authors wish to thank Dr. H. Klingenberg and Dr. D. Schürmann for their support of this work and for their critical comments.

REFERENCES

1. Ames, B.N., McCann, J., Yamasaki, E.: 1975, Mutation Research 31, pp. 347.
2. UBA, "Luftqualitätskriterien für ausgewählte PAH": 1979, Berichte 1/79, E. Schmidt Verlag, Berlin.
3. Schaefer-Ridder, M.: 1979, Nachr. Chem. Tech. Lab. 27, pp. 4.
4. Kaschani, D.T.: 1980, in: W. Bertsch et al., "Instrumental HPTLC", Proceedings of the First International Symposium of Instrumentalized High Performance Thin-Layer-Chromatography (HPTLC), Bad Dürkheim, pp. 185-208.
5. Kraft, J., Lies, K.-H.: 1981, SAE-Paper No. 810082, pp. 1-11.
6. Grimmer, G., Hildebrandt, A., Böhnke, H.: 1972, Erdöl und Kohle - Erdgas-Petrochemie 25, pp. 442.
7. Kraft, J., Hartung, A., Lies, K.-H.: 1981, Fourth International Symposium on Capillary Chromatography, Hindelang (West Germany), May 3-7, pp. 639-671.
8. Federal Register: 1980, Vol. 45, March.
9. Hartung, A., Kraft, J., Lies, K.-H., Schulze, J.: 1982, MTZ 6, pp. 263-266.

10. Kraft, J., Hartung, A., Lies, K.-H., Schulze, J.: 1982, SAE-Paper No. 821219, pp. 1-15.

11. Kraft, J., Hartung, A., Lies, K.-H., Schulze, J.: 1982, in: R.E. Kaiser, "Instrumental HPTLC", Proceedings of the Second International Symposium of Instrumentalized High Performance Thin-Layer-Chromatography (HPTLC), Interlaken, pp. 144-158.

12. Doerfel, K.: 1962, Analyt. Chem., 185, pp. 3.

13. Kraft, J., Postulka, A., Gring, H.: 1978, VW Research Report No. FMT 7802V/5, pp. 32-35.

ACTIONS OF THE EUROPEAN COMMUNITIES IN THE DOMAIN OF AIR
POLLUTION CAUSED BY HYDROCARBONS

F. Magdonelle

COMMISSION OF THE EUROPEAN COMMUNITIES
Brussels, Belgium.

The actions of the European Communities within this domain are
juridically based on the Action Programmes on the Environment
adopted by the Council of Ministers which is the decision-making
authority of the Community. The Commission of the European
Communities is the executive body and, as such, is charged
with the operation of Council Decisions.

The first Action Programme on the Environment was adopted by the
Council of Ministers on 22 November 1973 and was published in the
Official Journal (C 112) on 20 December 1973. A second programme
was adopted by the Council on 17 May 1977 and published on
13 June 1977 (C 139).

On 9 November 1981 the Commission submitted a draft action
programme to the Council for the period 1982-1986. This project
was published in the Official Journal on 25 November 1981 under
the number C 305. The project has not yet been adopted by the
Council and is not yet, therefore, in operation.

From the first Programme, and in the up-dates, the hydrocarbons
have been classified as pollutants of the First Category, always
with the restriction that they are to be considered as such if
they have known or probable carcinogenic effects. Also included
in the First Category are the organo - halogens and the organo-
phosphors as well as the photochemical oxidants, for which the
hydrocarbons - at least the non-methanes - are precursors and
intervene in the formation of ozone and PAN. Within the programmes
it is clearly specified that the first task to be undertaken is the
work necessary to normalise or standardise the measurement
methods.

D. Rondia et al. (eds.), Mobile Source Emissions Including Polycyclic Organic Species, 193–196.
© *1983 by D. Reidel Publishing Company.*

A mandate, although defined in very broad terms, was given to the
Commission to study the actions necessary to combat this type of
pollution. Given the relatively small availabilities of budget
and personnel - in the Commission there are only 3 people concerned
with air pollution, for example - and the enormity of the tasks,
priority was accorded to the principal pollutants of the First
Category, namely, sulphur compounds and suspended particulates
in the Air Quality Directive (80/779/EEC), nitrogen oxides, lead
in petrol and in the air, all of which were mentioned in the First
Action Programme as having priority within the First Category.

Actions on other pollutants in this category were, in the first
instance, partly preparatory, partly legislative but within a
restricted domain such that it was impossible to take complete
global action.

This is the case for hydrocarbons where our actions have been
relatively limited. Within our work we can distinguish two
aspects: preparations for legislative action on the one hand,
and specific legislation on the other.

Studies and preparations

A number of studies on motor-vehicle emissions have been realised
for the preparation of directives and have been formalised by
different Community Directives. Other studies on motor-vehicles
are destined to lead to a further stage in the reduction of emissions. It must be stated that these studies concern not only the
problems of hydrocarbon emissions but also the whole spectrum of
emissions; one cannot distinguish between different pollutants
and, apart from hydrocarbons, one must simultaneoulsy consider
nitrogen oxides, carbon monoxides as well as taking into account
economic aspects of energy consumption and prices. In these
studies no real distinction is made between the different types
of hydrocarbons, one considers total hydrocarbons.

Another type of study, entitled "Monitoring & analytical
requirements for environmental polycyclic aromatic hydrocarbons"
(Prof. Grimmer, 1978) was undertaken, but has not yet lead to
any legislative action.

In the preparation of actions on photochemical oxidants, the
Commission undertook a global study of the problem including non-
methanic hydrocarbons as precursors. This study will terminate
in 1983 and includes measurement of precursors, their methods of
control as well as proposals for possible actions.

Legislative actions

Legislative action in the European Communities is taken by the
Council of Ministers upon a proposition by the Commission. It
may take the form of Decisions, Directives or Recommendations.
The Decision is the most important and obliges the Member
States to put it into operation without leaving any choice in
the means. A Directive is also obligatory but leaves open the
means whereby it is put into operation. A recommendation is the
least restrictive and imposes no obligation. In the cases of
the Decisions and Directives the Member States are obliged to
transpose the requirements into their national legislation
within a specified time. Decisions, Directives and
recommendations are published in the Official Journal of the
European Communities at the time of their presentation to,
and their approval/adoption by, the Council of Ministers.
When published as an adopted act it is regarded as being in its
definitive form and acquires the force of law within the Member
States of the European Community.

The regulations for motor-vehicle emissions are contained in the
Directive 70/220/CEC of 20 March 1970 and were modified as
follows:

- 74/290/CEC of 28 May 1974
- 77/102/CEC of 30 November 1977
- 78/665/CEC of 14 July 1978

These modifications almost always appertain to an more rigourous
limitation on the emissions as a result of discussion within the
ECE at Geneva.

A further modification currently exists in the form of a proposal
and was submitted to the Council of Ministers on 14 April 1982;
it was published in the Official Journal C 181, 19 July 1982 and
is due for discussion by the Council within the next few months.
It modifies the preceeding legislation by specifying new test
values for motor-vehicles. As they are more than 100 pages
it is not possible for me to give you a reasonable résumé.

Requirements for an abatement policy for hydrocarbons in general and in particular for the preparation of a draft directive

1.   The first action is to study the healths effects of the
proposed compounds, including carcinogenicity, etc. Laboratory
studies on animals are insufficient and epidemiological studies
must be included to demonstrate any possible risks due to current
ambient concentrations or levels which may occur in the future.
These studies should provide some idea of threshold levels since
in defining levels zero is impossible.

2.   Additionally methods of sampling and measurement for each
compound are necessary, not only very elegant analytical methods
which may be usable in very few laboratories, but also simpler
or less-demanding methods for use in networks. A Directive/
legislation   without an adequate reliable control is
completely ineffective. Such methods must be conform to an
International Standard, preferably that of the ISO.

3.   The proposed legislation must also be supported by studies
on techniques for control and reduction of emissions thus
providing proof of the feasability of proposed reductions.
Such studies must consider the economic implications of
réductions. Such studies must consider the economic implications
as well as promoting networks, the present economic climate
underlining the importance of this aspect.

Stages in the adoption of legislation by the Council of Ministers:

1.   Draft by the Commission perhaps aided by contractants
(6 months to 1 year).

2.   Discussions with Government experts (1-2 years)

3.   Adoption by the Commission after approval by different
Services (2-3 months).

4.   Submission to Council of Ministers and publication in
Official Journal.

5.   Opinions of Economic and Social Committee and European
Parliament.

6.   Adoption at Council after Groups discussions which may
take several years.

This gives an overall view of the work necessary to obtain
European legislations; afterwareds it must be transcribed into
national legislation for which the delays are specified in the
different legislations.

Conclusions

Up to this time the work of the Commission of the European
Communities has been relatively limited and concerns, principally
motor-vehicles emissions.   However, the Commission is preparing
other actions relative to hydrocarbon emissions, notably within
the context of the reduction of photochemical oxidants.

COMPARISON OF NITRO-AROMATIC CONTENT AND DIRECT-ACTING MUTAGEN-
ICITY OF PASSENGER CAR ENGINE EMISSIONS

Marcia G. Nishioka, Bruce Petersen, Joellen Lewtas*

Battelle, Columbus Laboratories
*U.S. Environmental Protection Agency

Particulate emissions from internal combustion engines are
a very complex mixture of carbonaceous matter containing absorbed
and/or condensed organic components from the combustion of fuel
and lubricants. The organic solvent extractable fraction of these
particulates is extremely complex and has been reported to contain
hundreds of individual compounds. A number of studies have
reported that these solvent extracts are mutagenic as determined
by the Ames Salmonella microbial assay. The greatest activity
which has been found in certain compound class fractions within
these extracts does not require metabolic activation and is
called direct mutagenic activity. Chemical characterization
studies have been used in an attempt to identify the compound
classes or specific compounds present in these fractions.
Polynuclear aromatic hydrocarbons (PAH), particularly pyrene,
have been reported to readily react with nitrogen oxides to
form nitrated derivatives which are powerful direct-acting
mutagens. Both the PAHs and nitrogen oxides are present in
engine emissions and thus the formation of nitroaromatics may
indeed be possible. These results have prompted increased
attention to the characterization of particulate extracts to
identify the compounds or compound classes which are responsible
for the direct acting biological activity. Identification of
specific mutagens or classes of mutagens is important to determine
whether these components are present in the effluent which is
emitted to the atmosphere or are formed as a result of the
particulate collection and analysis procedures. This information
must be established before potential effects of these compounds
on the environment can be addressed.

This presentation will describe a study to identify and

197

D. Rondia et al. (eds.), Mobile Source Emissions Including Polycyclic Organic Species, 197–210.
© 1983 by D. Reidel Publishing Company.

quantify 23 nitroaromatic compounds in the extracts of particulate material from three light-duty passenger car diesel
engines and one gasoline engine. Mutagenic assay data were
also collected on these extracts using the Salmonella typhimurium TA98 bioassay without S9 metabolic activation. The results
of these two studies were compared to determine whether the
concentration of the nitroaromatics detected can fully account
for the direct-acting mutagenic activity indicated by the
bioassay data.

INTRODUCTION

The increasing number of automobiles and light-duty trucks
powered by diesel engines has generated concern over the emissions associated with these engines. Diesel engines have a
higher particulate emission rate than that of gasoline catalyst
engines. The extractable organics from both diesel and gasoline
particle emissions have been found to be mutagenic and carcinogenic[1]. Recently, nitro-substituted polynuclear aromatic
hydrocarbons (nitro-PAH, nitro-aromatics) have been identified in
diesel particle extracts[2]. Several of these nitro-aromatics
are very potent bacterial mutagens[3].

A study was carried out to identify and quantify nitro-
aromatic compounds in the extract of particulate material from
three different light-duty passenger car diesel engines and
one gasoline engine. The operating and sampling conditions
have been described elsewhere[4]. Mutagenic assay data were
also collected on these extracts using the Salmonella typhimurium
TA98 bioassay, without S9 metabolic activation. The results
of these two studies were compared to determine whether the
concentration of nitro-aromatics detected can fully account
for the direct-acting mutagenic activity indicated by the bioassay
data.

TECHNICAL APPROACH

Chemical

Each particulate extract was separated into four chemical
compound classes using open-bed silica gel column chromatography:
(1) aliphatic hydrocarbons--hexane elution, (2) polycyclic
aromatic hydrocarbons--hexane:benzene (1:1 v/v) elution,
(3) moderately polar neutrals--methylene chloride elution, and
(4) highly polar neutrals, primarily oxygenated compounds--
methanol elution. Fraction #2 was expected to contain the
mononitro-aromatics, fraction #3 was expected to contain the di-
and trinitro-aromatics, and fraction #4 was expected

to contain the oxygenated nitro-aromatics. The quantity of
silica gel and the volume of elution solvent used were adjusted
for the amount of total organic material available for fractionation.

A high resolution gas chromatography/negative chemical
ionization mass spectrometry (HRGC/NCIMS) method using on-column
injection was developed to analyze the particulate extracts. The
instrumentation consisted of a Finnigan Model 4000 GC/MS inter-
faced to a Finnigan/INCOS Model 2300 Data System. Methane was
selected as the reagent gas. A J&W DB-5 bonded, fused silica
capillary column was used. The on-column injection was made
at 40°C, and the GC was temperature programmed from 40-320°C
at 10°C/minute. Standard solutions and fractions #2, #3, and
#4 of each extract were analyzed by this method.

Standard solutions containing the following nitro-aromatic
compounds

1-nitronaphthalene	2,7-dinitrofluorene
2-nitrofluorene	2,7-dinitrofluorenone
9-nitroanthracene	1,6-dinitropyrene
3-nitrofluoranthene	1,3,6-trinitropyrene
1-nitropyrene	

were prepared over the concentration range 0.1-10 ng/μl for the
calibration curve. Perdeutero (d$_7$) nitronaphthalene was chosen
as the internal standard and it was spiked into all standards
and samples at the level of 2.5 ng/μl. The response factor for
each compound relative to the internal standard was calculated
over the calibration range. Nitro-aromatic compounds not present
in the standard which were detected in samples were quantified
using the response factor of the most similar compound from the
standard. In addition, separate solutions of 1,3- and 1,8-
dinitropyrene were analyzed to determine the elution order of
the three dinitropyrene isomers.

The relative retention time (relative to d$_7$-nitronaphthalene)
and mass spectrum of each standard compound were used to assign
specific isomer identification to a few of the nitro-aromatics
detected in the extracts. The other nitro-aromatic isomers
detected were identified by appropriate mass spectra and retention
times. The availability of other specific nitro-aromatic isomers
may allow assignments to be made of those compounds.

The methanol fractions (#4) of two of the light-duty diesel
particulate extracts were also analyzed by HRGC/MS and high
resolution mass spectrometry (HRMS) in an attempt to tentatively

identify the formulas of compounds in the most polar class fraction. From these data, structures and compound classes were tentatively assigned.

Biological

An aliquot of the total extract from each sample was tested in severe short-term bioassays as reported previously[1]. The slope of the dose response curve (rev/μg) in the Salmonella typhimurium TA98 plate incorporation assay[5] without metabolic activation (-S9) was utilized in this study to compare the nitro-aromatic content and direct-acting mutagenicity of each sample extract.

RESULTS

Twenty-three different nitro-aromatics were tentatively identified in the three light-duty diesel (LDDI, LDDII and LDDIII) engine extracts. However, only 1-nitropyrene was detected in the gasoline engine extract. In all cases, the 1-nitropyrene was the nitro-aromatic compound detected in greatest quantity, followed by the nitrophenanthrene/anthracene isomers. Quantities of the detected nitro-aromatics are given in Table 1 and vary over the range of 0.1 ppm to 589 ppm in the extracts.

Mono-nitro derivatives of the PAHs at molecular weight 228 (such as benz[a]anthracene and chrysene) and molecular weight 252 (such as benzopyrenes, perylene and benzofluoranthenes) were also detected. Two nitropyrenone isomers were tentatively identified in the LDDI and LDDIII samples and the three dinitro-pyrene isomers were identified in the LDDII sample.

The chromatograms from the NCI HRGC/MS analysis of the hexane:benzene and methylene chloride fractions of the LDDI diesel extract are shown in Figures 1 and 2, respectively. In spite of the relatively rapid temperature program rate, 10°C/minute, the excellent resolving power of the capillary column was maintained. As shown in Figure 1, analyses were complete in 30 minutes, achieving baseline separation for every significant peak. This resolving power enhanced the accuracy for quantifying closely eluting structural isomers.

In Figure 1, the peaks with intensity greater than 25% of the largest peak were identified as PAHs, with the exception of the peak at scan #1275. This peak was identified as 1-nitropyrene. The methylene chloride fraction was considerably more complex and included primarily carboxaldehyde and semi-quinone derivatives of PAH. The largest peak (scan #850) was identified as 9-fluorenone.

TABLE 1. QUANTIFICATION OF NITRO-AROMATICS DETECTED IN EXTRACTS OF ENGINE PARTICULATE

Compound	Concentration in Extract, wt ppm				Relative Retention Time(a)
	LDDI	LDDII	LDDIII	Gasoline	
1-nitronaphthalene	0.5	0.7	0.3	ND	1.003
Nitronaphthalene/azulene isomer	0.1	0.6	0.2	ND	1.031
Nitronaphthalene/azulene isomer	0.9	1.2	0.3	ND	1.042
Nitrofluorene isomer	ND(b)	1.0	0.5	ND	1.426
2-nitrofluorene	ND	ND	0.4	ND	1.478
Nitrofluorene isomer	1.9	1.4	1.1	ND	1.503
Nitrophenanthrene/anthracene isomer	98	285	63	ND	1.494
Nitrophenanthrene/anthracene isomer	2.7	33	25	ND	1.545
3-nitrofluoranthene	2.9	1.2	0.9	ND	1.804
Nitrofluoranthene/pyrene isomer	1.7	1.4	0.6	ND	1.813
1-nitropyrene	407	589	107	2.5	1.842
Nitromethylpyrene isomer	23	46	22	ND	1.843
Nitromethylpyrene isomer	1.9	3.3	1.9	ND	1.874
Nitro 228(c)	3.0	1.1	1.4	ND	1.881
Nitro 228	0.8	1.3	0.2	ND	1.938
Nitro 228	3.9	5.6	1.2	ND	1.977
Nitromethyl 228	2.6	ND	ND	ND	2.132

TABLE 1. (CONTINUED)

Compound	Concentration in Extract, wt ppm				Relative Retention Time (a)
	LDDI	LDDII	LDDIII	Gasoline	
Nitro 252(d)	29	10.0	1.1	ND	2.232
Nitropyrenone isomer(e)	4.8	ND	2.1	ND	1.901
Nitropyrenone isomer(e)	11.5	ND	3.9	ND	1.916
1,3-dinitropyrene	ND	0.6	ND	ND	2.090
1,6-dinitropyrene	ND	0.6	ND	ND	2.126
1,8-dinitropyrene	ND	0.4	ND	ND	2.149

(a) Relative to internal standard, d_7-nitronaphthalene.
(b) Not detected.
(c) PAH isomers of MW 228-benz(a)anthracene, chrysene, benz(c)phenanthrene, etc.
(d) PAH isomers of MW 252-B(e)P, B(a)P, perylene, benzofluoranthenes, etc.
(e) Tentative identification.

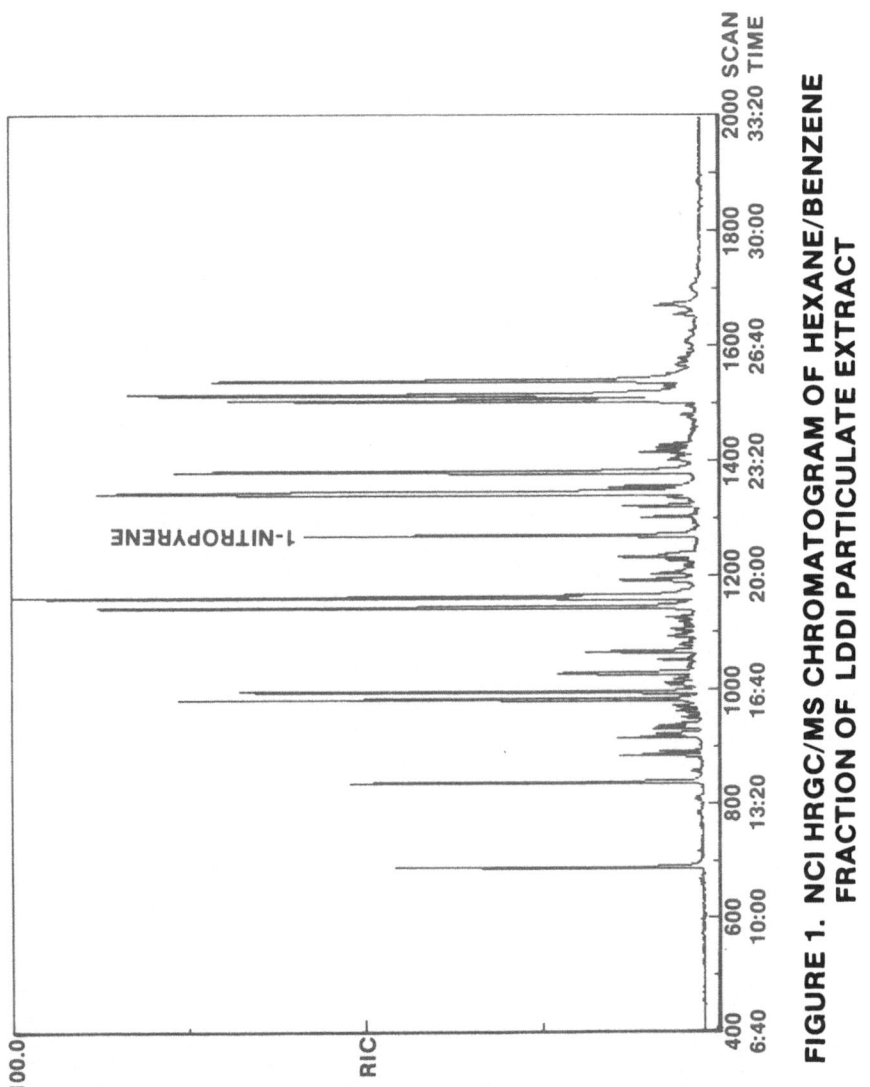

FIGURE 1. NCI HRGC/MS CHROMATOGRAM OF HEXANE/BENZENE FRACTION OF LDDI PARTICULATE EXTRACT

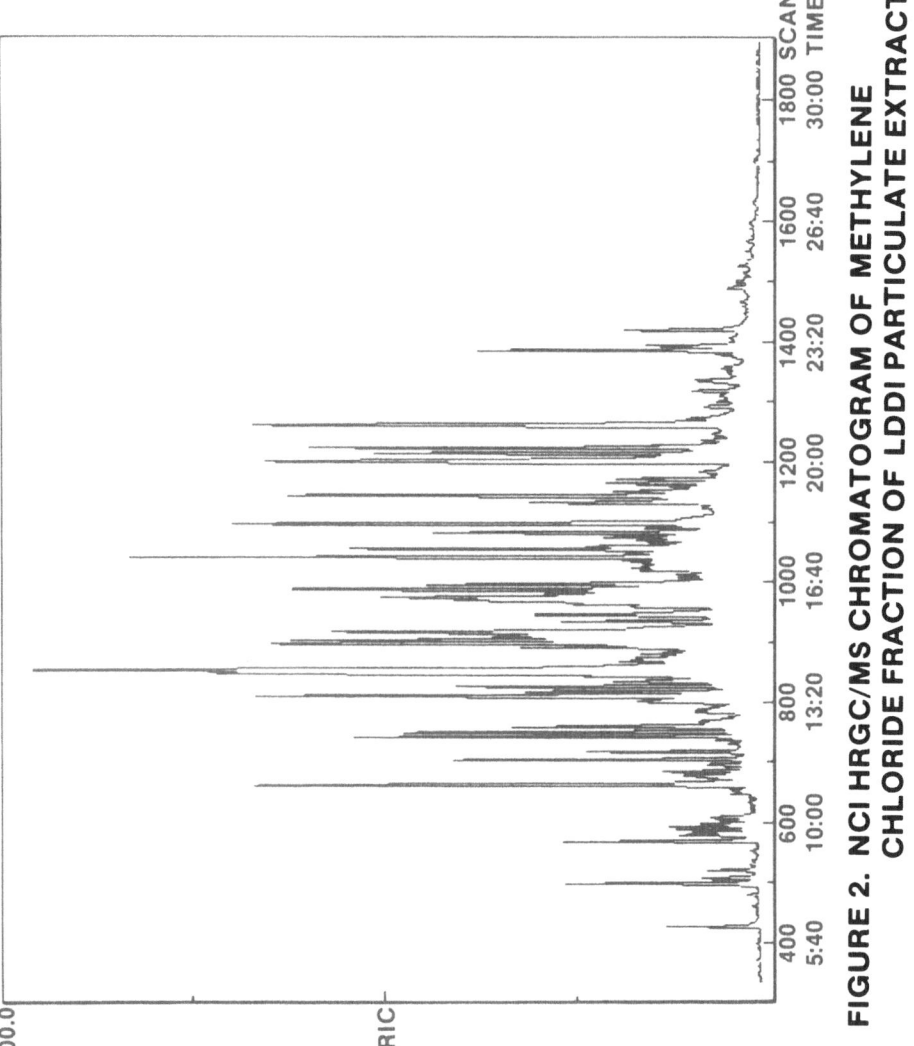

FIGURE 2. NCI HRGC/MS CHROMATOGRAM OF METHYLENE CHLORIDE FRACTION OF LDDI PARTICULATE EXTRACT

The mutagenic activity (rev/μg) by linear regression analysis for the engine emission extracts and selected nitro-aromatic compounds is given in Table 2. Note that the commercially available 1-nitropyrene appears to be contaminated with di-nitropyrenes.

Analyses by high-resolution mass spectrometry and electron impact HRGC/MS were used to determine that most of the compounds in the methanol fraction of the LDDI and LDDII extracts were quinones, ketones, and dicarboxylic acid anhydride derivatives of PAHs. Representative compounds tentatively identified in the methanol fractions are given in Table 3.

TABLE 2. MUTAGENIC ACTIVITY OF ENGINE EXTRACTS
AND PURE NITRO-AROMATICS

Sample	Mutagenic Activity, rev/μg TA98,-S9	Reference
LDDI	12.0	
LDDII	3.7	
LDDIII	5.8	
Gasoline	1.1	
1-nitronaphthalene	0.3	3
2-nitronaphthalene	1.2	3
2-nitrofluorene	65.9	3
3-nitrofluoranthene	22,500	6
1-nitropyrene (commercially available)	2,111	7
1-nitropyrene (99.99% pure)	918	7
1,3-dinitropyrene	496,000	8
1,6-dinitropyrene	629,000	8
1,8-dinitropyrene	870,000	8
nitrophenanthrene (nitration mix)	240	9
nitroanthracene (nitration mix)	1.8	9
nitrochrysene (nitration mix)	190	9
nitroperylene (nitration mix)	654	9

TABLE 3.   REPRESENTATIVE COMPOUNDS TENTATIVELY
           IDENTIFIED IN METHANOL FRACTIONS

Tentatively Identified Compound	Accurate Mass	Measured Mass Assignment Accuracy, ppm	Formula
Fluorenone	180.0574	-2.7	$C_{13}H_8O$
Acenaphthylenedione	182.0367	-3.4	$C_{12}H_6O_2$
Anthracenone	194.0730	0.5	$C_{14}H_{10}O$
Naphtho-pyrandione	198.0315	-0.2	$C_{12}H_6O_3$
Anthracendione	208.0522	-3.7	$C_{14}H_8O_2$
Pyrenedione	232.0523	0.4	$C_{16}H_8O_2$
Phenanthrenediquinone	238.0264	6.3	$C_{14}H_6O_4$
Cyclopenta(def)-chrysenone	254.0731	-1.1	$C_{19}H_{10}O$
Benzo(ghi)fluoranthenedione	256.0522	6.7	$C_{18}H_8O_2$
Pyrene-pyrandione	272.0471	3.8	$C_{18}H_8O_3$

CONCLUSIONS

Negative chemical ionization high-resolution gas chroma-
tography/mass spectrometry with on-column injection (NCI HRGC/MS)
provides sensitivity and selectivity for the detection of nitro-
aromatics. The limit of detection is approximately 50 pg for
the mono-nitro-aromatics and di-nitro-pyrenes in the full mass
scan data acquisition mode. The limit of detection can be
improved, if necessary by using selected ion monitoring
techniques.

The NCI on-column injection HRGC/MS method provides significant benefits for the analysis of nitro-aromatics over the
conventional EI GC/MS methods. It provides for injection into
the chromatographic system at a lower temperature in order to
avoid degradation of the thermally labile nitro-aromatics. The
chemical ionization technique is less energetic than electron
impact and thus enhances sensitivity for detection of a molecular
ion by decreasing the amount of fragmentation. The negative
chemical ionization technique is especially sensitive and
selective for the detection of nitro-aromatics. This is due
to the electro-negative nature of the nitro substitution which
is highly susceptible to attachment of a thermal electron from
the reagent gas plasma. Theoretically, the greater the nitro
substitution the higher the electronegativity and thus the
lower the limit of detection. This enhanced sensitivity has
been observed only when an extremely well deactivated GC
column was used. Column active sites presumably cause the
irreversible adsorption of nitro-aromatics and the magnitude of
the effect is greater for the di- and tri-nitro compounds than
for the mono-nitro compounds. In practice, sensitivity for
mono- and di-nitro compounds is approximately the same. For
the reasons listed above, the NCI on-column injection HRGC/MS
method is the most sensitive and selective method identified
to date for the analysis of nitro-aromatics.

For each extract, the percentage of measured mutagenic
activity due to 1-nitropyrene was calculated and is given in
Table 4. The value of 918 rev/µg was used as the mutagenic
activity of 1-nitropyrene for these calculations. In addition,
the percentage of measured mutagenic activity due to those
nitro-aromatics detected by mass spectrometry was calculated
for each extract using the specific activities listed in Table 2.

The nitration mixes produced synthetically were assumed to
be similar in isomer distribution to that found in the diesel
extracts. The mutagenic activity of each nitration mix was
applied to the appropriate isomers detected in the extracts.

Since the calculated contribution to mutagenic activity
from identified nitro-aromatics is well below 100 percent,
several conclusions should be considered: (1) nitro-aromatics
whose activities are unknown to date contribute significantly
to the direct acting activity, (2) compounds other than nitro-
aromatics, present in the extracts, are potent mutagens, or
(3) nitro-aromatics originally present in the extract oxidized
or decomposed with time.

The identification of compounds present in the methanol
fractions and the detection of the two nitropyrenone isomers
tend to support the hypothesis of oxidation occurring in the

TABLE 4.   PERCENTAGE CONTRIBUTION OF NITRO-AROMATICS
            TO DIRECT-ACTING MUTAGENIC ACTIVITY OF
            EXTRACTS

Sample	Measured Mutagenic Activity[a] rev/μg	Calculated Percentage Contribution to Mutagenic Activity due to	
		1- Nitro- Pyrene	Identified Nitro- Aromatics
LDDI	12.0	3.1	4.3
LDDII	3.7	14.6	37.5
LDDIII	5.8	1.7	2.9
Gasoline	1.1	0.2	0.2

a = TA98, -S9 assay, Reference 5

diesel engine extracts during storage. This hypothesis is
further supported by data from bioassays of the LDDI extract
which had been stored for two years. These recent bioassays
indicated much reduced mutagenic activity and higher toxicity
for the stored extract than measured originally in the fresh
extract. Additionally, quinones and keto-aromatics are
generally toxic but not mutagenic. It is possible that the
nitro-aromatics may have originally accounted for a larger
percentage of the activity, but with storage, many of the nitro-
aromatics may have been oxidized to quinones and keto-aromatics.

These observations may indicate that a greater concentration
of nitro-aromatics might have been originally present in the
extracts. This possibility was investigated by extracting a
particulate sample from LDDI which had been stored for two years.
Both a bioassay and analysis for nitro-aromatics were performed
on this freshly prepared extract. The mutagenic activity was
found to be 12.8 rev/μg and is virtually the same result as
was determined two years ago. Analytical results are presented
in Table 5 along with the results from the analysis of the
stored extract. Differences in the nitroaromatic concentrations
between the freshly extracted particulate and the stored extract
demonstrate that the nitroaromatics degrade substantially on
storage. It is important to note that the 1-nitropyrene
concentration is about a factor of four higher in the freshly

prepared extract that in the stored extract (1587 ppm vs 407 ppm).
Its proportion to the mutagenic activity of the extract thus
increases from 3.1 percent to 12.1 percent. These results
indicate that the nitroaromatics may be more stable when adsorbed
or condensed onto the soot particles than in a concentrated
organic solvent extract. Biological and chemical characterization
of particulates is very ddpendent on sample preparation and
therefore must be conducted simultaneously to assure that the
sample integrity is maintained. Standardization of methods for
analysis of nitroaromatics is currently under way.

TABLE 5.   CONCENTRATION OF NITRO-AROMATICS IN LDDI
PARTICULATE EXTRACTS

Compound	Concentration in Extract, wt ppm	
	Freshly Extracted Stored Particulate	Stored Extract
1-nitronaphthalene	1.4	0.5
Nitrofluorene isomer	18	1.9
Nitrophenanthrene	272	98
Nitroanthracene	8.8	2.7
3-nitrofluoranthene	7.0	2.9
Nitrofluoranthene/pyrene isomer	7.5	1.7
1-nitropyrene	1,587	407
Nitromethylpyrene isomer	233	23
Nitromethylpyrene isomer	11	1.9
Nitrated 228 MW PAH	16	3
Nitrated 228 MW PAH	7.2	0.8
Nitrated 228 MW PAH	31	3.9
Nitrated 242 MW PAH	2.8	2.6
Nitrated 252 MW PAH	0.4	ND[a]
Nitrated 252 MW PAH	1.2	ND
Nitrated 252 MW PAH	68	29
Nitropyrenone isomer[b]	15	4.8
Nitropyrenone isomer	35	12

(a) Not detected
(b) Tentative identification

REFERENCES

1. Nesnow, S., and Huisingh, J. L. Mutagenic and carcinogenic
    potency of extracts of diesel and related environmental
    emissions: Summary and discussion of the results. In:
    Health Effects of Diesel Engine Emissions, Vol. II. Edited
    by W. E. Pepelko, R. M. Danner, and N. A. Clarke, EPA-600/9-
    80-057b, U.S. Environmental Protection Agency, Cincinnati,
    Ohio, 1980, pp. 898-912.

2. Petersen, B. A., Chuang, C., Margard, W., and Trayser, D.
    Identification of mutagenic compounds in extracts of diesel
    exhaust particulates. In: Proceedings of the 74th Annual
    APCA Meeting, Philadelphia, Pennsylvania, 1981.

3. Rosenkranz, H. S., McCoy, E. C., Sanders, D. R., Butler, M.,
    Kiriazides, D. K., and Mermelstein, R., 1980. Nitropyrenes:
    Isolation, identification and reduction of mutagenic impuri-
    ties in carbon black and toners. Science 209:1038-1043.

4. Huisingh, J. L., Bradow, R. L., Jungers, R. H., Harris, R. D.,
    Zweidinger, R. B., Cushing, K. M., Gill, B. E., and Albert,
    R. E. Mutagenic and carcinogenic potency of extracts of diesel
    and related environmental emissions: Study design, sample
    generation, collection, and preparation. In: Health Effects
    of Diesel Engine Emissions, Vol. II, Edited by W. E. Pepelko,
    R. M. Danner, and N. A. Clarke, EPA-600/9-80-057b, U.S.
    Environmental Protection Agency, Cincinnati, Ohio, 1980,
    pp. 788-800.

5. Claxton, L. D. Mutagenic and carcinogenic potency of diesel
    related environmental emissions: Salmonella bioassay.
    In: Health Effects of Diesel Engine Emissions, Vol. II,
    Edited by W. E. Pepelko, R. M. Danner, and N. A. Clarke,
    EPA-600/9-80-057b, U.S. Environmental Protection Agency,
    Cincinnati, Ohio, 1980, pp. 801-807.

6. Austin, A., and Lewtas, J., private communication.

7. Austin, A., and Lewtas, J., private communication.

8. Mermelstein, R., Rosenkrantz, H. S., and McCoy, E. C.,
    Microbial Mutagenicity of Nitroarenes, In Genetoxic Effects
    of Airborne Agents. Edited by R. Tice, D. L. Costa and K. M.
    Schaich, Plenum Press, New York, 1982, pp. 369-396.

9. Austin, Ann, private communication.

THE INFLUENCE OF FUEL COMPOSITION ON PAH-EMISSION
A METHODICAL CONSIDERATION

Friedhelm Nunnemann, Daimler-Benz AG,
Stuttgart, Germany

ABSTRACT

The idea of this presentation was to present the special
problems of automobile emission measurement and other more or
less closely connected questions. The connection of fuel
composition and PAH emission is discussed.

INTRODUCTION

There are some good reasons to identify and determine engine
exhaust components.

Identification and Determination of Regulated
and Unregulated Automobile Exhaust Components

1. Regulated Pollutants

2. Nonregulated Components

- Components relevant to combustion processes
- Components relevant to air quality standards
- Components suspected to be carcinogens.

Not only regulated but also unregulated chemical compounds
are hints to what happens inside and round about the combustion
chamber and how they take part in ambient air processes.

D. Rondia et al. (eds.), Mobile Source Emissions Including Polycyclic Organic Species, 211–222.

The various emissions from an operating internal combustion engine are diagrammed in Figure 1.

EXHAUST GAS COMPONENTS			
INERT COMPONENTS	POLLUTANTS	IRRITANT TOXIC CARCINOGEN	
	REGULATED	UNREGULATED	
ORIGIN: COMBUSTION PROCESS			ORIGIN: ADDITIVES ABRASION
CARBONDIOXIDE WATER NITROGEN OXYGEN HYDROGEN	CARBONMONOXIDE HYDROCARBONS PARTICULATES (SMOKE) NITROGENOXIDES	ALDEHYDE/KETONES BENZENE PHENOLES POLYCYCLIC AROMATICS ODORANTS	LEAD SULFURDIOXIDE SULFATES METALOXIDES METALS

FIGURE 1. Classification of Exhaust Components.

We see many problems with the determination of so-called unregulated pollutants.

Problems with the Determination of
Unregulated Pollutants

1. Appropriate sampling system

    • Emission sample-airborne pollutants
    • Artifacts

2. Differential or total determination of a group of chemical compounds

    • Total sum of all species
    • Some representative compounds
    • Indicator substances

3. Poor correlation between different analytical methods

4. No comparisons of emissions data

5. Lack of an approved reference method

6. No reliable and comparable routine and
    on-line techniques.

These problems are increased by the interactions of the
emissions of different sources, the exhaust gas treatment, and
the sampling method as illustrated by the following figure.

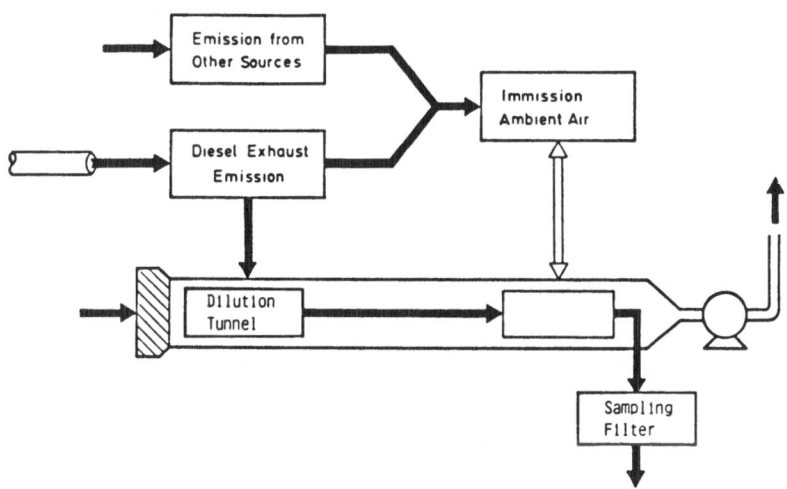

FIGURE 2.   Exhaust Gas–Ambient Air–Particulate Sample
            Different Sampling Sites–Different Samples

The emissions of a combustion engine depend on multifarious
engine parameters and it is a great task to investigate these
aspects.

Obviously there are some parameters like air-fuel ratio where
you can find definite and evident trends, and relate them to PAH
emissions.   Therefore, regulation of polycyclic aromatics is not
necessary since the methods to reduce the emission of carbon
monoxide and hydrocarbons by engine modifications and after
treatment devices also decrease the PAH emission.

The effect of fuel composition on PAH emission is a more
intricate and sophisticated question.

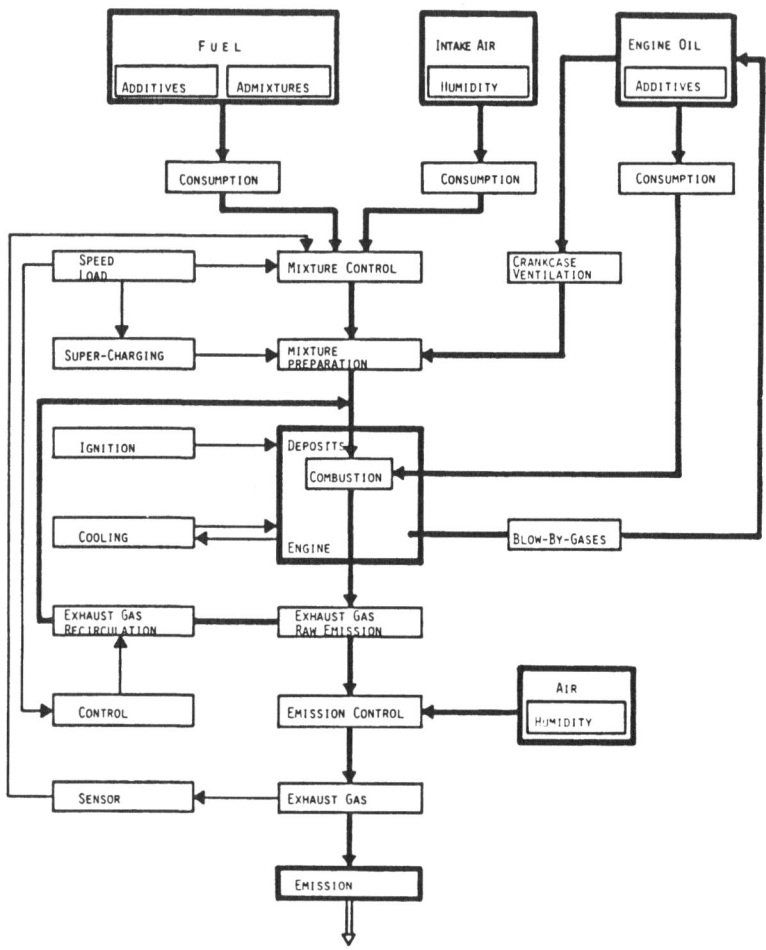

FIGURE 3.   Combustion and Emission Flow Scheme

   Designing a special program to investigate this, there are
some doubts, whether present available and practical analytical
procedures are sufficient to attack this problem.

EXPERIMENTAL

   First some information about the test program and the
analytical method are given in Table 1.

TABLE 1

SUMMARY OF ENGINE TESTS

Fuel No.	Isooctane (%)	Aromatics (%)	Lead (g/1)	Number of Tests	Test Run
1	100	0	0.0	3	1
2	100	0	0.21	3	5
3	100	0	0.44	3	9
4	80	20	0.0	3	2
5	80	20	0.21	3	6
6	80	20	0.44	3	10
7	65	35	0.0	3	3
8	65	35	0.21	5	7
9	65	35	0.44	3	11
10	50	50	0.0	3	4
11	50	50	0.21	3	8
12	50	50	0.44	3	12

Fuel Aromatics: Benzene, Toluene, Xylene (1:3:4)
Lead: Tetraethyl-lead and Scavengers
Balance: Isooctane
Cycle: 50 km/h, 10 min. cruise

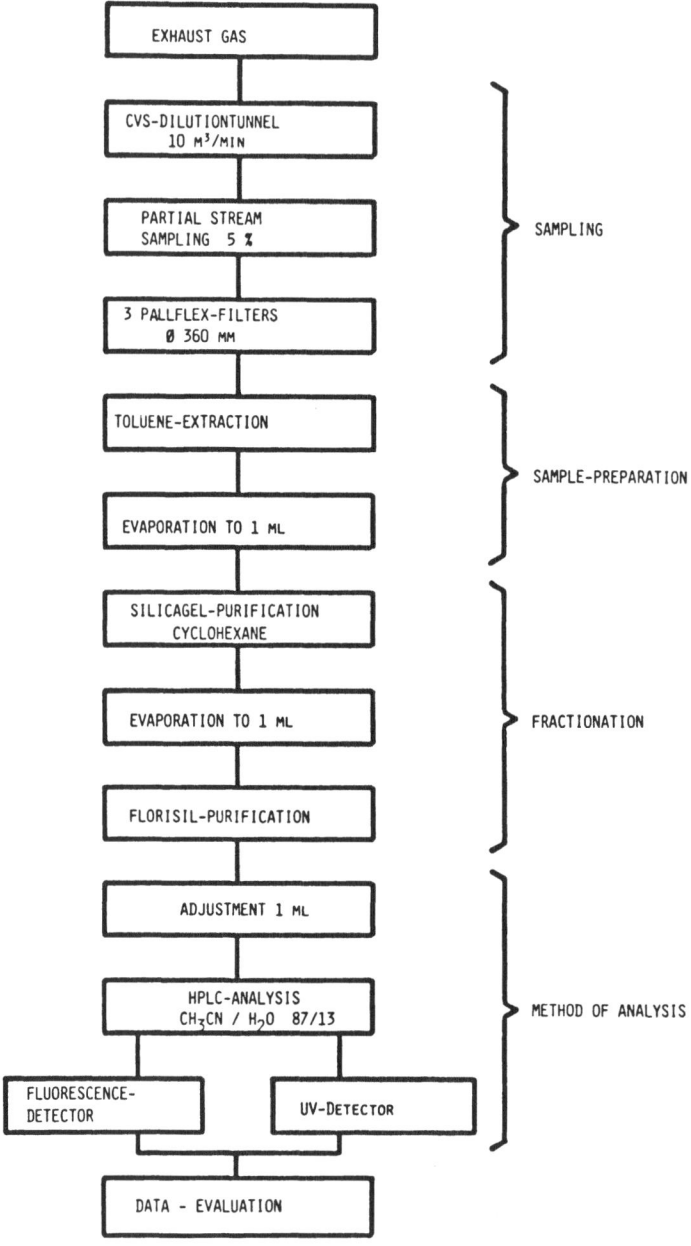

<u>FIGURE 4.</u> Flow Diagram of PAH Analysis Procedure.

The next figure (Figure 5) shows the lowest and highest rate of the nine PAH components. The single profiles are free of trends and independent of the various fuel compositions.

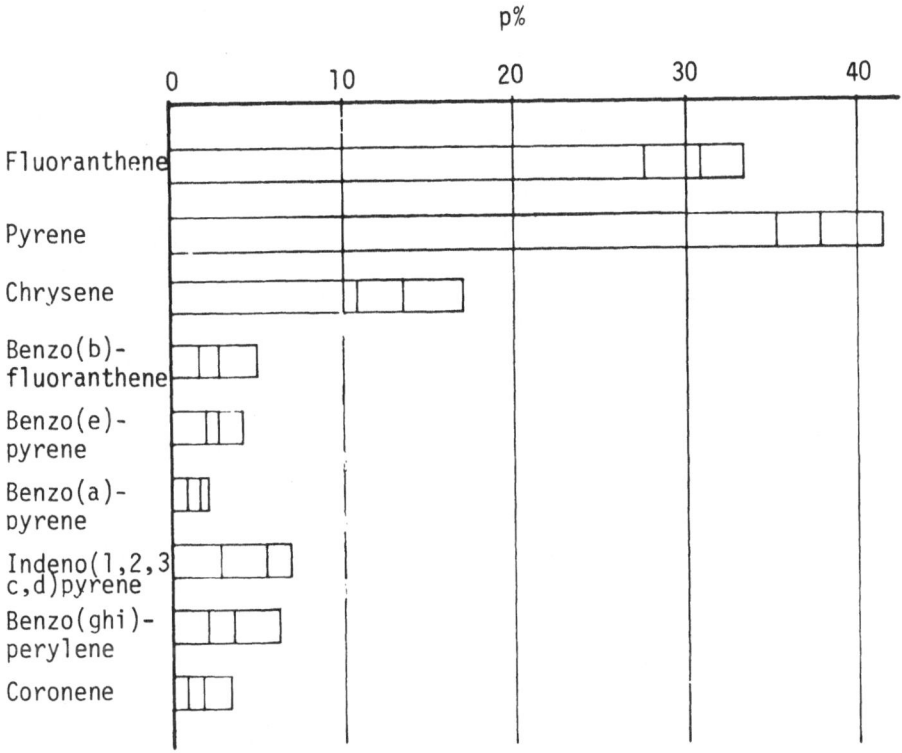

FIGURE 5. Percent Rate (Minimum and Maximum) of Individual Polycyclic Aromatics.

To reduce the amount of individual data, we only discuss in detail the sum of all measured compounds (Figure 6), and give some information about the sum of all polycyclic aromatics suspected to be carcinogens (Figure 7) and the benzo(a)pyrene emissions (Figure 8). At first sight you do not see any trends expected by apparent plausible theoretical considerations.

The next step was an analysis of variance described by R.E. Kaiser and G. Gottschalk. The result was that there are no significant effects both for aromatic and lead content. Possibly they are covered by the poor repeatability of the emission of these trace components and/or by the poor sampling and analytical methods. We obtained the same answer removing suspected outliers.

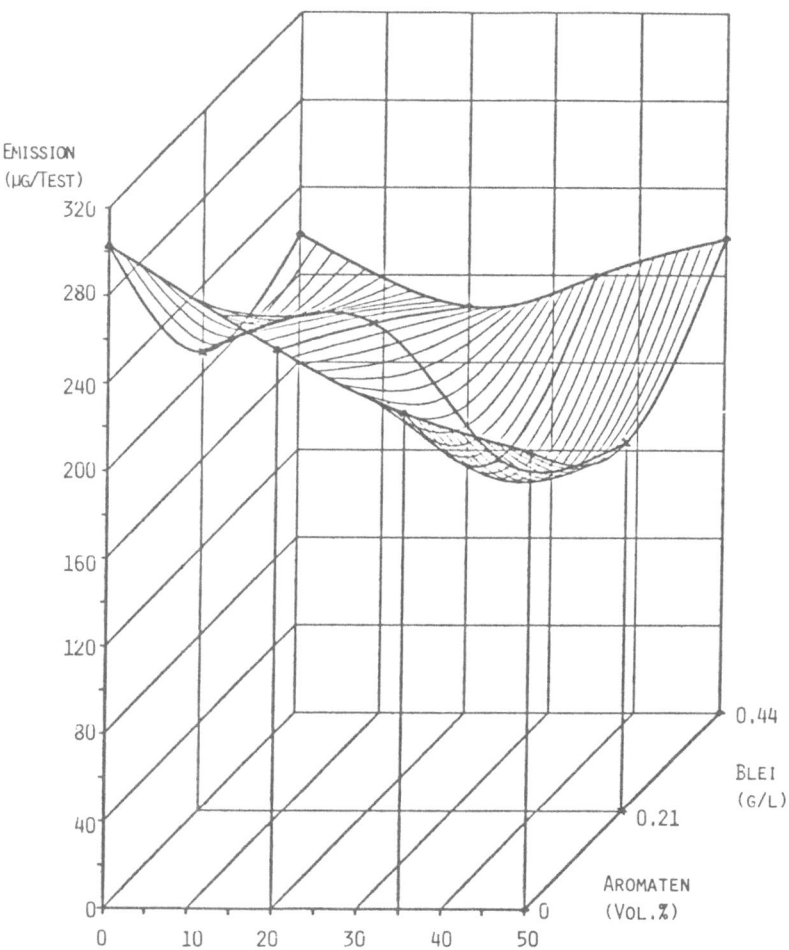

<u>FIGURE 6.</u>   Sum of PAH versus Lead and Aromatic Content
                  in Fuel Tests.

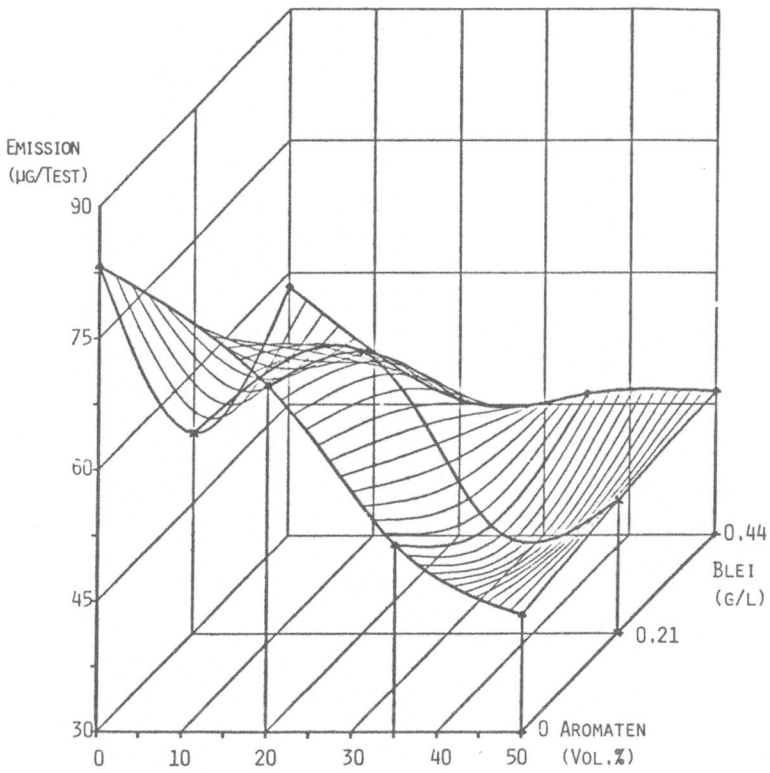

FIGURE 7. Sum of Carcinogenic PAH versus Lead and Aromatic Content in Fuel Tests.

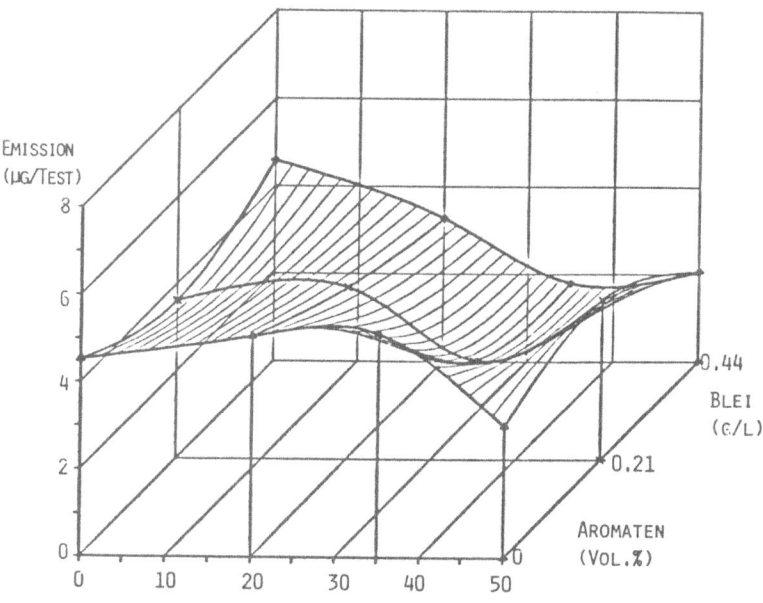

FIGURE 8. Benzo(a)pyrene Emissions versus Lead and Aromatic
Content in Fuel Tests.

TABLE 2

RESULTS OF ANALYSIS OF VARIANCE BY KAISER AND GOTTSCHALK

Aromatics (Vol%)

LEAD (G/L)	0Vol%.	20Vol%	35Vol%	50Vol%	Ki	x̄i	si	si²
0.00	302.55	255.50	226.28	208.91	4	248.33	40.96	1077.36
0.21	209.32	222.75	160.92	168.45	4	190.36	30.31	918.45
0.44	218.41	185.26	199.42	216.58	4	204.92	15.65	244.91
Lj	3	3	3	3				
x̄j	243.25	221.17	195.54	197.98		N = 12		
sj	51.40	35.15	32.85	25.86		x̿ = 214.53		
sj²	2642.33	1235.25	1079.27	668.72		s = 37.89		
						s_R = 25.85		

Number of Tests

k	1	2	3	4	5	6	7	8	9	10	11	12
1	315.02	213.99	254.63	307.41	228.68	167.41	265.12	167.36	229.93	221.77	163.46	228.05
2	289.60	216.12	197.95	187.92	244.45	221.73	185.20	179.00	201.83	195.22	150.46	205.11
3	303.04	197.84	202.64	271.18	195.12	166.63	228.52	170.63	166.43	209.73	191.43	-
4								168.43				
								119.20				
M	3	3	3	3	3	3	3	5	3	3	3	3
x̄	302.55	209.32	218.41	255.50	222.75	185.26	226.28	160.92	199.42	208.91	168.45	216.58
s_k	12.72	10.00	31.46	61.27	25.19	31.59	40.07	23.77	31.79	13.29	20.94	16.22
$\bar{s}_k$	26.52											

CONCLUSION

In part, the conclusion is that the analytical methods are
not good enough to investigate such special problems. Therefore
it is difficult to compare data from different sources, laboratories, and methods. In testing engines it is most important to
precondition the vehicle well, stabilize deposits from the engine
oil, combustion chamber, and exhaust pipe; and, design a good
test program.

ENVIRONMENTAL CARCINOGENS - SELECTED METHODS OF ANALYSIS:
IARC MANUAL SERIES INCLUDING POM MEASUREMENTS

Ian K. O'Neill

International Agency for Research on Cancer
150 cours Albert-Thomas
F69372 Lyon Cédex 08

ABSTRACT

Polycyclic aromatic hydrocarbons (PAH) in soot were the first
group of substances causally related to occupational cancer. The
International Agency for Research on Cancer published a Carcinogenic Risk Evaluation Monograph in 1973 on polycyclic organic
materials (POM) and a Manual in the title series in 1979. Two
IARC Working Groups will be held in 1983 concerning 39 POM and
the industries where POM exposure is experienced. It is proposed
to publish also an updated volume on POM analysis in the Manual
series. The work of the Manual series will be discussed in relation to its basis on the IARC Evaluation Monographs, the methods
of sampling and analysis selected for inclusion and the particular
interest in updating POM analysis.

In this presentation is given a description of the work of the
International Agency for Research on Cancer in preparing a series
of volumes in which analytical methods are provided for substances
that have been evaluated as having a carcinogenic risk to man.
Polycyclic organic material (POM), specifically PAH were first
evaluated at IARC in 1973 [1] (see Table 1) and continue to occupy
attention inasmuch as further work is planned for the next few
years. The participation of experts internationally is the basis
of IARC work and some attendees at this workshop have contributed
to this work.

D. Rondia et al. (eds.), Mobile Source Emissions Including Polycyclic Organic Species, 223–226.
© *1983 by D. Reidel Publishing Company.*

TABLE 1

POM EVALUATED IN IARC MONOGRAPHS SERIES IN 1973 [1]

Benz(a)anthracene (S)	Dibenzo(a,h)pyrene (S)
Benzo(b)fluoranthene (S)	Dibenzo(a,i)pyrene (S)
Benzo(j)fluoranthene	Dibenzo(a,l)pyrene
Benzo(a)pyrene (S)	Indeno (1,2,3-cd)pyrene (S)
Benzo(e)pyrene	Benz(C)acridine
Chrysene	Dibenz(a,h)acridine (S)
Dibenz(a,h)anthracene (S)	Dibenz(a,j)acridine (S)
Dibenzo(h,rst)pentaphene	7H-dibenzo(c,g)carbazole (S)
Dibenzo(a,e)pyrene	

(S) indicates sufficient evidence of
carcinogenicity in animals

IARC Evaluation of Carcinogenicity of Substances

It is necessary to briefly describe the IARC Monographs
Programme which provides the basis for selection of substances to
be covered in the Manual series. The Monographs Programme
convenes international working groups three times each year to
evaluate substances for which there is some information on carcinogenicity and for which human exposure is known or likely.
Prior to the meetings, there is an extensive bibliographic search
followed by preparation of draft documents. Each working group
comprises experts from the fields required to assemble a monograph
on each substance in relation to its occurrence and evaluation of
carcinogenicity; these include experts in epidemiology, mutagenesis, experimental carcinogenicity testing, industrial hygiene and
chemistry. The Working Group then meets in Lyon to discuss and
finalize the texts of the Monographs and to formulate the
evaluation on the degree of evidence of carcinogenicity in humans
(on available case reports and epidemiological studies) and in
experimental animals (based on relevant experimental data).

Each monograph deals with a substance, or an industrial
activity and working groups are planned for 1983 to evaluate 39
POM or POM-containing mixtures. Industries where effects of exposure to POM are not likely to be confused by workplace exposure
to other substances may be considered later in 1983. Information
on the substances and activities considered in the Monographs may
be obtained by consulting the latest volume [2].

The Manual Series

The Manual series aims to provide appropriate analytical
methodology that will permit improved data to be obtained, and
thus to better allow comparison in results for environmental
monitoring and for epidemiology. Methods of sampling and analysis
are provided for substances identified in the Monographs as potential or known human carcinogens, such that current exposure may
be monitored and in some cases may be appropriate for estimating
previous exposures for correlation with results in epidemiological
investigations.

Each volume of the IARC Manual series 'Environmental
Carcinogens - Selected Methods of Analysis' deals with a group of
related substances of known or suspect carcinogenicity. In addition to methods of analysis, introductory chapters cover
carcinogenicity, occurrence, sampling and other relevant background
material.

Each method of analysis is provided in a style similar to that
of the International Standards Organization, and deals with one
substance in a matrix or similar matrices where that substance is
particularly likely to be found. Amongst the many methods of analysis often described in the literature for each substance or matrix, one or two are chosen for each substance/matrix combination.
Where possible, methods are selected as having been validated by
collaborative/cooperative studies, although selection often must
rest on the experience of the expert Review Board where such validation has not been made. A main objective is to provide clearly
spelt-out methods that will yield comparable results in the hands
of the experienced and inexperienced alike.

Six volumes published or in press are :

Volume 1 : Volatile N-Nitrosamines in food [3]
Volume 2 : Vinyl Chloride [4]
Volume 3 : Polycyclic Aromatic Hydrocarbons [5]
Volume 4 : Aromatic Amines and Azo-Dyes [6]
Volume 5 : Mycotoxins [7]
Volume 6 : N-Nitroso Compounds [8]

Amongst those volumes planned is one on polycyclic organic
materials arising from a variety of sources, including coal and
oil combustion, coal liquefaction and the major industrial
fossil-fuel users. It is anticipated that sampling of air, water,
sediments and foods and vehicle exhausts will be covered. Already
volume 4 includes a description of the formation of highly mutagenic POM in cooking of food via the pyrolysis of protein.
Volume 3 listed above, after introductory texts on carcinogenicity,
nomenclature and sampling/separation approaches, provides eight

methods for determination of POM in a variety of matrices.
Included is description of sampling using the Europa drive cycle
and subsequent analysis for POM, validated by collaborative
studies.

The Manual series has been partially supported by the
United Nations Environment Programme and the Monographs series
is supported by the U.S. National Cancer Institute. All texts
and methods in the Manual have been provided gratis by leading
experts chosen by the Review Board of each volume, and editing
is performed jointly by IARC staff and experts in the field.

REFERENCES

1. IARC Monographs on the Evaluation of Carcinogenic Risk of the
Chemical to Man, 3 (1973) Certain polycyclic aromatic heterocyclic
compounds, IARC, Lyon.

2. IARC Monographs on the Evaluation of the Carcinogenic Risk of
Chemicals to Humans, 29 (1982) Some industrial chemicals and dye-
staffs, IARC, Lyon.

3. Environmental Carcinogens Selected Methods of Analysis, 1
(1978) Analysis of volatile nitrosamines in food (H. Egan,
R. Preussmann, M. Castegnaro, E.A. Walker and A.E. Wassermann,
eds) IARC Scientific Publications No. 18, Lyon

4. Environmental Carcinogens Selected Methods of Analysis, 2
(1978) Methods for the measurement of vinyl chloride in poly
(vinyl chloride), air, water and foodstuffs. D.C.M. Squirrel
and W. Thain, IARC Scientific Publications No. 22,
Lyon

5. Environmental Carcinogens Selected Methods of Analysis, 3
(1979) Analysis of polycyclic aromatic hydrocarbons in environmental samples (H. Egan, M. Castegnaro, P. Bogovski, H. Kunte
and E.A. Walker, eds) IARC Publications No. 29, Lyon

6. Environmental Carcinogens Selected Methods of Analysis, 4
(1981) Some aromatic amines and azo dyes in the general and
industrial environment (H. Egan, L. Fishbein, M. Castegnaro,
I.K. O'Neill and H. Bartsch, eds) IARC Publications No. 40, Lyon

7. Environmental Carcinogens Selected Methods of Analysis, 5,
(1982) Some mycotoxins, IARC Scientific Publications No. 45,
Lyon (in press)

8. Environmental Carcinogens Selected Methods of Analysis, 6
(1982) N Nitroso compounds, IARC Scientific Publications No 46,
Lyon (in press)

BIOLOGICALLY ACTIVE NITRO-PAH COMPOUNDS IN EXTRACTS OF
DIESEL EXHAUST PARTICULATE

Thomas C. Pederson

General Motors Research Laboratories
Biomedical Science Department
Warren, Michigan 48090 USA

ABSTRACT - Polycyclic organic matter extracted from diesel
exhaust particulate is active in bacterial mutation assays
without mammalian enzyme systems as required for activation of
polycyclic aromatic hydrocarbons. This direct-acting
mutagenicity is attributable to nitrated-PAH compounds activated
by bacterial nitroreductases. TLC and HPLC separations used in
combination with Salmonella mutation assays employing
nitroreductase-deficient strains evidence the involvement of a
variety of nitrated compounds. These include mononitro-PAH
derivatives such as 1-nitropyrene, but the larger proportion of
mutagenic activity is associated with more polar fractions which
likely contain further-substituted nitroarenes. Highly-
mutagenic derivatives of mononitro-PAH, such as the
dinitropyrenes, are evidently present in the particle extracts.
Similar products are formed by mammalian enzymes in the
Salmonella/S9 assay. The enhanced mutagenicity of 1-nitropyrene
with rat liver S9 enzymes is attributable to a metabolite formed
by P-450 monooxygenase activity, but the mutagenicity of this
product still requires activation by bacterial nitroreductases.
The intermediate metabolite, most probably an oxygenated
derivative of 1-nitropyrene, may be representative of the more
polar mutagens found in diesel particle extracts.

INTRODUCTION

     Because of the increasing use of light-duty diesel engines
in automobiles, there has been considerable interest in the
health effects which might be associated with exposure to diesel
exhaust emissions (1). This interest has centered on the

227

D. Rondia et al. (eds.), Mobile Source Emissions Including Polycyclic Organic Species, 227–245.
© *1983 by D. Reidel Publishing Company.*

particulate material in diesel exhaust which is submicron in size allowing it to penetrate and deposit in the deep lung. These particles contain a largely carbonaceous core and an adsorbed layer of organic material which can be extracted by solvents.   The extracted material contains various types of polycyclic organic matter.   Included in this mixture are a number of nitrated derivatives of the polycyclic aromatic hydrocarbons (nitro-PAH).

Diesel particle extracts have been repeatedly found to be mutagenic with the Salmonella assay in the absence of any mammalian enzyme system.   The components of the extract responsible for this activity are, therefore, referred to as direct-acting mutagens.   The first experimental evidence from our own studies to indicate a possible involvement of nitroaromatic mutagens was the finding that the mutagens in the particle extract were unreactive towards purified DNA which ruled out reactive compounds such as arene oxides (2).   On the other hand, nitroaromatic mutagens are not chemically reactive but direct-acting in bacterial assays because bacteria possess nitroreductase enzymes which catalyze the reductive activation of these compounds (3).

This article describes our subsequent studies which have demonstrated that the direct-acting mutagenicity of diesel particle extracts is largely attributable to nitroaromatic compounds, including both mononitro-PAH compounds and their multisubstituted derivatives.   These studies have also examined the processes by which mammalian enzymes in the rat liver S9 preparation modify the activity of these mutagens.   The use of normal and nitroreductase-deficient bacteria in the Salmonella assay has been instrumental in all of these studies.

COLLECTION AND EXTRACTION OF DIESEL EXHAUST PARTICULATE

The method commonly used for preparation of diesel particle extracts employs a Soxhlet extaction apparatus with dichloromethane as the extracting solvent.   This procedure has been found optimal for extraction of the mutagenic components adsorbed to the particulate (4).   Most of our studies have been conducted with an extract prepared from particulate collected by electrostatic precipitation, which allowed the collection of a moderate amount of material in a short period of time (5). Extracts were also prepared from diesel particulate collected by two other methods.   These extracts and the methods of particle collection are described in Table 1.   The baghouse filter apparatus was used to obtain a large quantity of particulate required for a skin-painting carcinogenesis assay (6).   This method involves collecting the particulate from hot undiluted

Table 1.

Diesel Exhaust Particle Extracts

Extract	Type of Engine	Operating Conditions[a]	Method of Particle Collection	Collection Duration	Extractable Mass
#1	1978 GM 5.7 L V8 diesel	1620 rpm 110 N·m 80 km/h	from undiluted exhaust at 100°C by electrostatic precipitation	30 min	6.4%
#2	1979 GM 5.7 L V8 diesel	1350 rpm 96 N·M 65 km/h	from undiluted exhaust at 100°C by baghouse filter	20 hrs	29%
#3	1982 GM 5.7 L V8 diesel	1100 rpm 125 N·m 72 km/h	from a 1:6.7 dilution at 22°C by Teflon-coated filter	20 hrs	40%

[a] Engines were run at steady speed and load simulating the indicated vehicle speeds.

exhaust for a prolonged period of time. The third extract was prepared from the particulate collected on Teflon-coated glass fiber filters from the animal exposure chambers being used for long-term exposure of animals to diesel exhaust (7).

FRACTIONATION OF EXTRACTS BY THIN LAYER CHROMATOGRAPHY

We have made extensive use of thin layer chromatography for separation of the components in diesel particle extracts (8,9). These separations, on normal silica gel TLC plates, distinguish one class of compounds from another based on their difference in polarity. Figure 1 shows the bands of material which are seen on the thin layer chromatograms of diesel particle extracts. The samples shown on these plates are the three particle extracts and a mixture of nitropyrenes. The two TLC plates contain identical samples, but exhibit two ways by which nitro-PAH compounds may be visualized. One plate is viewed under UV light with fluorescent paper behind, which allows both fluorescent and nonfluorescent/UV-absorbing components to be seen . The fastest migrating fluorescent band contains all the PAH compounds. Below this is a dark band of material which contains the monosubstituted nitro-PAH compounds. The UV-quenching is characteristic of all the nitro-PAH compounds. The lower half of the plate contains many fluorescent components. These are the more polar compounds and are most probably multisubstituted polycyclic organic compounds. The other plate shows how the same chromatogram looks after it has been sprayed with a reducing reagent containing sodium borohydride and a copper chloride catalyst (10). This reduces nitroaromatic compounds to their corresponding amines which are fluorescent. The band of material in the particle extracts attributed to mononitro-PAH becomes fluorescent as do the nitropyrene reference compounds. There are some other components in the polar region which also turn fluorescent, including material in the region between 2 and 3 cm from the bottom where the sample was applied to the plate. As shown in the following section, this fluorescence may be attributable to highly-polar nitroaromatic compounds which account for a considerable portion of the microbial mutagenicity exhibited by these extracts.

The reason for chromatographing the extracts on silica gel plates was to determine in what regions the mutagens were located. Successive bands of material were scraped from the plates, extracted with dimethylsulfoxide, and the extracts assayed for mutagenicity (9). Figure 2 shows the distribution of mutagenicity in the TLC chromatograms of all three particle extracts. The arrows at the top of the figure show the regions from which differing classes of reference compounds are recovered. The magnitude of the activity per fraction with each of the extracts has been scaled to represent the activity attributable to the material extracted from 1 mg of particulate.

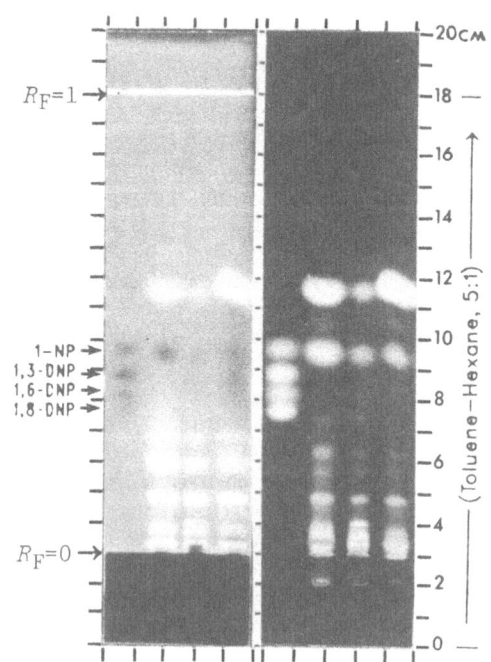

Figure 1. Thin layer chromatograms of diesel particle extracts viewed under UV light with a fluorescent background (Left) or after reduction by $NaBH_4/CuCl_2$ reagent (Right). The same samples are shown on both plates, beginning from the left: a reference compound mixture of 1-nitropyrene and the three dinitro-pyrenes which ascend in the order indicated; and particle extracts #1, #2, and #3 (250µg each). Samples were applied to the silica gel plates in the preabsorbant region on the bottom 3 cm of the plate.

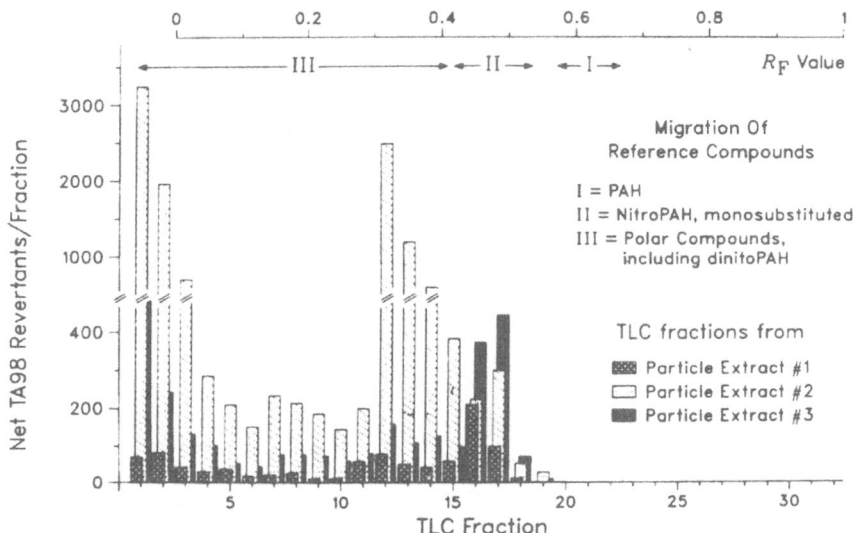

Figure 2. Mutagenic activity recovered in TLC fractions of the three diesel particle extracts described in Table 1.

All three extracts have a somewhat similar amount of activity in fractions 16 through 19, the fractions which would contain all the mononitro-PAH compounds. However, the major portion of activity in all three extracts was recovered from the more polar fractions. This is especially evident with extract #2 which has most of its mutagenic components in the regions comprised of fractions 1-3 and fractions 12-14.

CHARACTERIZATION OF MUTAGENS IN DIESEL PARTICLE EXTRACTS WITH NITROREDUCTASE-DEFICIENT BACTERIA

A comparison of mutagenic activity using normal and nitroreductase-deficient bacteria provides a method for detection and identification of nitroaromatic mutagens. But this idea is complicated by the existence of several bacterial nitroreductases with differing specificities (11). Three different nitroreductase-deficient derivatives of the normal tester strain TA98 were obtained from Dr H. S. Rosenkranz at Case Western University who has isolated and characterized these strains. Strain TA98NR was selected for its resistance to the nitrocompound niridazole and subsequently found to be resistant to the mutagenicity of niridazole, and also to that of the nitrofurans , nitronaphthalenes , and nitrofluorene (3). It is also resistant to 1-nitropyrene and a number of other nitro-PAH mutagens, but there are other nitroaromatic compounds, the dinitropyrenes in particular, which remain as mutagenic in this strain as in TA98. Therefore, another type of nitroreductase-deficient strain was derived by selecting for resistance to 1,8-dinitropyrene (11). These studies employ two of the dinitropyrene-resistant strains; strain TA98/1,8-DNP$_6$, derived from TA98; and strain TA98NR/1,8-DNP$_2$, derived from TA98NR.

The mutagenic activities of the three diesel particle extracts in normal and nitroreductase-deficient strains is shown in Table 2. As was evident in Figure 2, extract #2 is the most mutagenic of the three extracts. Its activity is decreased by about 60% in strain TA98NR and by nearly 90% in strain TA98/1,8-DNP$_6$. The mutagenicity of the other two extracts decreases by about 40% in TA98NR and by 50-60% in the dinitropyrene-resistant strain. Also shown in this table are the specific mutagenicities of the nitropyrenes. These activities are expressed in revertants/ng rather than revertants/μg. The most mutagenic member, 1,8-dinitropyrene, is one of the most potent mutagens known for the Salmonella mutation assay.

The contrasting specificities of bacterial nitroreductases is very evident in the activity of the nitropyrenes. The nitroreductase-deficiency in strain TA98NR effects a 90% reduction in the mutagenicity of 1-nitropyrene. The compounds 1,3-dinitro- and tetranitropyrene are also less mutagenic in this strain, but 1,6-dinitro-, 1,8-dinitro-, and trinitropyrene retain their activity. In the experiment recorded in Table 2, the activities of the latter three compounds were, in fact,

Table 2.

Specific Mutagenicity of Particle Extracts and the Nitropyrenes with Normal and Nitroreductase-Deficient Bacteria.

	TA98	TA98NR	TA98/1,8-DNP$_6$
	Revertants per μg per plate		
Particle Extract #1	32	20	16
" " #2	88	33	10
" " #3	13	7.6	5.6
	Revertants per ng per plate		
1-Nitropyrene	1.32	0.095	1.49
1,3-Dinitropyrene	230	40	5.6
1,6-Dinitropyrene	347	283	110
1,8-Dinitropyrene	618	556	19
1,3,6-Trinitropyrene	279	267	121
1,3,6,8-Tetranitropyrene	34	24	18

somewhat reduced in TA98NR, but they are usually of the same magnitude in this strain as in TA98 (12). There are variations from one experiment to another in the specific mutagenicities of nitro-PAH compounds which are probably due to differing metabolic competencies of the working cultures (9,11).

Strain TA98/1,8-DNP$_6$, exhibits a contrasting specificity for activation of the nitropyrenes. The mutagenicity of 1-nitropyrene is as great in this strain as it is in TA98, but the activities of all the multinitropyrenes are reduced. The relative response in the nitroreductase-deficient strains is a characteristic which can be used to identify the presence of these compounds in chromatographic fractions of diesel particle extracts. Figure 3 shows the way these strains were used with the TLC fractionation of diesel particle extract.

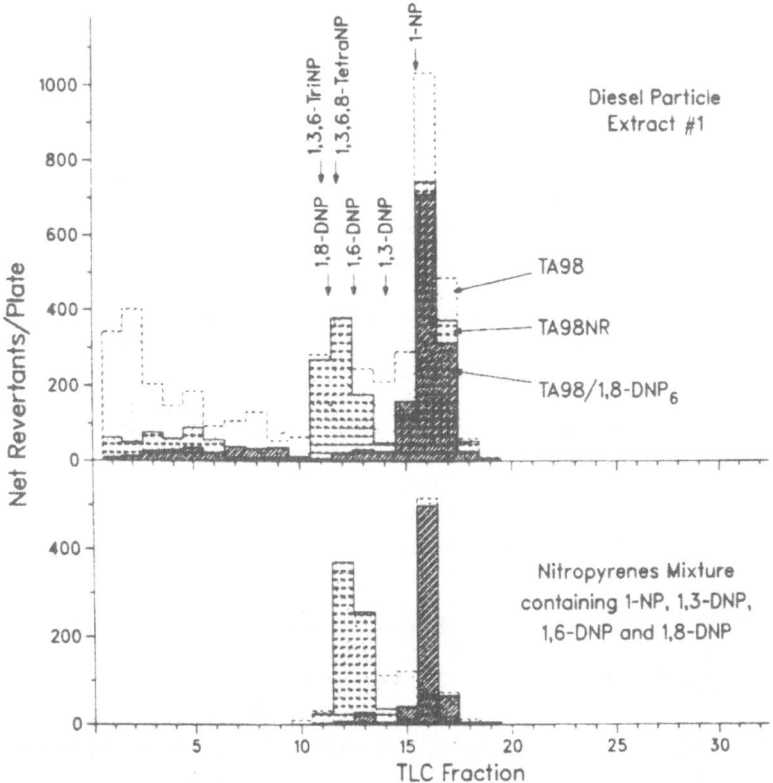

Figure 3. Mutagenic activity in TLC fractions of diesel particle extract (Upper) and a reference compounds mixture (Lower) assayed using tester strains: TA98; TA98NR; and TA98/1,8-DNP$_6$. The chromatographic positions of the nitropyrene reference compounds are shown for comparison purposes.

The upper half of Figure 3 shows the activity in TLC
fractions from particle extract #1 and compares this to the
distribution of activity seen with a mixture of nitropyrenes.
The mutagenic activiy in all the particle extract fractions is
reduced in at least one of the nitroreductase-deficient strains.
This implies that all the mutagenic components, even those in
the most polar fractions, are evidently some type of
nitroaromatic compound. There is also a similarity between the
profiles of activity from the nitropyrenes mixture and those
from the corresponding TLC fractions of the particle extract.
In fractions 11 through 13, the mutagens from the particle
extract retain their activity in strain TA98NR, but not in
TA98/1,8-DNP$_6$, as expected if attributable to multinitropyrenes,
in particular, the 1,6-dinitro- and 1,8-dinitro- isomers.
Activity similarly attributable to multinitropyrenes was also
evident with TLC fractions from the other two extracts.

IDENTIFICATION OF MULTINITROPYRENES IN DIESEL PARTICLE EXTRACTS

In order to achieve better chromatographic resolution, high
performance liquid chromatography was employed to determine
which of the multinitropyrenes may be present in the particle
extract. The TLC separation was still used as a preparative
technique to isolate a multinitropyrene fraction from the
particle extract. The extract was divided into five fractions,
with one fraction chosen to include the region of the
chromatogram believed to contain multinitropyrenes. Table 3
shows the amount of material and mutagenic activity recovered in
these fractions. Most of the mass of the extract was recovered
from either fraction A, probably containing paraffinic
hydrocarbons, or from fraction E. Fraction D, the
multinitropyrene fraction, contains only 2% of the mass but
about 25% of the mutagenic activity. This material was then
fractionated by HPLC using a cyano-derivatized silica column
eluted with a gradient of isopropanol in hexane.

Table 3.

Material and Activity Recovered in TLC Fractions of Diesel Particle Extract[a]

TLC Fraction	Characteristic Components	Fractional Weight	Fractional Mutagenicity
A, R_F = 0.63-1.0	Saturated hydrocarbons	58%	-
B, R_F = 0.54-0.63	Parent PAH Compounds	3.1%	-
C, R_F = 0.43-0.54	Nitro-PAH, monosubstituted	1.9%	39%
D, R_F = 0.25-0.43	Multinitropyrenes	2.0%	25%
E, R_F = < 0-0.25	Highly polar compounds	35%	36%

[a]TLC separation using 18 mg of extract #1

Although the nitropyrene reference compounds were clearly separated from each other, the multitude of components in fraction D obscured the small absorbance peaks which might be attributed to the multinitropyrenes. Thus, the only method suitable for detecting the presence of these compounds was the bacterial mutation assay. The column eluent was divided into fractions and assayed using the normal and nitroreductase-deficient strains. Figure 4 shows the activity detected in these fractions. Most of the activity is in the fractions which elute where the compounds 1,6-dinitro- and 1,8-dinitropyrene elute. There is also activity evident where 1,3-dinitro- and trinitropyrene would be found. The location of the activity coincides with the nitropyrene reference compounds and so does the comparative activity in the nitroreductase-deficient strains. Although this data does not constitute rigorous proof for the identity of these compounds, dinitropyrenes have also been detected by direct analytical methods in another diesel particle extract (13). It is therefore most probable that the more prominent mutagens in TLC fraction D of the diesel particle extract are multinitropyrenes, principally the compounds 1,6-dinitro- and 1,8-dinitropyrene.

The comparative distributions of mutagens shown in Figure 2 indicate considerable differences in the proportionate contribution of multinitropyrenes to the activity of these extracts. Extract #2 has about 35% of its total activity in the

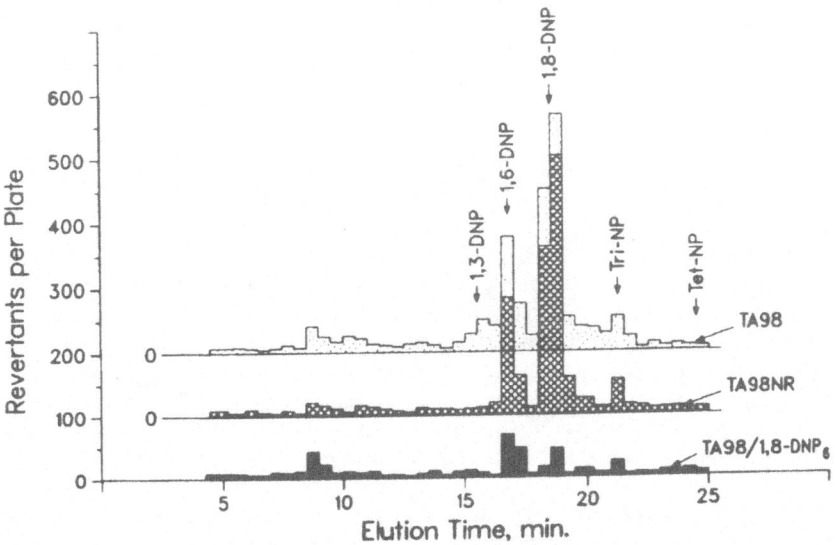

Figure 4. The mutagenic activity in HPLC fractions containing components from the multinitropyrene fraction of diesel particle extract described in Table 3.

fractions attributable to multinitropyrenes.   The remainder is
largely in the most polar fractions. The much higher content of
polar mutagens in this extract is probably the result of
chemical modifications which occur during collection of the
particulate (13,14).   The particulate for extract #2 was
collected from hot undiluted exhaust over a period of 20 hours.
Extract #3 was also collected over a period of 20 hours, but
from cool and diluted exhaust.   The specific activity of this
extract is much less than the extract of the baghouse filter
particulate, and only 17% of its activity is attributable to the
multinitropyrenes. All three particle extracts probably contain
nitro compounds formed during collection of particulate
material.   This collection artifact may be of considerable
significance for Extract #2, but it is unrealistic to assume the
nitroderivatives were formed only during the collection process.
Moreover, particulate material recovered from the lungs of
animals exposed directly to diesel exhaust were still found to
contain direct-acting bacterial mutagens (15).

ACTIVATION OR INACTIVATION OF NITRO-PAH BY RAT LIVER S9 ENZYMES

Since the direct-acting mutagenicity of nitro-PAH compounds
in the Salmonella assay is mediated by bacterial enzymes, an
obvious question is whether similar metabolic activations will
occur in mammalian tissues?   The addition of mammalian enzyme
systems to the Salmonella assay provides some indication of the
genotoxic potential of in vivo mammalian metabolism (16).   The
effect of S9 enzymes from Aroclor-treated rats on the mutagenic
activity of diesel particle extracts and on the nitropyrenes
have revealed some rather surprising results.

A problem encountered when using the Salmonella/S9 assay
with nitro-PAH compounds and diesel particle extracts is that
the S9 preparation will nonspecifically interfere with the
direct-acting mutagenicity and thus mask an activation catalyzed
by the S9 enzymes (17). To overcome this problem, the
nitroreductase-deficient strain TA98NR was used to reduce the
interference from direct-acting mutagenicity.   In addition, a
comparison was made between assays with overlay mixtures of
identical composition except for the absence or presence of
NADPH, the cofactor that supplies reducing equivalents required
for S9 enzyme activity.   The difference between assays with and
without NADPH is attributable to S9 enzyme activity.

Another experimental parameter of importance in these
experiments is the concentration of the S9 enzymes. Activation
of the diesel particle extract, and also of 1-nitropyrene, were
only evident when using low concentrations of S9 enzymes (17).
This is demonstrated by the results shown in Figure 5.   The

concentration of S9 enzymes most commonly used in the
Salmonella/S9 assay is in the range of 25-50 µL/plate (16). The
activation of benzo[a]pyrene is clearly optimal at the higher
concentrations. Yet, activation of the particle extract, or of
1-nitropyrene, is only observed when using much lower
concentrations.

In order to characterize the enzyme systems involved in
these NADPH-dependent activations, the soluble and membrane-
bound enzymes in the S9 preparation were separated by
centrifugation into cytosol and microsome fractions. Assays
using the separated fractions revealed that the enzymes
activating 1-nitropyrene reside in the microsomal fraction, but
activation of the particle extract is catalyzed by both
microsome and cytosol enzymes (17). It was therefore evident,
that the S9-catalyzed activation of diesel particle extract
involved more than one S9 enzyme and more than one component in
the particle extract.

The mutagenicity of the dinitropyrenes in the Salmonella/S9
assay was markedly reduced by NADPH-dependent enzyme activity.
However, when the cytosol and microsome enzymes in the S9
preparation were separated, the activity increased in assays
containing cytosol enzymes and decreased in the presence of

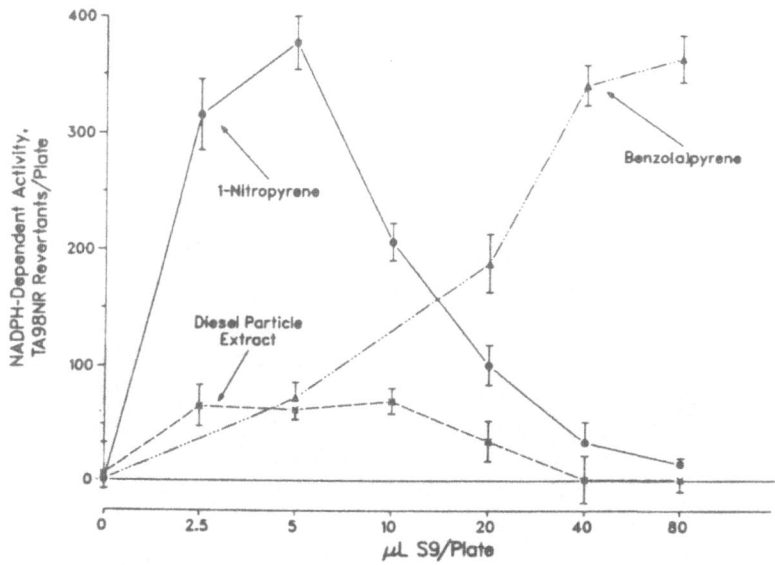

Figure 5. Effect of S9 enzyme concentrations on the activity of
diesel particle extract (Extract #1, 30µg/plate), 1-nitropyrene
(0.5µg/plate), and benzo[a]pyrene (5µg/plate).

microsomes (12). Figure 6 summarizes the results of experiments
comparing the effect of cytosol, microsome and unfractionated S9
enzymes on the activity of 1-nitropyrene, the dinitropyrenes,
and the three particle extracts. The cytosol activity has no
effect on 1-nitropyrene but markedly activates the
dinitropyrenes. NADPH by itself also increased the activity of
the dinitropyrenes, probably involving exogenous bacterial
enzymes, but the increase was much greater in the presence of
cytosol enzymes (12). Both cytosol and microsomal enzymes
contribute to the S9-dependent activation of all three particle
extracts, but not in equivalent proportions. The comparative
enhancement by cytosol and microsome enzymes can be correlated

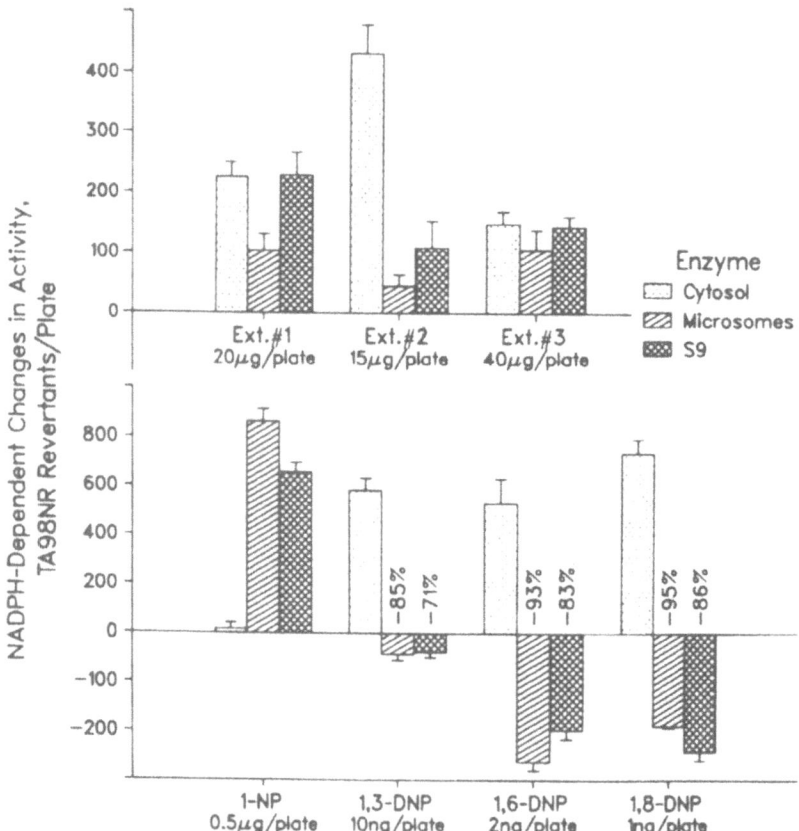

Figure 6. The effects of cytosol and microsome enzymes from a
rat liver S9 preparation on the mutagenicity of nitropyrenes and
diesel particle extracts. Loss of activity is also described as
a percent decrease in the residual direct-acting mutagenicity.

with the apparent contributions made by multinitropyrenes to the mutagenicity of these extracts. The ratio between cytosol and microsomal activations was greatest with extract #2 which has the largest proportion of activity attributable to multinitropyrenes. Conversely, particle extract #3 has the lowest ratio between cytosol and microsomal activations and the lowest apparent contribution from the multinitropyrenes.

MUTAGENIC ACTIVATION OF 1-NITROPYRENE BY THE COMBINED ACTIVITIES OF S9 AND BACTERIAL ENZYMES

The direct-acting mutagenicity of the dinitropyrenes was not reduced in strain TA98NR. Therefore, the dinitropyrene-resistant strains were also used to characterize the effects of S9 enzyme activities. Again, the mutagenicity of the dinitropyrenes was enhanced by cytosol enzyme activity and decreased by microsome activity (12). But surprisingly, the S9-enhanced mutagenicity of 1-nitropyrene was not detectable when using the dinitropyrene-resistant strains (18). As is shown in Table 4, a NADPH-dependent increase in the mutagenicity of 1-nitropyrene is very evident when using either strain TA98 or TA98NR, but there is no detectable activation when using the dinitropyrene resistant derivatives of the same two strains. The activity is actually decreased in strain TA98/1,8-DNP$_6$. Yet the activation of benzo[a]pyrene by S9 enzymes was observed with all four strains. This implied that the NADPH-dependent activity of 1-nitropyrene in the Salmonella/S9 assay was still dependent on bacterial nitroreductase activity.

Since the S9-enhanced mutagenicity of 1-nitropyrene appeared dependent on bacterial enzymes, an intermediate was presumably generated by one enzyme system and converted to the

Table 4

S9-Enhanced Mutagenicity in Nitroreductase-Deficient Strains.

Net Revertants/Plate

	TA98	TA98NR	TA98/18-DPN$_6$	TA98NR/18-DNP$_2$
1-Nitropyrene[a]				
-NADPH	284±22	26±5	265±12	27±6
+NADPH	970±15	623±15	60±5	34±6
	+686	+597	-205	–
Benzo(a)pyrene[b]				
+NADPH	478±9	490±17	457±32	331±16

[a] 0.5μg 1-NP and 2.5μL S9 per plate;   [b] 5μg BaP and 40μL S9 per plate.

ultimate mutagen by the other. The existence of such an intermediate was demonstrated by incubating 1-nitropyrene with S9 enzymes in the absence of bacteria and recovering the accumulated products by extraction with dichloromethane. The material recovered in these extracts was assayed for direct-acting mutagenicity using strains TA98, TA98NR, and TA98/1,8-DNP$_6$. The results of this experiment are shown in Figure 7. During the first thirty minutes of incubation, the extracts became increasingly mutagenic towards strains TA98 and TA98NR but not towards the dinitropyrene-resistant strain. The activity towards strain TA98/1,8-DNP$_6$ actually decreased, demonstrating the depletion of 1-nitropyrene in the reaction mixture. In the final hour of the two hour incubation, the activity of the reaction extract declined, indicating that the mutagenic metabolite is either unstable or further metabolized.

The experiments demonstrating the S9-enhanced mutagenicity of 1-nitropyrene and formation of an extractable intermediate were all conducted under aerobic conditions. Thus, the possible involvement of the P450 monooxygenases or other oxygenase activity was investigated. Experiments were conducted to determine whether formation of the highly mutagenic intermediate product was dependent on oxygen or inhibited by any of the well-known inhibitors of P450 activity. As shown in Table 5, the mutagenic product was formed in the reaction mixtures incubated

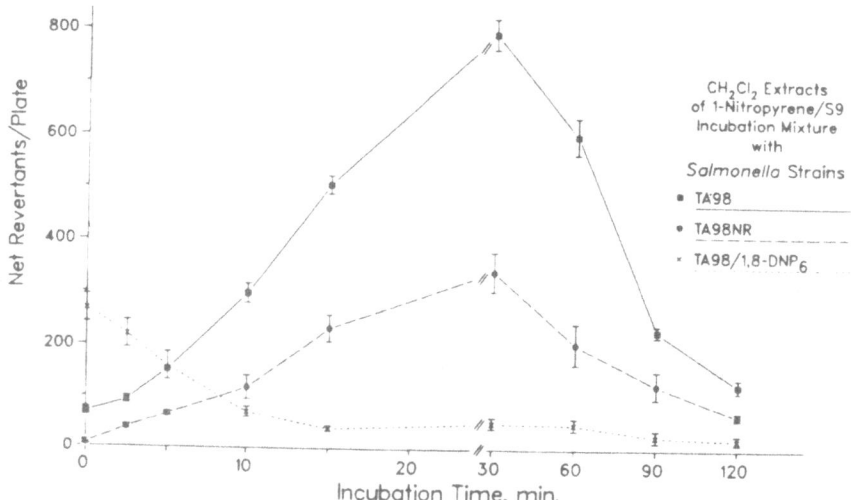

Figure 7. Accumulation of a mutagenic metabolite in an incubation mixture containing 1-nitropyrene (2μg/mL) and S9 enzymes (10μL/mL). The activity is attributable to the extracted material from 0.1 mL of the incubation mixture with TA98 and TA98NR or from 0.4 mL with TA98/1,8-DNP$_6$.

Table 5.

Mutagenicity of Metabolites from 1-NP/Microsome Reaction

Incubation Mixture	Atmosphere	Net TA98NR Revertants/Plate
Complete	Air	645±15
"	N_2	18±2
"	80%N_2/20%O_2	670±38
"	80%CO/20%O_2	65±5
+ 7,8-Benzoflavone, 1μg/mL	Air	381±32
+ 9-Hydroxyellipticine, 1μg/mL	Air	41±10

under air or under a mixture of oxygen and nitrogen, but not in the anaerobic incubation mixture. The reaction was also inhibited by carbon monoxide, 7,8-benzoflavone, and 9-hydroxyellipticine, which are all inhibitors of P450-catalyzed reactions. Thus, the mutagenic metabolites of 1-nitropyrene formed in these reaction mixtures are most probably the product of microsomal P450 monooxygenase activity. Since the microsomal enzymes form an intermediate product, the ultimate mutagen in the Salmonella/S9 assay is evidently formed by bacterial enzymes, presumably a reductive reaction catalyzed by the nitroreductase which activates dinitropyrenes.

ISOLATING THE MUTAGENIC METABOLITE OF 1-NITROPYRENE

When the metabolites of 1-nitropyrene extracted from the S9 incubation mixture were separated by thin layer chromatography, the mutagenic activity was recovered from fractions which would contain compounds considerably more polar than 1-nitropyrene. This TLC separation was performed using a much more polar solvent than was used for separation of the particle extracts. Consequently, polar compounds migrate much further. As shown in Figure 8, one region, comprised of fractions 13 and 14, contained over 98% of the activity recovered from all the fractions. Also shown in this figure are the positions on the chromatogram in which the reference compounds 1-nitropyrene and 1-aminopyrene were located. There was no detectable 1-aminopyrene in the reaction mixture extract.

Visual inspection of the chromatogram revealed that the region on the TLC plate in which the active metabolite was located contained at least three partially-resolved bands of material. These components were recovered and their UV-visible

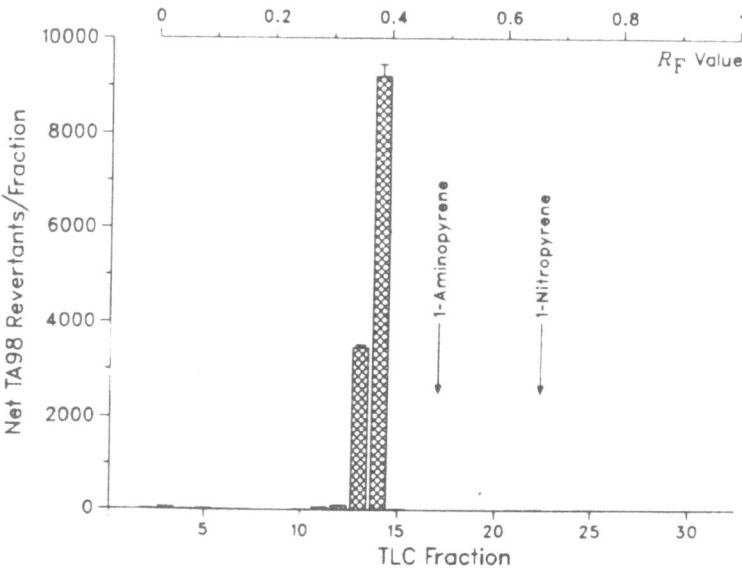

Figure 8. The location of the mutagenic products extracted from the incubation of 1-nitropyrene with S9 enzymes after TLC separation on silica gel using the solvent mixture: toluene/dichloromethane/methanol (9:1:0.4).

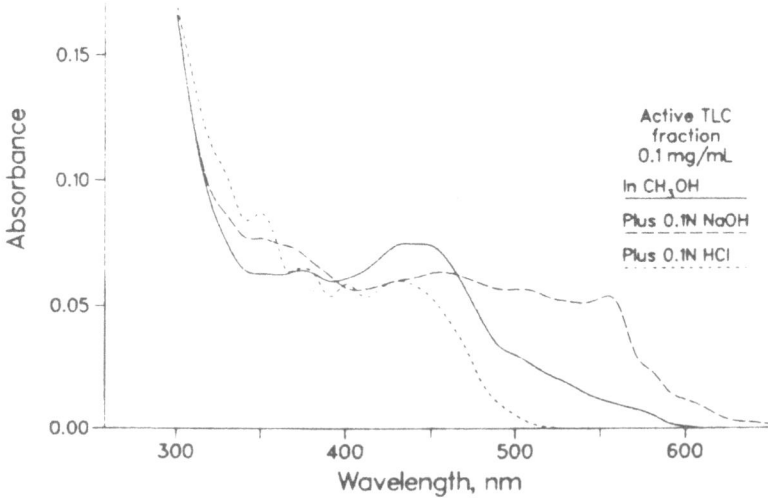

Figure 9. The absorption spectra of the material recovered from TLC fractions containing the mutagenic metabolite(s) of 1-nitropyrene.

absorbance spectrum is shown in Figure 9. The absorbance spectrum was measured under both alkaline and acidic conditions, because the active metabolite could very probably be a ring-hydroxylated derivative of 1-nitropyrene. The change in absorbance maxima to longer wavelengths under alkaline conditions, as is shown by these spectra, is characteristic of phenolic compounds. We do not, however, know that the material producing the absorbance spectra is in fact the mutagenic metabolite. Continued efforts are being made to isolate and identify this product.

SUMMARY COMMENTS

This article has described the results of our studies which were undertaken to determine what type of chemical components in the extracts of diesel exhaust particulate account for their direct-acting mutagenicity in the Salmonella assay. These findings, along with those of other investigators, have demonstrated that this activity can be ascribed to a variety of nitroaromatic compounds, principally, the nitrated derivatives of polycyclic aromatic hydrocarbons. We have also studied the way in which mammalian enzyme activity can affect the mutagenic properties of these compounds. There are clearly future research needs in both of these areas.

Considerable progress has already been made in identifying the monosubstituted nitro-PAH compounds found in diesel particle extracts. Certain other nitro-compounds, such as the dinitropyrenes, have also been identified, but the identities of the highly polar compounds contributing to the mutagenicity of these extracts are unknown. It is also important to determine if the particle extracts being used for chemical and biological studies are sufficiently representative of the material being emitted from diesel-powered vehicles, because it is known that chemical modification of the extractable material can occur during the collection of these particulates.

These studies have shown that if mammalian enzyme systems are included in bacterial mutation assays with nitro-PAH compounds, careful attention must be given to both experimental methods and the interpretation of results. The complexity of the metabolic processes occurring in the Salmonella/S9 assays is perhaps indicative of the complexity to be encountered in determining the in vivo metabolic fate and genotoxic action of these compounds in animals.

REFERENCES

1. US National Academy of Sciences: 1981, "Health effects of exposure to diesel exhaust", the report of the Health Effects Panel of the Diesel Impact Study Committee, National Research Council-National Academy of Sciences USA, Washington DC.

2. Pederson, T.C.: 1980, In "Health Effects of Diesel Engine Emissions", Proceedings of an International Symposium, EPA-600/9-80-057a, (W.E. Pepelko, R.M. Danner and N.A. Clark, eds.) 1, pp. 481-497.

3. Rosenkranz, H.S. and Poirier, L.A.: 1979, J. Nat'l. Cancer Inst. 62, pp. 873-892.

4. Siak, J-S., Chan, T.L., and Lee, P.S.: 1981, Environmental Internat'l. 5, pp. 243-248.

5. Chan, T.L., Lee, P.S., and Siak, J-S.: 1981, Env. Sci. Technol. 15, pp. 89-93.

6. Depass, L.R., Peterson, L.G., Weil, C.S., and Chen, K.C.: 1981, EPA 1981 Diesel Emissions Symposium, Oct.5-7, Raleigh, NC, Extended Abstracts.

7. Schreck, R.M., Soderholm, S.C., Chan, T.L., Smiler, K.L., and D'Arcy, J.B.: 1981, J. Appl. Toxicology 1, pp. 67-76.

8. Pederson, T.C. and Siak, J-S.: 1980, American Chemical Society, Div. Environ. Chem., Extended Abstracts 20, pp. 533-535.

9. Pederson T.C. and Siak, J-S.: 1981, J. Appl. Toxicology 1, pp. 54-60.

10. Johnson H. and Sawicki, E.: 1966, Talanta 13, pp. 1361-1373.

11. Rosenkranz, E.J.,McCoy, E.C., Mermelstein, R., and Rosenkranz, H.S.: 1982, Carcinogenesis 3, pp. 121-123.

12. Pederson, T.C. and Siak, J-S.: 1982, In "Polynuclear Aromatic Hydrocarbons: Physical and Biological Chemistry", (M. Cooke, A.J. Dennis and G.L. Fisher, eds.) Battelle Press, Columbus, pp. 623-640.

13. Schuetzle, D., Riley, T.L., Prater, T.J., and Salmeen, I.:
     1982, In "Analytical Techniques in Environmental Chemistry
     II",(J. Albaiges,   ed.)   Pergamon Press, Oxford, pp.
     259-280.

14. Gibson, T.L., Ricci, I., and Williams, R.L.: 1981, In
     "Chemical analysis and Biological Fate: Polynuclear
     Aromatic Hydrocarbons", (M. Cooke and A.J. Dennis, eds.)
     Battelle Press, Columbus, pp. 707-717.

15. Siak, J-S. and Strom, K.A.: 1981, The Toxicologist 1, p. 75
     Abstract.

16. Ames, B.N., McCann, J., and Yamasaki, E.: 1975, Mutation
     Res. 31, pp. 347-364.

17. Pederson, T.C. and Siak, J-S.: 1981, J. Appl. Toxicology 1,
     pp. 61-66.

18. Pederson, T.C. and Siak, J-S.: 1982, The Toxicologist 2, p.
     49 Abstract.

VEHICLE EMISSION CONTROLS AND ENERGY - THE ROLE OF AROMATICS AND LEAD COMPOUNDS

R. PERRY, A.E. McINTYRE, J.N. LESTER & A. CLARK

Public Health Engineering Laboratory, Imperial College, London SW7 2BU

1. INTRODUCTION

The last decade has seen considerable advances within the field of environmental control engineering where in particular, changes relating to vehicle emission controls have been significant. Modifications made to vehicles have been both innovative and beneficial and much can be learned from the experience of the last few years with regard to future policy in areas of the world where, to date, few changes have been made.

Legislation introduced in some countries however has also represented a balance between political and scientific opinion and there has been some inflexibility in modifying initial policies in relation to new constraints.

Earlier decisions relating to vehicle emission controls were heavily influenced by the Californian smog situation and policies implemented to alleviate the photochemical problems of this area were frequently accepted as a requirement in other areas, despite significant differences in meteorological conditions and no clear definition of necessary air quality standards.

Environmental controls associated with any combustion process necessarily involve energy penalties and it is clear that this balance between environmental concern, on the one hand, and the definition of what is acceptable in terms of energy conservation on the other, needs to be re-examined again at the commencement of this present decade. Recent trends in Europe and North America have shown significant changes in fuel composition where,

247

D. Rondia et al. (eds.), Mobile Source Emissions Including Polycyclic Organic Species, 247–258.
© 1983 by D. Reidel Publishing Company.

in particular, there has been a decrease in the use of lead alkyl antiknock agents and an appropriate increase in the use of aromatic components to compensate for this change. These factors, that can lead to increased emissions of aromatic and polycyclic aromatic hydrocarbons (PAH), need to be critically evaluated with regard to any change in fuel policy with its associated environmental and energy considerations.

It is fortunate that many developing countries who are evaluating the possibility of legislation, can still weigh this delicate balance carefully in relation to their own very different national priorities, taking into account the. required energy constraints and the air quality objectives required.

2. VEHICLE EMISSION CONTROLS

Currently, there are four main test procedures being used in the world for vehicle testing and certification involving specific test cycles and subsequent exhaust gas analysis. These methods include those introduced as the 1973 and 1975 US procedures, the European ECE 15 cycle and the Japanese test procedures. Controls instituted have largely related to three exhaust gas, parameters namely, total hydrocarbons (HC), oxides of nitrogen (NO_x) and carbon monoxide (CO). Typical global trends in the control of these emissions can be assessed from Figure 1 where it can be seen that the US (the 49 states and California) together with Japan have to date legislated for controls that involve 90% or more reduction.

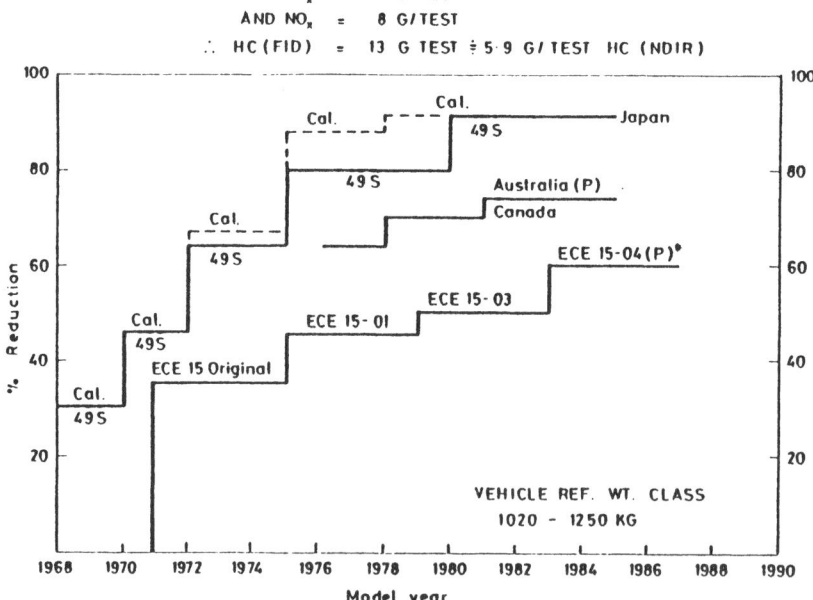

Figure 1. Hydrocarbon controls.

The technology employed by the car manufacturers in these countries to meet such stringent controls is based upon the use of
oxidation or 3-way noble metal catalysts. Associated with this
is the requirement to use unleaded gasoline that has in turn
necessitated reversion to engine compression ratios of pre 1970
values with their associated increased fuel usage.

In other countries all the other standards are currently being
met with largely preventitive techniques as defined in Figure 2,
with the exception of manifold air injection methods.

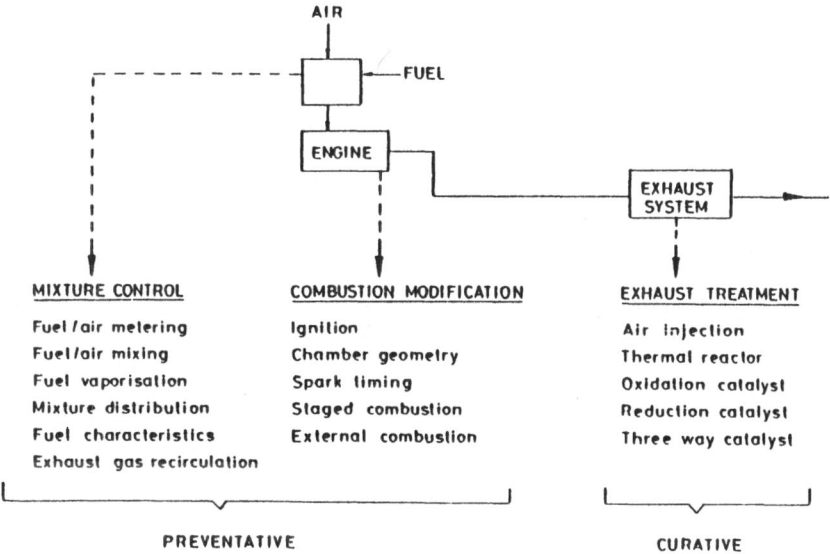

Figure 2. Approaches to the control of engine emissions.

Such a high degree of emission control is costly to achieve and
there is still controversy both in the definition of the air
quality standards required as related to both photochemical
activity and health requirements as well as a continuing debate
concerning the relevance of the individual source contribution to
the overall air quality of the areas involved.

3. ENERGY IMPLICATIONS OF EXHAUST EMISSION CONTROLS

A typical relationship between exhaust emission standards and the
associated energy penalties is shown in Figure 3. This is based
largely upon experience gained in the United States over the last
decade. Calculations relating to the use of unleaded fuel are
dependent upon the change in compression ratios that have been
necessary to accommodate this situation. This enables the energy
consumption of vehicles operating without catalysts and utilizing
leaded gasoline to be compared with those of vehicles operating

with catalysts and on unleaded gasoline under increasingly stringent emission standards.

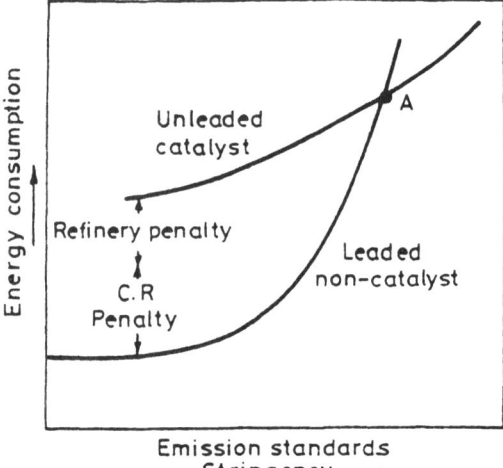

Figure 3. Energy relationships.

Compression ratio reductions (Figure 4) of 1.2 by the year 1978
in the United States accounted for an equivalent increase of fuel
consumption of 6.7%.   At the 1978 consumption levels of 7.4
million barrels of motor gasoline a day, if all vehicles were
operating in this mode, additional fuel requirements would be
equivalent to an additional 496,000 barrels of gasoline per day
or 2.6% of total US oil requirements.

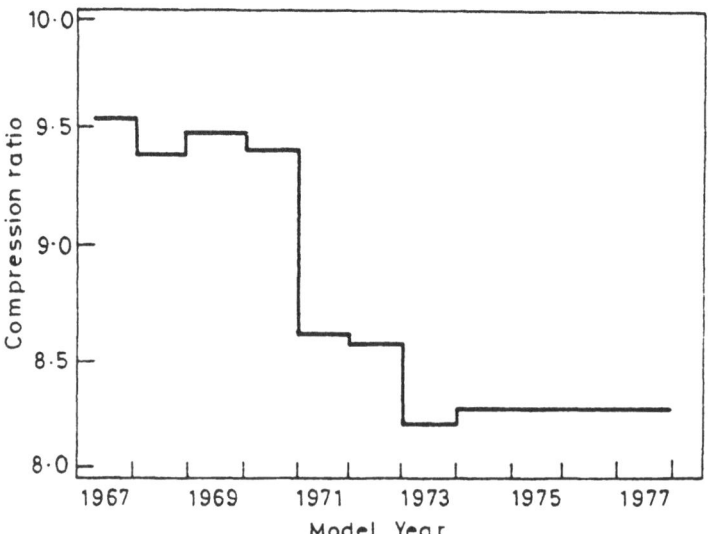

Figure 4. Trend compression ratio of US cars
(weighted average).

Clearly, the situation is complex and much depends upon decisions taken in terms of operating compression ratios for vehicles involved. Utilization of higher compression ratio engines with their associated improvement in thermal efficiency and fuel consumption is dependent however upon the availability of high octane fuels.

In Europe the average premium grade octane quality has increased from a research octane number of around 82 in the early 1950's to around 98-99 by the mid-1960's. The improvement in octane quality has been accompanied by a corresponding increase in the compression ratios of engines. The improvement in engine effi-ciency and its associated fuel consumption benefits has been well documented and is illustrated in Figure 5, where it can be seen that each unit change in compression ratio is associated with a fuel consumption change of about 5% by volume.

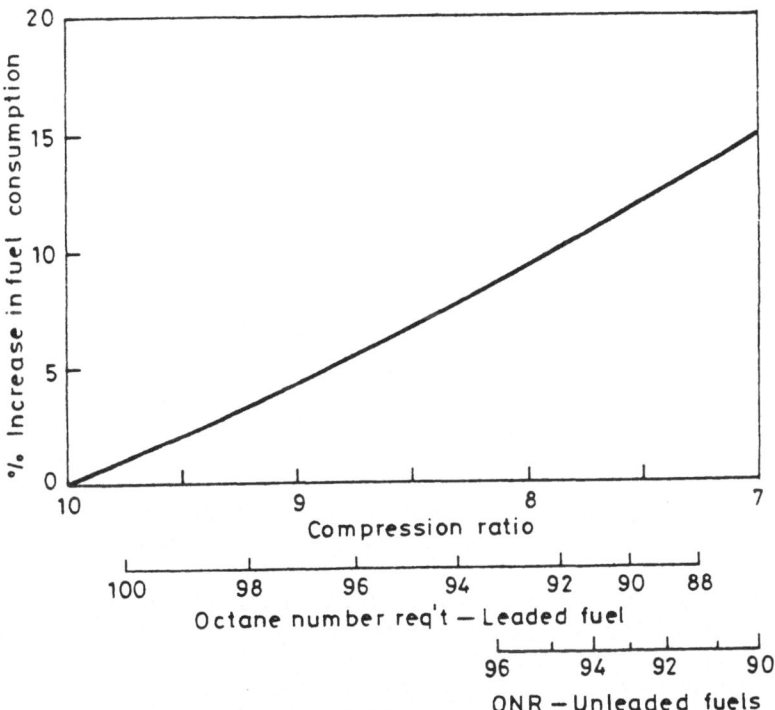

Figure 5. Effect of compression ratio on fuel consumption

European legislation requiring about 60% reduction in unburned hydrocarbons and carbon monoxide from the vehicle exhaust and a more modest reduction of oxides of nitrogen has been achieved to date without significant fuel consumption penalties. More re-cently, however, legislation has been passed for the reduction of lead in gasoline in the United Kingdom and this, of course, will carry with it an associated energy penalty for refineries. Similar legislation has been in operation for some time in both Germany and Sweden although the full effects of such changes are still to be evaluated. In Sweden and now in the United Kingdom a phased reduction in the use of lead down to 0.15 g l^{-1} has been planned and, clearly, if compression ratios are to be maintained the octane requirement will have to be met by alternative refinery practices. Inevitably this will require higher percentages of aromatic compounds in fuels and, with the complex combustion processes involved will undoubtedly lead to higher concentrations of benzene and associated aromatic materials in urban air.

The changing situation in Europe with respect to the lead content of gasoline and its associated aromatic content is shown in Table I.

Table I. Lead and aromatic content of European gasolines

Country	Lead Content (g Pb/litre)		% Aromatics (v/v)	
	1980	1981	1980	1981
Italy	0.64	0.38	29.0	32.6
France	0.5	0.39	33.1	36.8
UK	0.45	0.38	35.1	34.5
West Germany	0.15	0.14	42.7	43.1

From this data it can be seen that for 1980 gasoline sold in West Germany with 0.15 g Pb l^{-1} contained some 32% more aromatics than the average content marketed in the other three countries.

Three of the major aromatics present in gasoline are benzene, toluene and xylene (o-, m- and p-). Although the benzene content is generally limited in many countries either by law or by agreement, because of the well-documented health effects associated with this material, there is need for a detailed study of combustion changes that occur with the other aromatics leading to formation of benzene.

It has been estimated by Häsänen et al. (1981) that approximately 3-4% of aromatics are expelled unburnt with exhaust gases. In addition, benzene emissions linearly dependent on the benzene content of fuel up to 15% v/v have been claimed during ECE 15 cycles. Other workers have suggested that the total aromatic content present in the fuel capable of dealkylation in the combustion process is the critical factor in defining benzene emissions (Black et al., 1980; Morris and Dishart, 1971). Further consideration must also be given to evaporative losses which under certain circumstances can account for up to 20% of the benzene concentration of street air.

A preliminary investigation carried out by the Public Health Engineering Laboratory of Imperial College utilized three fuels of similar RON and benzene content (Table II). Benzene exhaust emissions were monitored using varying aromatic contents in the input fuels balanced with differing lead alkyl additions to maintain the same octane requirements.

A regularly serviced test car was used for this study and was maintained at a constant speed of 90 k hr^{-1} on a dynomometer system. Samples of exhaust gas were collected utilizing a constant volume sampling system with exhaust gas samples being collected in clean dry preconditioned Tedlar bags. Analyses of the samples were carried out by direct injection of the diluted

Table II. Composition of the test fuels compared with West German and UK aromatic contents

Lead Content of 97 RON fuel g l^{-1}	Benzene Content % vol/vol	Toluene Content % vol/vol	Xylene Content % vol/vol
0.64	4.33	14.95	15.44
0.4	4.59	19.47	18.14
0.15	4.43	21.85	21.75
West Germany (0.15 g l^{-1}) August 1980	5.23	13.7	10.09
UK (0.4 g l^{-1}) January 1981	3.2	10.7	7.73

exhaust gas samples into a portable photoionization gas chromatograph (Photovac model 10A10) equipped with a 6 ft × 1/8 in column packed with 1.5% OV17 + 1.95% QFI on Supelcoport 100/120. The detector utilized in this system offered considerable sensitivity towards aromatic materials and a typical benzene, toluene, xylene exhaust analysis is shown in Figure 6.

Exhaust gas analyses with respect to fuels used are presented in

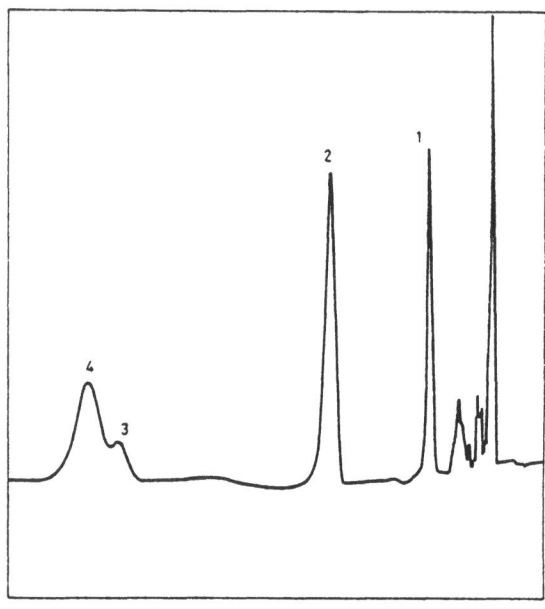

Figure 6. A typical exhaust sample taken from Tedlar bag (see text). Carrier gas flow rate = 20 ml min^{-1}, attenuation 100, chart speed = 1 cm min^{-1}.
1=benzene, 2=toluene, 3=ethyl benzene, 4=xylene.

Tables II and III and graphically in Figure 7.

Decreasing lead contents of fuels balanced by increased utiliza- tion of xylene in particular leads, under certain circumstances, to increased levels of benzene emissions. Although this data presented is of a preliminary nature, it clearly indicates the need for detailed study of the changing hydrocarbon pattern of emissions associated with changing compositions of fuels, parti- cularly where regulation involving lead reduction is involved and aromatic materials are used to balance octane requirements. Clearly, such studies should not only involve dealkylation pro- cesses but should take into account the more complex aromatic blending components that could now be involved.

Apart from the more simple emissions of aromatic compounds such as benzene, previous work has shown that the fuel aromatic con- tent itself strongly influences the emissions of PAH (Candelli et al., 1974 and Gross, 1973). Fuel aromatic contents influence particle associated PAH emissions almost linearly and a 10% increase in aromatics has been shown to increase emissions of benzo(a)anthracene, benzo(a)pyrene and benzo(ghi)perylene by approximately 20% (Pedersen et al., 1980). As typical increases

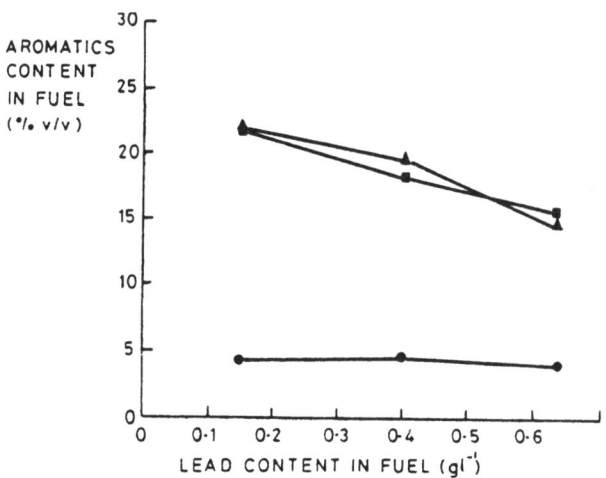

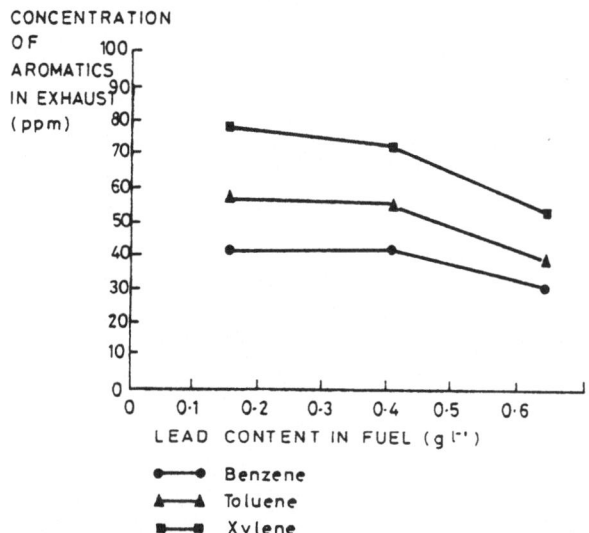

Figure 7. Aromatic content in fuel and exhaust for the three test fuels.

in aromatic contents of fuels are of the order of 15% when decreasing the lead content from 0.64 to 0.15 g l^{-1}, consequent increases in PAH emissions could be expected if 97 RON fuels are to be maintained.

With the present emphasis on energy conservation, considerable effort is now being devoted to the improvement of efficiency of spark ignition engines operating at high compression ratios. Such engines including those operating at compression ratios of up to and even exceeding 14:1, utilizing 'lean-burn' conditions

Table III. Concentration of aromatics in exhaust gas

Fuel	Compound	Exhaust Concentration in ppm (mg m^{-3} in parenthesis)		
0.64 gl^{-1} lead content	Benzene	30.2 33.5 31.3	$\bar{x}$ = 31.7 (103 mg m^{-3} SD = 1.37 RSD = 4.3%		
	Toluene	38.0 41.7 38.1	$\bar{x}$ = 39.3 (150 mg m^{-3}) ·SD = 1.72 RSD = 4.4%		
	Xylene	54.6 54.6 54.6	$\bar{x}$ = 54.6 (241 mg m^{-3}) SD = 0 RSD = 0%		
0.4 g l^{-1} lead content	Benzene	47.7 42.2 38.6	$\bar{x}$ = 42.8 (139.1 mg m^{-3}) SD = 3.7 RSD = 8.7%		
	Toluene	62.0 55.0 51.2	$\bar{x}$ = 56.1 (215 mg m^{-3}) SD = 4.47 RSD = 8.0%		
	Xylene	83.0 72.8 65.0	$\bar{x}$ = 73.6 (325.1 mg m^{-3}) SD = 7.37 RSD = 10%		
0.15 g l^{-1} lead content	Benzene	40.0 39.3 46.6	$\bar{x}$ = 42 (136.5 mg m^{-3}) SD = 3.29 RSD = 7.8%		
	Toluene	54 54.8 65	$\bar{x}$ = 57.9 (221.9 mg m^{-3}) SD = 5.01 RSD = 8.64%		
	Xylene	72.8 78 88.4	$\bar{x}$ = 79.7 (352 mg m^{-3}) SD = 6.48 RSD = 8.1%		

look most promising and current indications are that the efficiency of these engines over complete load ranges will be satisfactory.

It is clear that this type of development is very much linked to
a requirement to maintain the use of high octane quality gasoline
and it is in this situation that the energy penalty and environmental consequences of utilizing gasolines with reasonable levels
of lead additives or with higher aromatic contents need to be

carefully evaluated in the context of the current energy situation and realistically defined air quality standards.

It is perturbing that, at present, very little air quality data relating to vehicle emission problems are available in many developing countries. Automatic adoption of legislation from Western countries, implemented prior to the awareness of the current energy problems would be relatively more costly in energy terms and would be premature prior to detailed cost benefit analysis of the situation as it applies to each different country in its own right.

The necessity for acquisition of data at this stage can only be emphasized by a consideration of the differences in meteorological conditions that apply, where for instance, in many Far Eastern countries the high humidities and in other countries, the high dust levels, are bound to significantly affect the level of photochemical activity. Compounded with different types of industrial processes and natural sources of emission this can only further stress the importance of data acquisition and evaluation prior to institution of optimized air quality parameters.

4. CONCLUSIONS

As with many environmental controls the benefits of restricting emissions from vehicles need to be quantified with respect to the energy penalty involved. Although it is evident that certain air quality objectives can be met by modest control of vehicle exhaust emissions it is equally clear that stringent emission controls lead to high energy penalities.

Detailed data are generally not available in many developing countries and it is clear that if realistic air quality objectives are to be defined the necessary monitoring work and cost benefit studies need to be undertaken in order to determine the contributions made to air quality by vehicle as well as industrial and natural sources.

It would be ludicrous at this stage to impose the Californian standards upon countries having major energy problems particularly where the climatic conditions are significantly different with respect to photochemical activity.

Changes of fuel composition involving either lead compounds or other hydrocarbon materials leading to the production of increased levels of aromatic materials within the atmosphere is a situation that needs close scrutiny in that the alternative environmental problems associated with many of these materials are well documented and disturbing.

REFERENCES

Black, F.M.; High, C.E. and Lang, J.M. (1980) Composition of automobile evapourative and tailpipe hydrocarbon emissions. Journal of Air Pollut. Control Assoc., 16, 67.

Candelli, A.; Mastrandrea, V.; Arteconi, M. and Sezzi, F. (1974) Carcinogenic air pollutants in the exhaust from a European car operating on various fuels. Atmos. Environ., 8, 693.

Gross, G.P. (1973) Gasoline composition and vehicle exhaust gas polynuclear aromatic content. Publication No. PB-233 529.

Häsänen, E.; Karlsson, V.; Leppamaki, E. and Juhala, M. (1981) Benzene, toluene and xylene concentrations in car exhausts and in city air. Atmos. Environ., 15, 1755.

Morris, W.E. and Dishart, K.T. (1971) Influence of vehicle emission control systems on the relationship between gasoline and vehicle exhaust hydrocarbon composition. Effect of automative emission requirements on gasoline characteristics. ASTM. STAND. 487, 63.

Pedersen, P.S.; Ingwersen, J.; Neilson, T. and Larson, E. (1980) Effects of fuel, lubricant and engine operating parameters on the emeeions of polycyclic aromatic hydrocarbons. Environ. Sci. Technol., 14, 71.

MECHANISMS OF PAH FORMATION AND DESTRUCTION IN FLAMES
RELATION TO ORGANIC PARTICULATE EMISSIONS

G. Prado[*] and J. Lahaye[***]

[*] Université de Haute Alsace, 2, rue des Frères Lumière
68093 MULHOUSE CEDEX France

[***] Centre de Recherches sur la Physico-Chimie des Surfaces
Solides, C.N.R.S.
24, Avenue du Président Kennedy, 68200 MULHOUSE France

ABSTRACT

The formation and destruction of polycyclic aromatic hydrocarbons (PAH) in flames have been extensively investigated during
the last 20 years. Although many questions remain to be answered,
some important steps have been elucidated. Profiles of PAH concentration in premixed and diffusion flames indicate that formation of PAH occurs very early in the process and is followed by
PAH destruction either by burn-out or through formation of soot
particles.
PAH emission in exhaust gases results from competition between
those two processes (formation and destruction), and are discussed with special emphasis on the following points :
- Formation of PAH from aliphatic and aromatic fuels.
- Role of aromatic and acetylenic species in PAH formation.
- Distinction between PAH as stable by products, PAH as reactive
  intermediates and PAH containing heteroelements (N,O,S).
- Relation between PAH and soot.

INTRODUCTION

Organic particulates emitted from practical combustion devices
are very complex mixtures. Broadly speaking, they consist of

259

D. Rondia et al. (eds.), Mobile Source Emissions Including Polycyclic Organic Species, 259–275.
© 1983 by D. Reidel Publishing Company.

hydrocarbons more or less adsorbed on solid soot particles. For
analytical purpose, these two different phases are usually separated through Soxhlet extraction.
Present legislations focus on carbonaceous particulate emissions,
without considering separately soot particules and extractible
matter. However, mutagenic properties of these emissions are
essentially related to the extractible fraction of the organic
particulates. Although they generally occur together, overall
particulate emissions and extractible loading are not always
well correlated, as illustrated later in this paper.
The extractible fraction itself contains hundreds of compounds,
and only a limited number can be routinely analysed. Among these
molecules, the family of polycyclic aromatic hydrocarbons has
been by far the most extensively studied, in terms of mechanism
and properties. PAH are indeed quite easily produced in laboratory flames and can be routinely analysed, at least for molecules
with masses lower than 300 a.m.u. During the workshop, several
papers have discussed in detail the analysis of polycyclic organic
matter. However, as a brief introduction, it is useful to define
the different compounds considered.

1. ANALYSIS OF POLYCYCLIC ORGANIC MATTER

One can consider two classes of PAH
- PAH containing only C and H, no heteroatoms
- PAH containing heteroatoms, mainly O, N, S.
The first class of PAH is the only one observed in most laboratory
flames, where all fuel passes through the flame zone.
Fig. 1 illustrates a chromatogram of PAH produced from a turbulent diffusion flame of kerosene (1). Table 1 lists the compounds
identified with gas chromatography mass spectrometry. It is striking that very similar chromatograms are obtained when analyzing
extractible matter resulting from burning many different fuels,
aliphatic or aromatic, on premixed or diffusion flames, laminar
or turbulent. The nature of the molecules produced, and often
their relative amounts, are little influenced by the system investigated. With regards to health effect, much of the mutagenic
activity in the AMES, or-similar tests, can be accounted for by
8 compounds : phenanthrene, alkyl phenanthrene, alkyl anthracene,
fluoranthene, pyrene, cyclopenta (cd) pyrene and benzo(a) pyrene
(2).
The PAH containing heteroatoms are not observed in studies of
laboratory flames. They constitute however an important fraction
of extractibles from emissions of practical systems. So far, they
have been essentially related to diesel engine emission, although
they are also produced by other systems such as coal burning equipment and domestic oil burners. In the case of diesel engine emissions, the identification and testing of oxygenated, polar compounds are an active area of research as reflected by the contri-

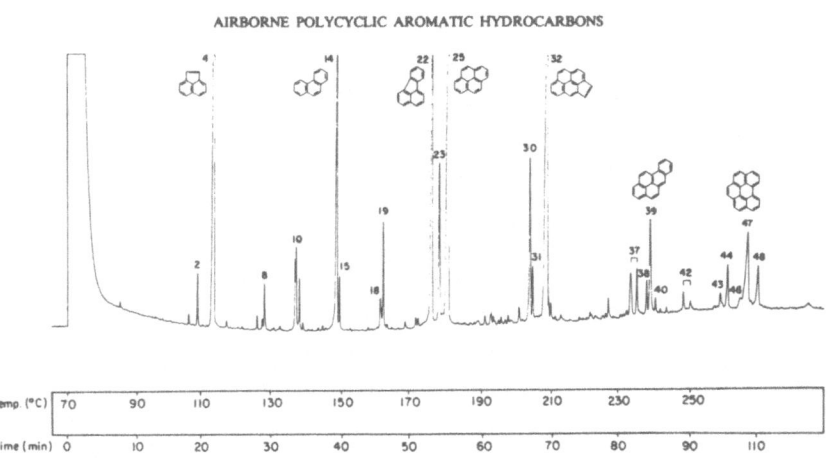

Figure 1. Capillary-column gas chromatogram of the PAH
fraction of kerosene combustion. Key : Table 1.

butions to this workshop.
Also it is worth noting that the first category of PAH shows mutagenic activity only in the presence of oxidative enzyme activator
(Aroclor PMS), whereas the second category is equally, or more
active, in the absence of PMS (3).
All compounds mentioned so far have molecular masses in the 100-
300 a.m.u. range. The vapor pressure of PAH with molecular masses
larger than 300 are too low for such compounds to be chromato-
graphed in the gas phase. They are however present in many samples, as indicated by a recent work of Peaden et al. (4), who
published analysis of compounds from mass 300 (7 rings) to 600
(11 rings), by using HPLC on a reversed phase C_{18} column. Separated fractions were collected and analysed by mass spectrometry
and fluorimetry for identification.

Table 1. PAH identified by g.c.m.s.

Peak No.	Compound	Peak No.	Compound	Peak No.	Compound
1	Methylnaphthalene	17	Propyldibenzofuran[b]	33	Benz[a]anthracene
2	Biphenyl	18	Methylphenanthrene[c]	34	Chrysene
3	Ethylnaphthalene[a]	19	4H-cyclopenta[def]phenanthrene	35	Methylchrysene[e]
4	Acenaphthylene	20	Methyl-4H-cyclopenta[def]phenanthrene	36	Methylcyclopenta[cd]pyrene[f]
5	Methylbiphenyl	21	Ethylphenanthrene[a]	37	Benzofluoranthene
6	Dibenzofuran	22	Fluoranthene	38	Benzo[e]pyrene
7	Propylnaphthalene[b]	23	Benz[e]acenaphthylene	39	Benzo[a]pyrene
8	Fluorene	24	Benzo[def]dibenzothiophene	40	Perylene
9	Methyldibenzofuran	25	Pyrene	41	Methylbenzopyrene[e]
10	$C_{14}H_8$[h]	26	Ethyl-4H-cyclopenta[def]phenanthrene[a]	42	$C_{21}H_{12}$ (unknown)
11	Methylfluorene	27	Methylfluoranthene[d]	43	$C_{21}H_{12}$ (unknown)
12	Ethyldibenzofuran[a]	28	Benzo[a]fluorene	44	Indeno[1,2,3-cd]pyrene
13	Dibenzothiophene	29	Benzo[b]fluorene	45	Dibenz[a,h]anthracene
14	Phenanthrene	30	Benzo[ghi]fluoranthene	46	Dibenz[a,c]anthracene
15	Anthracene	31	$C_{18}H_{10}$ (unknown)	47	Benzo[ghi]perylene
16	Ethylfluorene[a]	32	Cyclopenta[cd]pyrene	48	Anthanthrene

[a] Could be dimethyl.
[b] Could be trimethyl or ethylmethyl.
[c] Could be methylanthracene.
[d] Could be methylpyrene.
[e] Could be methylbenz[a]anthracene.
[f] Could be methylbenzo[ghi]fluoranthene.
[g] Could be methylbenzofluoranthene.
[h] Probably cyclopent[bc or fg]acenaphthylene, see Ref. 17.

The mechanisms discussed in the subsequent sections refer essentially to PAH containing only C and H, as very little work has been published on the formation of PAH containing heteroelements in flame conditions (5).

2. GENERAL TRENDS OF PAH FORMATION.

A striking feature of PAH formation in flames is the rapidity of the process. Indeed, PAH appear very early in the process, at a point where only a fraction of the initial fuel has reacted. This is especially the case with aromatic fuels (6) as illustrated in fig. 2.
In this case, a toluene flame was sampled at different heights with a quartz microprobe, producing yield data that showed a dramatic, sharp peak at approximately 2 mm above the burner. By comparison, the bottom of the yellow, luminous region occured at 1.0 mm above the burner. From the peak, total yield rapidly decreased by two orders of magnitude as distance from the burner

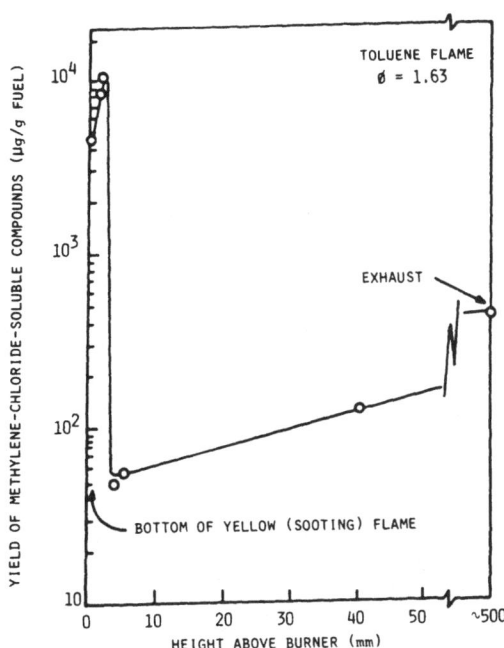

Fig. 2.   Effect of distance from burner surface upon
          yield of extractable compounds.

increased ; then the yield slowly increased again.
Similar profiles have been reported by different authors (7, 8),
in the case of premixed flames of aromatic and aliphatic fuels.
In the case of aliphatic fuel, however, the maximum does not
appear to be as sharp, and in some cases, a continuous increase
through the process, with no maximum, has been reported (9).
Another example of the high rate of PAH formation is given by
reactions occuring in the fuel zone of a diffusion flame (10).
Fig. 3 represents such a flame, methane burning in air. Optical
techniques, such as laser fluorescence and scattering allow the
precise determination of the zone of PAH and soot formation, respectively. Profiles of light scattering and fluorescence above
the burner along the axis of the flame are plotted in fig. 4. It
is clear that PAH are formed in a zone where the temperature is
relatively low ($\approx$ 500°C), and in the absence of oxygen. Similar
results have been reported by d'Alessio et al.(11).

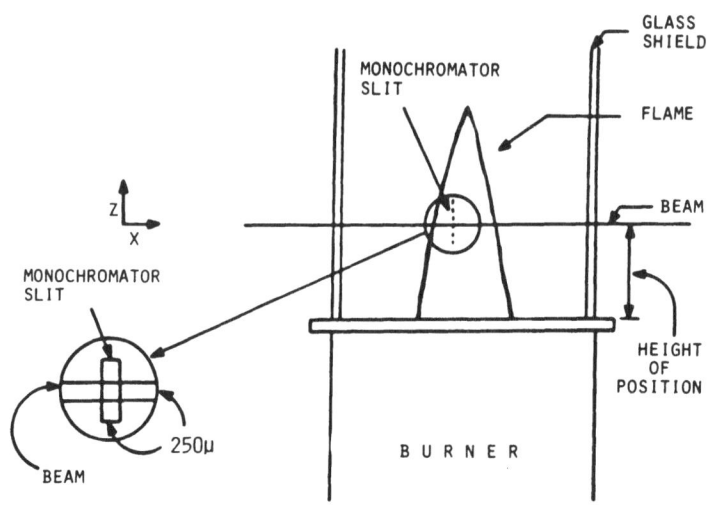

Fig. 3. Diffusion Flame methane - air.

In all these examples, it is also clear that PAH formation occurs
before soot formation, the decrease of PAH concentration being
associated with the beginning of soot formation. PAH are, therefore, possible intermediates of soot, and their destruction might
occur either through their incorporation into a solid phase
(soot), or through direct burnout (or both).

3. MECHANISMS OF FORMATION

The different routes proposed for the soot formation have been
listed by Palmer et al.(12), Lahaye et al. (13) and Haynes et al.
(14). Today, there is little doubt that large polyaromatic
hydrocarbon (PAH) molecules are the precursors of soot nuclei in
usual combustion devices.
PAH in flames are formed within a few milliseconds. Different
mechanisms are mentioned in the literature :
. cyclisation of polyacetylenes
. direct addition of cycles in the case of aromatic fuels
. ions may have a major role in the production of soot precursors as well as soot onset.
The aim of the present paper is not to present an exhaustive
review of works carried out in that field but to illustrate the
actual knowledge by some recent results. For more information,

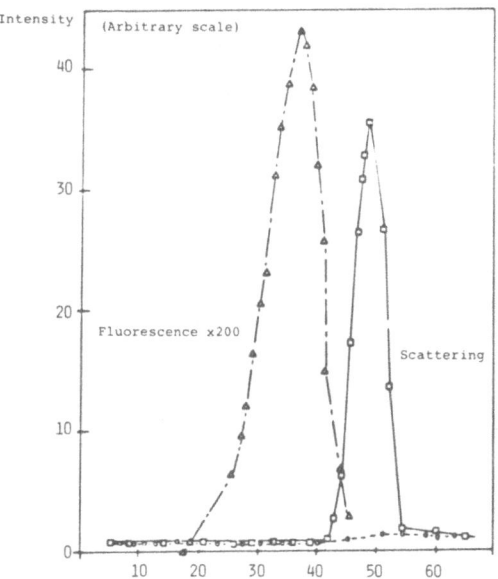

Fig. 4. Axial profiles of scattering (-) and fluorescence
         intensities (— - —)

review articles (12-14)and books (15-16) can be consulted.
One of the most relevant works published recently in that field
is concerned with intermediate species produced during combustion
of benzene (17). We begin to examine PAH formation from aromatic
fuels.

3.1. Formation of PAH from aromatic fuels

   Primary reactions. When an aromatic hydrocarbon is submitted
to high temperature in the presence of oxygen, hydrogen atom(s)
can be stripped from the cycles which can subsequently react with
species present in the flame or rearrange into other cyclic molecular form. The initial aromatic cycle can also be destructed into
aliphatic fragments.
Bittner et al. (17) studied a premixed flame of benzene, oxygen
and argon. The flame is produced at a burner chamber pressure of
2.67 kPa. Reported results are referring to near-sooting and sooting conditions corresponding to fuel equivalence ratios φ of 1.8

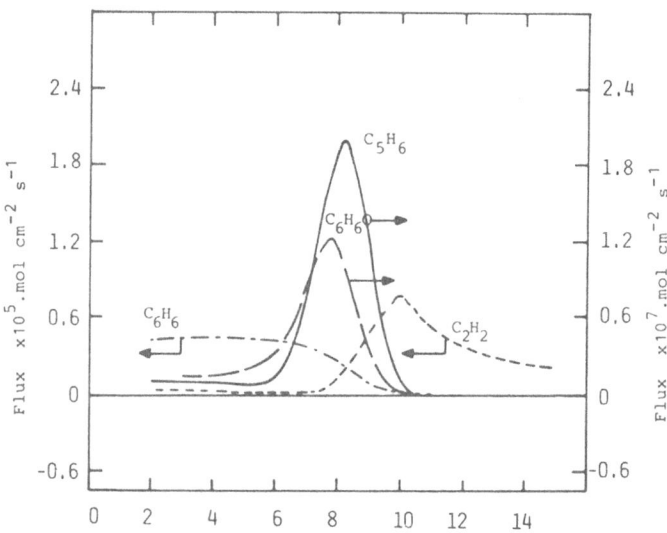

Fig. 5. Molar fluxes of C_6H_6, C_6H_6O, C_5H_6, C_2H_2 vsi distance
from burner in a near-sooting ($\emptyset = 1.8$) benzene
(13.5 m%) – oxygen (56.5 m%) – argon (30.0 m%)
flame. Cold gas velocity = 0.5 m/s. Pressure =
2.67 kPa (20 torr)(17).

and 2.0, respectively. A molecular beam mass spectrometer system
has been developped to measure stable and radical species ; ion
currents have been carefully calibrated.
The early appearance of C_5H_6 along with C_6H_6O in a zone where
benzene decay and CO production begin (fig. 5) suggests that the
initial attack into C_6H_6O is followed by production of CO and
C_5H_6. The reaction of OH (rather than O) with benzene may be the
primary source of C_6H_6O. Acetylene (C_2H_2) is produced after an
induction period which might be the result of the balance
of C_2H_2 production and destruction by OH. Actually, C_2H_2 is
essentially produced at less than about 10 mm above burner, where
benzene is almost depleted and the OH mole fraction rises drama-
tically : therefore its destruction by OH is slow.
When benzene is depleted, the carbonaceous molecular species present in the system are mainly aromatic radicals and aliphatic species such as C_2H_2, C_4H_2, C_4H_4 etc. The question which arises
immediately is, what are, among these species, the

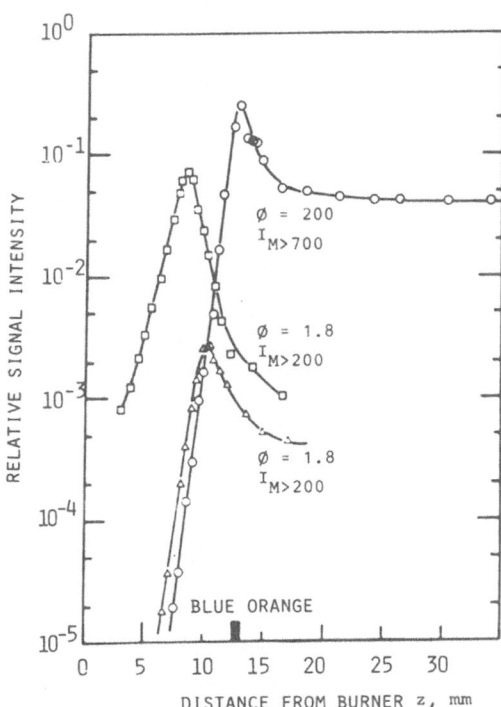

Fig. 6. Relative intensities of signals vs. distance from
burner. ( ) species larger than mass 200,
$I_{M>200}$; and ( ) species larger than mass 700,
$I_{M>700}$; in near sooting ($\emptyset$ = 1.8) benzene (13.5 m%)
- oxygen (56.5 m%) - argon 30.0 m%) flame.
(o) benzene (14.7 m%) - oxygen (55.3 m%) - argon
(30.0 m%) flame. Pressure = 2.67 kPa (20 torr), cold
gas velocity : 0.5 m/s for both $\emptyset$ = 1.8 and $\emptyset$ = 2.0
flames. Shaded region at 13 mm designates blue-
orange boundary in sooting ($\emptyset$ = 2.0) flame(17).

intermediate species of PAH formation?

Build_up of PAH. Bittner et al. have measured the signal
intensities I of high-mass species ; the fluxes are classified
into two categories of masses (i) $I_{M> 200}$ a.m.u. and (ii)
$I_{M>700}$.
The mole fraction of high-mass species (fig. 6)maximizes between
8 and 10 mm above the burner. In sooting flame $I_{M>200}$ maximizes

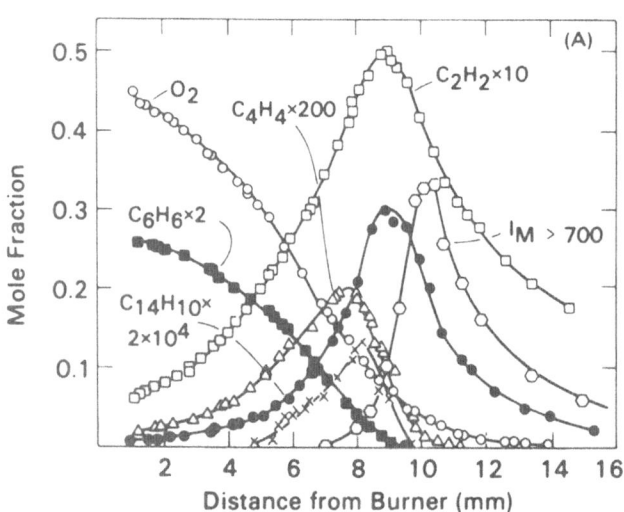

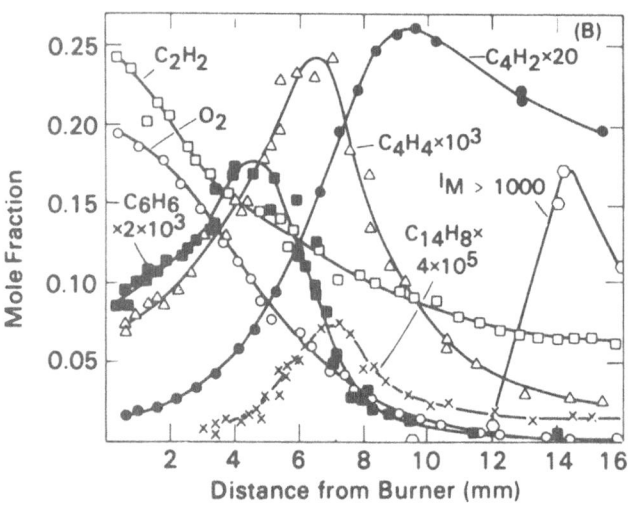

Fig. 7. Mole fractions of several species and high mass
signals ($I_{M>m}$, scale arbitrary) vs. distance
from the burner in (A) a near-sooting
($\emptyset = 1.8$) benzene (13.5 mol %) – oxygen (56.5 mol %)
– argon (30.0 mol %) flame and (B) a sooting ($\emptyset = 3.0$
acetylene (52.9 mol %) – oxygen (44.1 mol %) –
argon (3.0 mol %) flame. Pressure = 2.67 kPa (20
torr), cold gas velocity = 0.5 m/s for both flames (17

at 8.5 mm above burner while $I_{M > 700}$ maximizes at 10.5 mm.
Though   part of low molecular PAH is oxidized, it is clear
that a fraction leads to high molecular PAH. Since mass numbers between 200 an 300 a.m.u. correspond to aromatic structures,
it is believed that this high mass material is aromatic rather
than polyacetylenic.
The structure of the lower molecular weight PAH formed in the
early stages of growth as well as the profile of aliphatic
species indicated that they are products of reactions of aromatic species with C_2, C_3 and C_4 species. From the results of
Bittner et al. and   from the work of Smith on toluene
pyrolysis (18), it is clear that the production of higher-
molecular-weight PAH results also from reaction between mono-
cyclic aromatic species such as benzene, phenyl radical,
toluene and benzyl radicals and non-aromatics such as C_2H_2,
C_4H_2, C_4H_4, C_4H_3, C_2H_2 etc. The role of the initial
ring in this process may be to provide
a structure that is capable of stabilizing adducts formed from
addition of non aromatic species by ring closure, more rapidly
that it can decompose back to reactants. Thermochemical considerations allow the determination of the possible reactions
for PAH formation ; the results obtained suggest that the combined presence of aromatics and non aromatics like acetylene
or vinylacetylene promotes the rapid growth of PAH.

3.2. Formation of PAH from aliphatic fuels.

The formation of PAH by combustion and by pyrolysis of aliphatics has been studied for a long time. In 1973, Crittenden
and Long (7) in an article on the formation of PAH in rich
premixed acetylene and ethylene flames (19) give a good description of the state of art.
What is clear is that PAH can be produced during the combustion
of aliphatic fuels. The respective routes of formation of PAH
and polyacetylenes as well as their role in soot production
have been the object of many "scientific conflicts" for twenty
years. Homann and Wagner (19) considering that since acetylene
is always formed in considerable amounts prior to solid carbon
in flames of ethylene, methane and propane, chose acetylene
as a model substance for aliphatic fuels. In rich premixed
acetylene flames burning at low pressure, concentrations of
polyacetylenes go through a maximum while the concentrations
of PAH increase steadily behind the oxidation zone without
going through any maxima. This would be a general characteristic of all polycyclic aromatics, up to molecular masses of
about 300, in flames of acetylene. The authors conclude that
PAH cannot be important intermediates or "nuclei" for carbon
formation in acetylene flames : otherwise, its rate would not
go down to zero while the concentrations of the compounds
are still increasing.

On the opposite Crittenden and Long (17), also using premixed
acetylene flames burning at low pressure, found that PAH
concentrations maximize in the oxidation zone. Chain lengthen-
ing of acetylene leads to the formation of unsaturated
C_4, C_6, C_8 etc. radical species which can be stabilized as
polyacetylene or either by formation of a bran-
ched radical with ring-closure or by cyclisation, forming
an aromatic ring.

As noticed by Bittner et al. (17), these seemingly conflicting
results may be reconcilied if it is assumed that two different
mechanisms of production of PAH are operating in the oxidation
zone and in the post flame zone. In the oxidation zone where
radical concentrations are high the rapid production and
decline of PAH is free radical in nature. In the post flame
zone radical mole fractions are much lower and PAH might be
formed by molecular mechanisms or heterogeneous mechanism
on the surface of soot particles. Such PAH formation mechanisms
would be very sensitive to the hydrocarbon composition in
the post flame zone, which in turn is very sensitive to
changes in Φ. On the other hand, the intermediate hydrocarbon
and radical species in the oxidation zone are relatively
insensitive to changes in Φ. Thus, at $\Phi = 4.5$ the amount of
PAH formed in the post flame zone may be high enough to
obscure the much smaller amount formed in the oxidation zone.

Hence, the agreement between Bittner results and those of
Crittenden and Long and the apparent disagreement with
Homann and Wagner may be explained by the similarities and
difference in Φ.

Bittner et al. (17) compared (fig. 7) the structure of a near-
sooting $\Phi = 1.8$ benzene flame and the sooting $\Phi = 3.0$
acetylene flame.

In both flames, the mole fractions of the relatively small
PAH (represented by $C_{14}H_{10}$ in the benzene flame and $C_{14}H_8$
in the acetylene flame   maximize in the presence
of benzene, near the maximum C_4H_4 mole fraction and prior
to the maxima of the high mass   signals.

Although the high mass signals ($I_{M} > 700$ and $I_{M > 1000}$) are
not directly comparable since   different
low mass cutoffs have been used in the two flames, there is
more overlap between the profiles of the relatively small
PAH and the high mass signals in the C_6H_6 flame than in the
C_2H_2 flame. This may be due to the   much higher C_6H_6
and PAH mole fractions early in the benzene flame
and their sudden disappearance near 10 mm. Since the C_2H_2
flame is much richer, C_6H_6, C_4H_4, $C_{14}H_8$ and other PAH
survive well into the   burned gas region.

The above results suggest that the mechanism of PAH formation
proposed by Bittner and Howard in the case of aromatic fuels may
be occurring in flames of non-aromatic fuels.
The details of molecular reactions responsible for the formation
of PAH and polyacetylene is not relevant to the present review.
However details about the role of radical species can be obtained
in recent articles such as those of Homann et al. (20) and
Warnatz (21).

3.3. Role of ions.

Whether ions are important or not in PAH and in soot formation
remains unclear.
Measurements of ions (22,23,24) in premixed flames of benzene and
acetylene show a bimodal distribution with height above burner,
of both positively and negatively charged particles, the first in
the flame front and the second just beyond the point where soot
emission is first observed visually. $C_3H_3^+$ ions dominate in non-
sooting flames with larger species becoming increasingly more
important when approaching sooting limit.
The role of ions in the mechanisms of soot formation is still being
discussed. According to Calcote (25), primary flame ions produced
by chemi-ionization react by rapid ion molecule reactions with
neutral flame species, such as acetylene, polyacetylene, and free
radicals, to produce larger ions which rapidly rearrange to produce even larger polycyclic aromatic ions. Some of these ions are
neutralized by recombination with electrons produced in the primary flame ion reaction and become neutral incipient soot particles ;
others grow to produce charged soot particles. These particles
then grow by surface addition as well as by coagulation to form
larger soot particles. As these particles reach a critical size
at a high enough temperature, their ionization potential becomes
sufficiently low that the particles are thermally ionized. The
charge on these particles determines their rate of agglomeration
and produces chains of individual particles distinct from aggre-
gates.

4. OVERALL RELATIONSHIPS BETWEEN PAH AND SOOT.

The environmental concern is mainly related to PAH. However, legislation, as well as most exhaust measurements focus on solid particulates. For example, most field tests consist of smoke number
measurements without considering separately soot and extractibles.
It is therefore important to consider the overall relationship
between soot and PAH.
Soot and PAH are closely related in many ways. The mechanisms of
formation of soot and of PAH are closely intertwined. In addition,
many PAH are found adsorbed on soot particles. However, as pointed before, overall particulate emissions and PAH loading are not

Table 2. Residential oil burner tests.
          (Flame retention head)

Bacharach Smoke No.	% Excess Air	Soot mg/g fuel	Extract mg/g fuel (1)	Bacterial Mutagenic Activity % of BaP (2)
1	24	0.006	0.003	6
5	17	0.06	0.0005	2
9	13	0.5	0.0007	0.05
5(cyclic)	-	0.11	0.01	1.0
Diesel	-	3.0	0.6	∿1.0

(1) The extractable fraction contains oxygenated hydrocarbons.

(2) With Aroclor (PMS). Extrats were considerably more active
    without activation.

always well correlated. Let's illustrate this point by considering
emissions from a residential oil burner (2), commercially available, consuming 1 gallon p. hour of fuel.
Soot emission from these burners is controlled by increasing excess
air to reduce the Bacharach smoke number. In the field, most burners operate in a smoke number range of 1-4. As shown in Table 2
decreasing excess air from 24 to 17 % increases the smoke number
from 1 to 5 and increases the soot production from .006 mg/g fuel
to .06, a factor of 10. A further decrease to 13 % excess air increases soot output by another factor of 10. Production of methylene chloride soluble extract, however, decreases when excess air is
decreased below 24 % and, of considerable interest, the mutagenicity decreases drastically as soot production increases so that
adjusting a mildly smoking burner (smoke number = 5) to a smoke
number of 1 greatly increases the atmospheric loading of mutagenic
material. These measurements were made under steady state conditions ; however, the normal mode is cyclic. Results of a 5 minutes on - 10 minutes off series of tests where excess air was adjusted for a steady state smoke number of 5 are also shown. The
particulates and extractibles loadings are significantly increased
over those found for steady state operation and the specific mutagenic activity was increased by a factor of 5.
Clearly, future tests should consider soot and PAH separately,

and move away from smoke number measurement.

CONCLUSION

PAH emitted from practical combustion systems consist of two
classes of molecules:
-PAH containing only C and H
-PAH containing heteroatoms (O,N,S,)
Most of the studies so far have focused on the first category,
and major results refer only to this group of compounds.
PAH profiles measured in flames indicate a rapid formation of PAH,
followed by destruction, with in several systems a subsequent
increase of PAH concentration in the exhaust.
Mechanistic studies carried out at low pressure, by using molecular beam sampling, indicate that PAH build-up occurs through addition of aliphatic fragments to aromatic rings. Consequently, the
presence of aromatic fractions in the fuel promotes formation of
PAH.
PAH destruction occurs via transformation into soot and by direct
burnout mainly due to attack by OH radicals. Effect of flame parameters on these important processes remains largely unknown, and
deserve further studies.

REFERENCES

(1) Lee, M.L., Prado, G.P., Howard, J.B. and Hites R.A. : 1977,
Biomedical Mass Spectrometry, vol. 4, n° 3, pp. 182.
(2) Longwell, J., Soot in Combustion Systems and its Toxic
Properties. J. Lahaye and G. Prado editors. Plenum Press,
New-York, London, (In. Press).
(3) Thilly, W.G., Soot in Combustion Systems and its Toxic
Properties. J. Lahaye and G. Prado editors. Plenum Press,
New-York, London, (In. Press).
(4) Peaden, P.A., Lee, M.L., Hirata, Y. and Novotny M. : 1980,
Analytical Chemistry, 52, pp. 2268-2271.
(5) Kausch, W.J. Jr., Clampitt, C.M., Prado, G., Hites, R.A.
and Howard, J.B. : 1981, 18th Symposium (International) on
Combustion. The Combustion Institute, pp. 1097.
(6) Center for Health Effects of Fossil Fuel Utilization : 1980,
2nd Annual Progress Report, Massachusetts Institute of Technology.
(7) Crittenden, B.D. and Long, R. : 1973, Combustion and Flame,
20, pp. 359.
(8) Wenz, H. : 1978, Bestimmung der Konzentrationsprofile. Poly-
zyklischer Aromatischer Kohlenwasserstoffe an einer Russenden
Propan-Sauerstoff-Flamme. Diplomarbeit - Technischer Hochschule
Darmstadt.
(9) Haynes, B.S., Jander, H. and Wagner H. Gg. : 1980, Ber.
Bunsenges. Phys. Chem., vol. 84, p. 585.
(10) Prado, G., Garo, A., Lahaye, J. and Sarofim, A., (to be
published).
(11) D'Alessio, A., Personal communication.
(12) Palmer, H.B. and Cullis, C.F. : 1965, Chemistry and Physics
of Carbon, vol. 1, (P.L. Walker Jr. Ed.) Marcel Dekker Inc.,
New-York, p. 265.
(13) Lahaye, J. and Prado, G. : 1978, Chemistry and Physics of
Carbon, vol. XIV, (P.L. Walker Jr. and P.A. Thrower ed.), p. 167.
(14) Haynes, B.S. and Wagner, H. Gg. : 1981, Progr. Energy
Combust. Sci., 7, p. 229.
(15) Particulate Carbon - Formation during Combustion : 1981,
D.C. Siegla and G.W. Smith editors. Plenum Press, New-York-London.
(16) Soot in Combustion Systems and its Toxic Properties : 1983,
J. Lahaye and G. Prado editors, Plenum Press, New-York-London,
(In Press).
(17) Bittner, J.D. and Howard J.B.
. 18th Symposium (International) on Combustion : 1981, (The
  Combustion Institute, Pittsburgh, Pa), p. 1105.
. Particulate Carbon - Formation during Combustion : 1981,
  D.C. Siegla and G.W. Smith editors, Plenum Press, New-York-
  London.
. Soot in Combustion Systems and its Toxic Properties : 1983,
  J. Lahaye and G. Prado editors, Plenum Press, New-York -
  London.

(18) Smith, R.D. : 1979,
. Combustion and Flame, 35, 179
. J. Phys. Chem. 83, 1553.
(19) Homann, K.H. and Wagner H. Gg. : 1967, 11th Symposium
(International) on Combustion. The Combustion Institute,
Pittsburgh, Pa, p. 371.
(20) Homann, K.H. and Schweingurth, H. : 1981, Ber. Bunsenges.
Phys. Chem. 85, p. 569.
(21) Warnatz, J. : 1983, Soot in Combustion Systems and its
Toxic Properties. J. Lahaye and G. Prado editors. Plenum Press,
New-York-London.
(22) Michaud, P., Delfau, J.L. and Barassin, A. : 1981,
18th Symposium (International) on Combustion. The Combustion
Institute, Pittsburgh, Pa, p. 443.
(23) Olson, D.B. and Calcote, H.F. : 1981, 18th Symposium
(International) on Combustion. The Combustion Institute,
Pittsburgh, Pa., p. 453.
(24) Homann, K.H. : 1979, Ber. Bunsenges. Phys. Chem., 83, p.738.
(25) Calcote, H.F. : 1983, Soot in Combustion Systems and its
Toxic Properties. J. Lahaye and G. Prado editors. Plenum Press,
New-York-London.

PAH EMISSION FROM VARIOUS SOURCES AND THEIR EVOLUTION OVER THE
LAST DECADES

Thomas Ramdahl, Ingrid Alfheim and Alf Bjørseth

Central Institute for Industrial Research
P.O. Box 350, Blindern, Oslo 3, Norway

ABSTRACT

Anthropologic sources for polycyclic aromatic hydrocarbons (PAH)
and their emission factors are discussed. The major sources are
divided into two categories, stationary sources (such as industrial sources, residential heating, power and heat generation,
incineration and open fires) and mobile sources (such as cars,
trucks and airplanes). The emission factor for each source of
each category is established. Uncertainties in the emission factors are introduced by limited data from each source, different
sampling methods, and different methods for PAH analysis.

Using USA, Sweden and Norway as examples, the major known sources
of PAH were residential combustion of wood, the aluminium industry, forest fires, coke manufacturing, oil fired intermediate
commercial/industrial boilers, mobile sources and residential
combustion of bitumious coal.

Data from several countries over the past 20-30 years indicate a
substantial decrease in PAH emissions. However, emissions of
other copollutants have also changed. In view of the plausible
formation of potent transformation products, no firm conclusion
can be made about changes in potential health hazards of PAH
and their derivatives.

INTRODUCTION

As long as man has had fire, he has had PAH in his environment.
Since Percival Pott discovered (1) that chimney sweepers in

D. Rondia et al. (eds.), Mobile Source Emissions Including Polycyclic Organic Species, 277–297.
© 1983 by D. Reidel Publishing Company.

Britain often developed cancer in the scrotum, there has been
an awareness of the harmful effects of soot and tar. Considerable work has been done in order to isolate and identify the constituents of soot, tar and pitch, causing the carcinogenic effect.
Starting with two tons of gas-work pitch, Hieger and coworkers (2)
undertook a large scale isolation of a fluorescent carcinogenic
constituent resulting in the identification of benzo(a)pyrene
(BaP), a carcinogenic PAH. By 1976, more than 30 parent PAH compounds and several hundred derivatives of PAH were reported to
have carcinogenic effects (3,4).

Because of the many sources of PAH, the compounds are widely distributed in the environment (5). Potential hazards from the occurrence of PAH in the environment have been noted in the drinking
water standards set forth by the World Health Organization's
Committee on the Prevention of Cancer (6) as well as by several
national agencies concerned with PAH in food, working atmospheres,
and effluents from industries and mobile sources (7,8).

In this paper we are attempting to estimate the emission factors
for the known sources of PAH and establish their relative importance. Furthermore, by the historical development of PAH emissions
as revealed by urban ambient air, PAH levels will be estimated.

FORMATION

PAH are formed by incomplete combustion and pyrosynthesis (3).
Badger's model (9) of the free-radical pyrosynthesis of BaP and
other PAH, in which free radicals combine to condensed systems, is
widely accepted. Some ring systems are more easily formed than
others, giving rise to characteristic PAH profiles (10). Thus,
every process involving combustion or strong heating of organic
materials will, if not very carefully controlled, result in PAH
emission.

SOURCES

Although there are natural sources of PAH (11,12), i.e. volcanic
activity and biosynthesis, the anthropogenic sources are predominant and by far the most important to air pollution (9).

Table 1 lists the main sources, divided into two categories:
The stationary sources include industrial sources, power and heat
generation, residential heating, incineration and open fires. The
second category is the mobile sources, which include gasoline
engined automobiles, diesel engined automobiles, trucks, airplanes and sea traffic.

Stationary sources

1) Residential Heating
    -Furnaces, fireplaces and
    stoves (wood and coal)
    -Gas burners

3) Power and heat generation
    -Coal and oil fired power
    plants
    -Wood and peat fired power
    plants
    Industrial and commercial
    boilers

2) Industry
    -Coke Production
    -Carbon black production
    -Petroleum catalytic cracking
    -Asphalt production
    -Aluminium smelting
    -Iron and steel sintering
    -Ferro-alloy industry

4) Incineration and open fires
    -Municipal and industrial
    incinerators
    -Refuse burning
    -Forest fires
    -Structural fires
    -Agricultural burning

Mobile sources

5) -Gasoline engines automobiles
    -Diesel engined automobiles
    -Rubber tire wear

    -Airplanes
    -Sea traffic

Table 1.   Sources of PAH.

In all the listed processes, organic material is burned or strongly heated, and will, in the absence of sufficient oxygen, result
in emission of PAH.

EMISSION FACTORS

The amount of PAH released from any process is largely dependent
upon raw materials and the combustion technology. Therefore, a
given process with a fixed set of process parameters, may produce
a specific and invariable amount of PAH. This amount can be correlated to the process parameters whereby a relative emission factor for the process can be established.

Industrial emissions are mostly determined by the production level
of the factories emitting PAH. The emission factor can be expressed as weight of PAH per ton of product. In the case of mobile
sources, the consumption of fuel could be chosen as the emission
determining unit giving emission factor expressed as microgram
PAH per kilogram of fuel consumed.

In spite of the numerous publications dealing with PAH, very few
studies link the emission to any process parameter. These studies
are therefore not suited for emission factor development, as such.
When studies are suited for calculating emission factors, another
problem arises. For different reasons benzo(a)pyrene (BaP) has
been measured more frequently than any other PAH. BaP can in
these cases help judging how a given source contributes to the
total PAH emission by comparing it to values of BaP emission
from other sources. However, BaP is only a minor component, usually less than five percent of the total amount of PAH.

In studies where more PAH than BaP are measured, the number of
PAH analyzed may vary, depending upon the analytical technique
employed. Hangebrauck, von Lehmden and Meeker (13) analyzed ten
different PAH in emission samples by column chromatography and
ultraviolet-visible spectrophotometry, whereas modern glass capillary gas chromatography can separate and quantify more than
forty PAH in comparable samples (14). The term "Total PAH
Emissions" therefore strongly depends on number of compounds determined. Disregarding this fact may result in misleading conclusions. Because measurements of the same source often give large
differences, the emission factors presented consist of a minimum,
maximum and an intermediate value when possible. The emission of
bicyclic aromatic compounds are not included in the calculation
of PAH emission factors.

Few publications have been dealing with emission factors for PAH
as such. The Environmental Protection Agency in the US has published a comprehensive report (15). Most of the emission factors

in this report were developed from data reported by Hangebrauck, von Lehmden and Meeker in their relatively comprehensive study from 1967 (13). A literature search conducted by EPA in order to update this information resulted in very few additional data.

In this work we have attempted to update the emission factors where new results have been available. In particular, we have included new results from several Scandinavian studies. A letter requesting additional data to environmental offices in nine different countries, including Japan and USA, gave little information.

EMISSION FACTORS FOR SPECIFIC SOURCES

Residential Heating

Wood and coal fueled stoves and wood-burning fireplaces are frequently used for residential heating. The combustion of wood and coal is often incomplete due to slow, low-temperature burning with insufficient access to air at the burning surface. This results in formation of large amounts of PAH, which may be released directly to the atmosphere.

For wood stoves two Scandinavian studies yielded emission values between 1 and 36 mg PAH (22-26 compounds) per kg dry fuel (16,17) while an American study (18) measured 270 mg per kg (26 compounds). These differences may reflect variations in design, sampling methodologies, testcycles, etc., but do also illustrate the difficulty of calculating a single emission factor for a given source. In Table 2 the emission factor is given, with an intermediate value of 40 mg PAH per kg dry wood.

The emission factor for residential coal firing is 60 mg/kg (13,19), but can vary over 5 orders of magnitude with respect to the coal used (20).

Residential heating with oil varies with the size of the burner. A 100, 000 Btu/hr (30 kW) burner has been found to emit 150 μg/k oil (13), whereas a smaller 25,000 Btu/hr (7.5 kW) unit has been found to emit 10 mg/kg oil (21). More measurements are needed to establish good emission factors for residential oil heating.

Industrial Sources

Aluminium Production. Primary aluminium production is based on two different technologies basically using Söderberg or prebaked anodes. PAH are emitted from the carbon electrodes containing tar and pitch as binder (22). The emissions from the prebaked electrodes are usually 1-10% of that of the Söderberg electrodes. Only one major study has been undertaken to determine the total

PAH emission from an aluminium reduction plant (23). The PAH
emission to air was determined to 35 metric tons PAH at a plant
with an annual production of 83,500 tons aluminium using the
Söderberg process. The PAH in this study represent the sum of
some 25 compounds.

Based on these data, assuming that the Söderberg and the prebaked
process each cover 50% of the worldwide production, and that the
prebaked process emits 5% of that of the Söderberg process, an
emission factor of 235 g per ton produced aluminium is established for this source, as shown in Table 3.

Coke Production. Coke ovens are major sources of PAH emissions.
The leaks from the oven doors are found to be the major source
(24), giving an emission factor of 15 g per ton coal charged.

Petroleum Catalytic Cracking. The catalytic cracking process is
used to upgrade heavy petroleum fractions by breaking up long-
chain hydrocarbons to produce high octane gasoline. An emission
factor weighted by the cracker population in the US is 25 mg
per m^3 fresh feed without any emission control (13,15,25). Assuming a density of 900 kg per m^3 fresh feed, this yields an
emission factor of 28 mg per ton. When the regenerator exhaust
gas is recirculated and burned in waste heat boilers, the emission factor is 1 mg per ton (13,15,25).

Other Industrial Sources. Carbon black is widely used as pigment
i.e. in rubber tires. Literature data indicate an emission factor
of 300 mg per metric ton of product (15,26,27).

Asphalt production and modification involve various processes.
For instance, air blowing has an emission factor of 50 mg per
metric ton and shingle saturators 5 mg per ton (13,15).

Iron and steel sintering is a source of significant air emissions.
An emission factor of 17 mg BaP per ton has been reported (15,28).
Assuming this to be 5% of the total PAH, this source has an emis-
sion factor for PAH of 340 mg per ton.

In the ferro-alloy industry the PAH emission is due to contact
between hot metal and tarry products in electrodes and shutters.
Only data for PAH emissions to water has been found for this
source (29). However, assuming equal emission to water and air
(50% efficiency in the wet scrubbers), this source has an emission factor of 10 g PAH per ton alloy. In a study in a Norwegian
iron work, emissions to water were determined (29). PAH emission
is due to the use of Söderberg electrodes. Making the same assumptions as above with the scrubber efficiency, the iron works emit
60 g PAH per ton produced iron.

SOURCE	Benzo(a)pyrene		PAH		Ref.
	Range µg/kg	Typical µg/kg	Range mg/kg	Typical mg/kg	
Wood stoves	1 - 10,000	500	1 - 370	40	16-18
Fireplaces		700		29	18
Coal furnaces	2.0 - 4.4	1500	1 - 1200	60	13,19
Oil (30 kW)		2.2	0.006 - 0.75	0.15	13
Oil (7.5 kW)			0.9 - 21.6	10	21

Table 2. Emission factors for residential heating.

SOURCE	Benzo(a)pyrene		PAH		Ref.
	Range mg/ton	Typical mg/ton	Range mg/ton	Typical g/ton	
Aluminium smelting	10,000-38,000	15,000	$2 \cdot 10^5 - 6.5 \cdot 10^5$	235	23
Anode baking		66		50	23
Coke production				15	24
Petroleum catalytic cracking					
- No control	0.03-0.71	0.41	1.4-3200	0.03	13,15,25
- CO Waste Heat Boiler	0.007-0.16	0.04	0.32-2.4	0.001	13,15,25
Carbon Black production			220-490	0.3	15,26,27
Asphalt production					
- Air blowing	0.16-750	1	2.8-4100	0.05	13,15
- Shingle saturators	0.08-0.4	0.001	1-50	0.005	13,15
Iron and sttel sintering	0.6-1100	17		0.34	15,28
Ferrous founderies				7.7	27
Ferro-alloy industry				10	29
Iron works				60	29

Table 3. Emission factors for industrial sources.

In ferrous foundries the shake-out process of casting metal
emits PAH, resulting in an emission factor of 7.7 g per ton
casting produced (27).

Power and Heat Generation

In Table 4 a survey of the emission factors for power generation
is given. Each source is discussed below.

Coal and Oil. There are several types of coal fired power plants,
such as vertically, front-wall, tangentially and cyclone fired
plants and plants using a moving grate. In an emission st··dy a
weighted emission factor of seven types of coal fired power
plants was found to be 19 μg per kg coal charged (13,15,30).
Coal fired industrial boilers are usually smaller units. However, they generally give a higher emission relative to produced energy. A weighted emission factor for six types of boilers
was found to be 41 μg per kg fuel (13,15,30).

Three coal fired power plants in Scandinavia (25-600 MW) have a
PAH emission less than 0.5 μg per kg. Thus, when properly operated,
these power plants appear to give a very low contribution to the
overall PAH emission (31). Two other Swedish coal fired power
plants (350-400 MW) have been measured to emit 30 μg PAH per kg
coal (32). Improperly operated, the emissions may be higher.

A small fluidized bed coal fired power plant (2 MW) emitted 36 mg
per kg coal (31). It was claimed that the plant was not properly
operated at the time of sampling, and this clearly demonstrates
the importance of optimalization of the operation. A few such
heavy emission periods may overwhelm the otherwise low and steady
emission, as far as total emission per year is concerned.

In a recent study at a well tuned 500 MW oil fired power plant, no
detectable amount of PAH could be found (33). Detection limit was
0.05 μg of each component per sample. Recalculated BaP emission
is less than 0.01 g per hour. Other oil fired power plants in
Sweden have been estimated to emit 10 μg PAH per kg oil (32). Oil
fired intermediate boilers (6-8 MW) have an emission factor of
23 μg per kg (13,15).

Gas. Gas fired intermediate boilers (1.2 MW) heated by premix
burners have an emission factor of 1 mg per kg gas burned (13,15).

Biomass. Wood and peat are studied in several countries as a potential fuel for power generation. In a study of a small 2 MW hot
water boiler, the combustion of wood and peat gave emission
factors of 2 mg and 15 mg PAH per kg fuel, respectively (34).
Other studies have shown much lower emissions for biomass combustion. As for the other sources the combustion technology is of

SOURCE	Benzo(a)pyrene		PAH		Ref.
	Range µg/kg	Typical µg/kg	Range µg/kg	Typical µg/kg	
Pulverized coal fired power plants	1.1-2.7	1.6	0.5-32	19	13,15,30,31
Coal fired fluidized bed		150		36,000	31
Coal fired industrial boilers		0.93		41	13,15,30
Oil fired industrial boilers		1.1	5.3-100	23	13,15
Oil fired commercial boilers		40		820	13,15
Gas fired intermediate boilers		10	490-1100	1000	13,15
2 MW hot water boiler					
- Wood	10-36	22	1180-3390	2000	34
- Peat				15,000	34

Table 4. Emission factors for power and heat generation.

main importance. In some cases (35,36) no PAH could be detected in the emissions.

Incineration and Open Fires

In Table 5 a survey of the emission factors established for in- cenerators and open fires is given.

Incinerators. There are many different types of incinerators with different combustion systems. A geometric mean for large incine- rators with emission control systems is 17 μg per kg refuse charged (13,15,37). Smaller commercial indicators burning 2 to 5 tons per day have an emission factor of 6.8 mg per kg (13,15).

Open Burning. Open burning of automobile tires emits 240 mg per kg and municipal refuse 1.4 mg per kg (13,15,38).

Burning of coal refuse banks results in large emissions of PAH (3), but no emission factor can be established.

Forest fires are potential sources of PAH emission. An emission factor of 20 mg per kg were calculated from laboratory studies (39).

Agricultural burning is suspected to emit large quantities of PAH. Although no data for PAH emission from this source has been found, a rough estimate of 20 mg per kg, similar to that of forest fires, is suggested.

Structural fires will by burning of organic material emit PAH, but no emission factor for this source could be established.

Mobile Sources

In Table 6 a survey of the emission factors developed for mobile sources is given.

Gasoline fueled Automobiles. Few studies suitable for determina- tion of emission factors for total PAH are available. Grimmer et al. (40) characterized the PAH (59 compounds) emitted from two different vehicles and developed an emission factor of 10 mg per kg for noncatalytic cars. Catalytic converters for the exhaust reduce this figure significantly. An American study reports the emission factor for BaP to be 4.6 μg per kg without catalyst and only 0.36 μg BaP per kg with catalyst, a reduction of 92% (41). Approximately 50 per cent of American cars now have catalytic converters.

Other factors like age of engine, fuel aromaticity, driving mode, oil consumption, cold start, choking, etc., also have an important impact on the emission (40,42,43).

SOURCES	Benzo(a)pyrene		PAH		Ref.
	Range µg/kg	Typical µg/kg	Range mg/kg	Typical mg/kg	
Incinerators (USA, geometric mean)		0.13		0.017	13,15,37
Incinerators (2-5 tons/day		200		6.8	13,15
Open burning					
- Automobile tires				240	13,15,38
- Municipal refuse				1.4	13,15,38
Forest fires	17-140	100	3.5-31.5	20	39
Agricultural burning				20	

Table 5. Emission factors for incineration and open fires.

SOURCE	Benzo(a)pyrene		PAH		Ref.
	Range µg/kg	Typical µg/kg	Range mg/kg	Typical mg/kg	
Automobiles					
- Gasoline, no catalyst		50	5-50	10	40
- Gasoline, no catalyst		4.6			41
- Gasoline, catalyst		0.36			41
- Diesel				10	
- Diesel	0.2-31.7 µg/ mile				44
Trucks, diesel				5	45
Two-stroke engines		3600			3
Airplane, jet	2-10 mg/ min				46

Table 6. Emission factors for mobile sources.

Recent results indicate that engine design developments reduce
fuel consumption and PAH emissions considerably and that the
emission factor given above may be adjusted when new studies are
published.

Diesel Engines. Although diesel emissions have been carefully
studied the last few years (44), few publications are suitable
for the purpose of establishing relative emission factors. BaP
emissions from diesel engines can vary by two orders of magnitude depending on engine design alone (44). BaP emissions from·
the diesel engine having lowest emissions are the same order of
magnitude as those from gasoline engines with catalytic converters (44). An emission factor of 10 mg per kg is estimated,
based on available data.

PAH emissions from heavy duty diesel trucks have been estimated to
5 mg PAH per kg (45).

Two-Stroke Engines. Motorcycles, snow-scooters, lawn-mowers,
chain-saws, etc., all have two-stroke engines with low fuel-
efficiency. An emission factor for motorcycles of 3.6 mg BaP per
kg has been reported (3). No estimates are available on other two-
stroke engine emissions.

Rubber Tire Wear. Degradation of automobile tires releases carbon
black particles to the atmosphere. Carbon blacks are used in tires
and contain PAH. No direct emission factor could be established,
but the NAS study estimated the annual BaP emissions to be 11 tons
per year in the US (3). The total US mileage is approximately
$2 \cdot 10^{12}$ km per year which gives a BaP emission of 5 µg per km.

Aircraft and Ship Traffic. Shabad (46) reports that a jet air-
plane emits 1-10 mg BaP per minute. No other studies were available, but a recent report from Sweden shows a decreasing amount
of "fall out" PAH at increasing distance from the runway (47). No
data are available with respect to sea traffic.

TOTAL EMISSIONS

Another important parameter for the total emission estimate is the
total production or consumption of the various sources. It is the
product of the relative emission factor and the production or consumption capacity that finally show the emission severity of the
source. Based on statistical data and the relative emission factors given previously, total PAH emissions are calculated for
the US, Sweden and Norway (Table 7). In the following, these data
will be discussed.

In the US vehicular traffic seems to be a major source emitting approximately 2100 tons PAH per year, 24% of the total emission.

Another major source is the aluminium industry. Based on the emission factor from Table 3, this industry emits 1000 tons PAH annu-
ally. However, this figure is very uncertain and must be considered only as a very rough estimate, as the emission factor is
developed on the basis of one single aluminium reduction plant.
Iron and steel work are estimated to emit 1860 tons PAH and coke
manufacturing 630 tons PAH per year. In total the industrial sources are responsible for 41% of the total emission. Residential
heating is the third largest source, 16% of the total emission.
Residential combustion of wood has gained greater popularity in
the US recently, due to rising oil prices (48,49), giving rise
to large local emissions. Of particular importance is the potential use of relative emission factors in areas near major sources
to determine possible environmental impact as a function of geographical and meteorological factors.

In Table 7, similar data for Sweden and Norway are also included.
As revealed in this table, the relative importance of the different sources varies depending on the industrial profile of each
country. It must be stressed that the figures in Table 7 are very
uncertain, caused by the uncertainties in the emission factors,
and they should only be considered as estimates. Nevertheless, the
ranking and magnitude of the sources should be representative for
the real situation.

VARIATION OF PAH EMISSION WITH TIME

Several studies have shown that PAH emissions undergo both seasonal and long term variations (50). When discussing variations of
PAH emissions with time, this has to be taken into account. In
particular, when the dominating sources are residential heating or
car driving in cold climate, the winter values may be considerably higher than the summer values. However, when the dominating
sources are related to industrial activities, the seasonal variations will be smaller. Relatively few studies are published demonstrating the variations of PAH emissions over long terms. Furthermore, some of these studies may be difficult to evaluate. Over the
past decades there has been an introduction of improved cleaning
technologies as well as improved combustion techniques, resulting
in reduced emissions. These effects vary for different sources,
and an evaluation of each source is not feasible. Therefore, a
better impression is obtained when studying the urban ambient air
PAH levels. However, comparing results over a long period requires
that both sampling, techniques and analytical methodologies, sampling time and the seasonal variations are taken into account. Only
a few of the published studies fulfill these requirements. In
most of the studies, only one or a limited number of PAH compounds

Source	Norway		Sweden		USA	
	Metric tons/year	o/o	Metric tons/year	o/o	Metric tons/year	o/o
Residential combustion:						
- Wood, coal	48	16	96	19	450	5
- Oil, gas	14.5	5	36	7	930	11
Industrial production:						
- Coke manufactoring	5.1	2	18	4	630	7
- Carbon black			< 0.1	< 0.1	3	< 0.1
- Asphalt production	0.1	< 0.1	0.3	< 0.1	4	< 0.1
- Aluminium production	160	54	35	7	1000	12
- Iron and Steel works	34	12	258	51	1860	22
- Ferroalloy industry	3.5	1	1	0.2		
- Petroleum cracking			< 0.1	< 0.1		
Power generation:						
- Coal & Oil fired power plants			< 0.1	< 0.1	1	< 0.1
- Peat, wood, straw	0.1	< 0.1	6.5	1		
- Industrial boilers	1.2	0.4	6.5	1	400	5
Incineration						
- Municipal incineration	0.3	< 0.1	2.2	0.4	50	0.6
- Open burning	0.4	< 0.1			100	1
- Forest fires	7	2	1.3	0.2	1000	12
- Agricultural burning	6	2				
Mobile sources						
- Gasoline automobiles	13	4	33	6	2100**	24
- Diesel automobiles	7	2	14	3	70	0.8
- Air traffic	0.1	< 0.1	< 0.1	< 0.1		
Total	295		510		8600	

Table 7.   Estimated PAH emission.*

* Numbers are encumbered with great incertainties. To be used as an indicator of order of magnitude.

** Approximately 50 per cent of American cars have catalytic converters. This is not corrected for in this number.

have been analyzed. In the following, we will only discuss BaP data.

A decline in the BaP concentration has been observed in London, U.K. The annual average BaP concentration in London decreased from 46 ng/m^3 in 1949-51 to 14 ng/m^3 in 1957-64 and to 4 ng/m^3 in 1972-73 (51). In these studies, the same sampling and analytical techniques have been used over the whole period. In total, the data show a 90% reduction over the past 25 years. By using selected PAH ratios, the contribution of BaP to the air from coal burning was shown to decline considerably.

Decline in the PAH level has been published in the Federal Republic of Germany (52). Fig. 1 shows concentrations of the PAH level in three different German cities in the period from 1969 to 1974. With one exception it is seen that the PAH level has a constant decline (52).

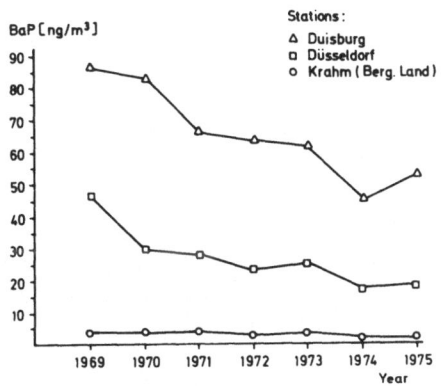

Fig. 1. Decline in BaP concentrations in German cities, 1969-1975.

The most comprehensive monitoring of BaP was performed under the National Air Surveillance Network Programme in the US (53). Data are available for 240 stations for the period 1966 to 1970, and for 40 stations for the period 1971-1976. The data from the NASN suggest a considerable decline in the BaP concentration for the urban atmosphere during the period 1966-1975. In Fig. 2 the four quarter moving average of percentile values also are given. The annual average BaP concentrations decline from 3.2 ng/m^3 in 1966 to 2.1 ng/m^3 in 1970, and to 0.5 ng/m^3 in 1975. This decline is believed to be due primarily to the decrease in coal consumption for house heating as well as improved disposal of solid waste and restrictions on open burning (53,54). It is worthwhile mentioning, however, that a relatively high number of urban sampling sites were located in areas where coke ovens were present, and that the decline in BaP concentrations may have resulted from reduced emis-

sions from these sources. The quarterly BaP from the NASN also
illustrate the seasonal variations in BaP concentrations. The
figure presents the seasonal dependence in the 50th-90th percentile
in BaP concentrations for 34 NASN urban sites. The figure shows
more pronounced seasonal variations from 1966 to 1969 when a remarkable dumping in the amplitude is noted. From 1973 the seasonal
variations are again minor when compared to the 1966-1969 period.

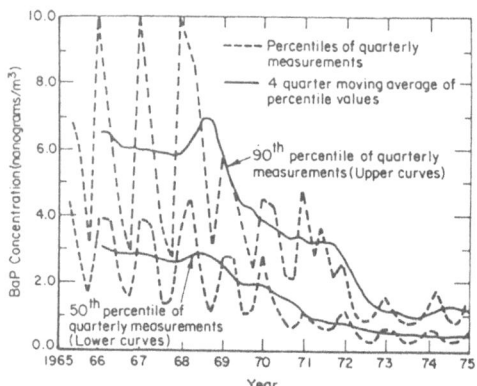

Fig. 2. Trends in BaP concentrations for 34 NASN urban sites.

It has recently been realized that PAH may undergo reactions with
copollutants and that some of these transformation products may
represent a considerable health hazard. It is therefore worthwhile
also to look into the emission of some of the copollutants. Recent
data from the Netherlands were published by Bos et al. (55) on the
change with time of atmospheric concentrations of NO_x and SO_x. The
data show that the concentration of SO_2 in Delft was 300 $\mu g/m^3$ in
1961 and this had dropped to 30 $\mu g/m^3$ in 1975. The opposite trend
was found for NO_x. In 1970 the NO_2 concentration was 30 $\mu g/m^3$ and
this had increased to 50 $\mu g/m^3$ in 1977. Other studies show a
dramatic increase in the nitrate component of acid rain (56).
Higher concentrations of NO_2 may result in higher concentrations
of nitro-PAH, some of which are potent mutagens (57). Therefore,
in evaluating the potential health hazards of PAH emissions, the
emissions of copollutants and potential formation of transformation
products also have to be taken into account.

CONCLUSION

For several reasons emission factors always require a certain degree of caution and a certain degree of consideration. These requirements are vital for the following reasons:

Many of the emission factors are obtained from a relatively small
number of measurements, in some cases even only one. Thus, it is

not always certain that they are representative for the exact
average of the prevailing cases.

Emission factors are meant to be representative of average values
and thus they will not correspond to the true values for each of
the individual cases. Therefore, an emission factor will not only
depend on management and state of maintenance, but also on the
special measures which have been taken to limit the extent of the
emissions.

The lack of standardization of analytical procedures and the
number of PAH reported, make the term "Total PAH" very uncertain.
In order to reduce the uncertainties, more measurements are needed for many sources like wood and coal residential heating, aluminium smelting, fluidized bed coal fired power plants, air and
sea traffic, steel industry, petroleum cracking, coke production,
and municipal incinerators.

ACKNOWLEDGMENT

Financial support from the Nordic Council of Ministers through
the MIL-2 project is gratefully acknowledged.

REFERENCES

1. Pott, P., Chirurgical Observations Relative to the Cataract,
    the Polypus of the Nose, the Cancer of the Scrotum, the
    Different Kinds of Ruptures, and the Mortification of the
    Toes and Feet: 1775, L. Hawes, W. Clarke and R. Collins,
    London.

2. Cook, J., Hewett, C.L. and Hieger, I.: 1933, J. Chem. Soc.
    p. 395.

3. Particulate Polycyclic Organic Matter: 1972, National
    Academy of Science, Washington, D.C.

4. Dipple, A., in C.E. Searle (ed.), Chemical Carcinogens,
    ACS Monograph 173: 1976, Am. Chem. Soc., Washington D.C.

5. Suess, M.J.: 1976, Sci. Total Environ. 6 p. 239.

6. World Health Organization, International Standards for
    Drinking Water, 3rd Edition: 1971, Geneva.

7. Walker, E.A., Pure & Appl. Chem.: 1977, 49 p. 1673.

8. Occupational Safety and Health Administration, Department
    of Labor, Federal Register: 1976, 41 p. 46741.

9. Badger, G.M., Nat. Cancer Inst. Monogr.: 1962, 9 p. 1.

10. Imissionsmessungen von polycyclishen aromatischen Kohlen-
    wasserstoffen (PAH), Untersuchung im Auftrag des Ministers
    für Arbeit, Gesundheit und Soziales des Landes Nordrhein-
    Westfalen: 1981, Essen.

11. Blumer, M., Sci. Amer.: 1976, 234 p. 35.

12. Andelman, J.B. and Suess, M.J., Bull. World Health Org.:
    1970, 43 p. 479.

13. Hangebrauch, R.P., von Lehmden, D.J. and Meeker, J.E.,
    Sources of Polynuclear Hydrocarbons in the Atmosphere,
    U.S. HEW.: 1967, Public Health Service, AP-33, PB 174-706,
    Washington, D.C.

14. Bjørseth, A. and Eklund, G., Anal. Chem. Acta: 1979,
    105 p. 119.

15. Energy & Environmental Analysis Inc., Preliminary Assess-
    ment of the Sources, Control and Population Exposure to Air-
    borne Polycyclic Matter (POM) as Indicated by Benzo(a)-
    pyrene (BaP): 1978, Report to EPA under contract # 68-02-2836.

16. Rudling, L., Ahling, B. and Löfroth, G., Statens Naturvårds-
    verk PM 1331: 1980, Solna (In Swedish).

17. Ramdahl, T., Alfheim, I., Rustad, S. and Olsen, T. Chemos-
    phere: 1982, 11 p. 601.

18. De Angelis, D.G., Ruffin, D.S., Peters, J.A. and Reznik, R.B.,
    Source Assessment: Residential Combustion of Wood, EPA-
    600/2-80-042b: 1980, Research Triangle Park.

19. De Angelis, D.G. and Reznik, R.B., Source Assessment: Resi-
    dential Combustion of Coal, EPA-600/2-79-019a: 1979,
    Research Triangle Park.

20. Beine, H., Staub - Reinhalt. Luft: 1970, 30 p. 334.

21. Herlan, A., Combust. Flame: 1978, 31 p. 297.

22. Bjørseth, A., Bjørseth, O. and Fjeldstad, P.E., Scand. J.
    Work Environ. & Health: 1978, 4 p. 212.

23. Bjørseth, A. and Wickstrøm, L., Unpublished results.

24. Trenholm, A.R. and Beck, L.L., Assessment of Hazardous Orga-
    nic Emissions from Slot-Type Coke Oven Batteries, Internal
    EPA Report: 1978, March 16, Durham.

25. The Oil and Gas Journal, Worldwide Directory: Refining and
     Gas Processing 1977-1978: 1977.

26. Vandegrift, A.E., Shannon, L.J. et al., Handbook of
     Emissions, Effluents, and Control Practices for Stationary
     Particulate Pollutant Sources, Report of NAPCA
     Contract No CPA 22-69-104: 1970.

27. Vena, F., Environment Canada, Personal communication.

28. Louis, R.St., Pennsylvania Dept. of Environmental Resources
     Emission Test Results: 1977, Harrisburg.

29. Berglind, L. and Gjessing, E., Norwegian Institute for Water
     Research, Report No A3-25: 1980 (In Norwegian), Oslo.

30. Supernant, N. Hall, et al., R., Preliminary Emissions Assess-
     ment of Conventional Stationary Combustion Systems,
     Vol.II; EPA-600/2-76-046b: 1976.

31. Bergström, J., Emissions from combustion of coal and oil in
     powerplants, Studsvik Technical Report EK-81/103: 1981,
     Studsvik, Sweden (in Swedish).

32. Erikson, K., PAH,NO_x- and metal emissions, etc., from coal-
     and oil combustion. ÅF-Energikonsult 1980-06-04: 1980. (In
     Swedish).

33. Report to Södertälje (Sweden) Community, Swedish Steam Users
     Assosiation: 1980, (in Swedish).

34. Alsberg, T. and Stenberg, U., Chemosphere: 1979, 8 p. 487.

35. SSVL Stiftelsen Skogindustriernas Vatten- och Luftvårdsforsk-
     ning. Emissions of polycyclic organic compounds from bark
     combustion. SSVL 99:1:1980. (In Swedish).

36. Eklund, G., Determination of heavy metals and organic com-
     pounds in emissions from Växjö woodfired power plant.
     Studsvik Arbetsrapport E2-80/159: 1980. (In Swedish).

37. Davies, I.W., Harrison, R.M., Perry, R., Fatanyaka, O. and
     Wellings, R.A., Environ. Sci. Technol.: 1976, 10 p. 451.

38. EPA, Compilation of Air Pollution Emission Factors, 3. ed.,
     AP-42: 1977.

39. McMahon, C.K. and Tsoukalas, S.N. in Jones, P.W. and
     Freudenthal R.I. (eds.), Carcinogenesis, Vol. 3: Polynuclear
     Aromatic Hydrocarbons.: 1978, Raven Press, New York.

40. Grimmer, G., Böhnke, H. and Glaser, A., Erdöl & Kohle: 1977, 30 p. 411.

41. Gross, G.P., Third Annual Report on Gasoline Composition and Vehicle Exhaust Gas Polynuclear Aromatic Content, CRC APRAC Project No CAPE-6-68, EPA Contract No 68-0400-25, APTD 1560, PB 218-873: 1972, Detroit.

42. Begeman, C.R. and Burgan, J.G., Polynuclear Hydrocarbon Emission from Automotive Engines, Soc. of Automotive Engineers, SAE Paper 700469: 1970, Detroit.

43. Pedersen, T.S., Ingwersen, J., Nielsen, T. and Larsen, E., Environ Sci. Technol.: 1980, 14 p. 71.

44. National Research Council, Impacts of Diesel-powered Light-Duty Vehicles; Health Effects of Exposure to Diesel Exhaust: 1981, National Academy Press, Washington, D.C.

45. Santodonato, J., Basu, D. and Howard, P., Health Effects Associated with Diesel Exhaust Emission; Literature Review and Evaluation, EPA-600/1-78-063: 1978, Research Triangle Park.

46. Shabad, L.M. and Smirnov, G.A., Atmos. Environ.: 1972, 6 p. 153.

47. Colmsjö, A., Report to the health authority of Stockholm (Sweden) Community: 1980, Stockholm.

48. Waldrop, M.M., Science: 1981, 211 p. 914.

49. Wood Heating Alliance, Proceedings Documents for Wood Heating Seminars 1980/1981: 1981, Washington, D.C.

50. Sawicki, E. Analysis of atmospheric carcinogens and their cofactors, in: Environmental Pollution and Carcinogenic Risks. C. Rosenfeld and W. Davis (eds), Ed Inserm (IARC Sci. Publ. 13): 1976, Paris, p. 297.

51. Lawter P.J. and Waller, R.E., Coal fires, industrial emissions and motor vehicles as sources of environmental carcinogens: 1976, INSERM, 52 p. 27.

52. Bundesamt Umweltberichte 1/79. Luftqualitätskriterien für ausgewählte polyzyklische aromatische Kohlenwasserstoffe: 1979, Erich Schmidt Verlag, Berlin, p. 96.

53. Faoro, R.B. and Manning, J.A., J. Air Pollut. Contr. Assoc.: 1981, 31 p. 62.

54. Faoro, R.B., J. Air Pollut. Contr. Assoc.: 1975, 25 p 638.

55. Bos, R., Goudena, E., Guicherit, R., Hoogeveen, A. and
     de Vreede, J., Atmospheric precursors and oxidants concen-
     trations in The Netherlands: 1978, TNO-Gravenhage.

56. Brimblecombe, P. and Stedman, D.H., Nature: 1982, 298 p. 460.

57. Pitts Jr., J.N., van Cauwenberghe, K.A., Grosjean, D.,
     Schmid, P., Fitz, D.R., Belzer, W.L., Knudson, G.B. and
     Hynds, P.M., Science: 1978, 202 p. 515.

THE IDENTIFICATION AND POTENTIAL SOURCES OF NITRATED POLYNUCLEAR AROMATIC HYDROCARBONS (NITRO-PAH) IN DIESEL PARTICULATE EXTRACTS

D. Schuetzle, M. Paputa, C. M. Hampton, R. Marano,
T. Riley, T. J. Prater, L. Skewes and I. Salmeen

Ford Motor Company
Chemical Analysis Research
Engineering and Research Staff - Research
Dearborn, Michigan 48121

ABSTRACT

HPLC procedures have been used to fractionate a diesel particulate extract into two non-polar, five moderately-polar and one polar chemical fraction. The moderately-polar fractions accounted for approximately 10% of the total extract mass and more than 60% of the total recovered extract Ames Salmonella TA98 mutagenicity. Approximately 200 nitro-PAH species were found to be present in the moderately-polar fractions as determined using GC^2/nitrogen selective detection (ND) and GC^2/MS procedures. Absolute isomer identification for approximately 25 of these compounds have been made by using GC retention times and low and high resolution mass spectra of synthesized standards as criteria. The chemistry of nitro-PAH formation and loss during combustion, in the tailpipe and during sampling is described.

INTRODUCTION

Previous studies have shown that most of the direct-acting Salmonella typhimurium mutagenic activity (>50%) in diesel particulate extracts is concentrated in chemical fractions which contain compounds of moderate polarity (1-4). These moderately-polar fractions have been found to consist

299

D. Rondia et al. (eds.), Mobile Source Emissions Including Polycyclic Organic Species, 299–312.
© 1983 by D. Reidel Publishing Company.

primarily of polynuclear aromatic hydrocarbon (PAH) derivatives (3,5-7). Analyses of these samples using nitroreductase deficient strains of Salmonella typhimurium have indicated that most of the extract mutagenicity is probably due to nitro-PAH (4,8). These results have been confirmed by recent work which has shown that seven nitro-PAH account for up to 40% of the total extract mutagenicity (9).

Mass spectrometry/mass spectrometry (MS/MS)(10) and high resolution mass spectrometry (HRMS) (3,7,10) have been used to show that a large number of nitro-PAH are potentially present in these extracts. However, only a few specific isomers have been positively identified. The purpose of this paper is to describe techniques for identifying a large number of nitro-PAH which may be present in these samples.

Nitro-PAH have been shown to be formed during sampling (9,11). However, the extent of these reactions have not been fully evaluated. Another purpose of this paper is to present new information on possible formation of nitro-PAH during combustion, in the tailpipe, and during sampling.

EXPERIMENTAL

A light-duty diesel particulate sample was collected from a dilution tube and extracted with dichloromethane and fractionated using HPLC as previously described (12). The HPLC fractionation scheme was slightly modified from that previously reported in order to elute the moderately-polar fraction with a greater degree of resolution. The programming scheme is shown in Figure 1. Capillary column (GC^2/ND) was undertaken (13) using a 30m x 0.25mm fused-silica capillary column bonded with DB-5 (obtained from J and W Scientific). The column operated with a temperature profile initially held at 120°C for one minute, followed by a temperature of 8°C/min to 310°C and an eight-minute hold at 310°C. Sample introduction was undertaken using a Grob splitless injector operating at 275°C. The nitrogen-phosphorous detector operated at 310°C under 4.4 cm^3/min air and 22 cm^3/min helium auxiliary flow.

RESULTS AND DISCUSSION

Sample Fractionation - Figure 1 shows the HPLC chromatogram for a diesel particulate extract and elution times for some standards. The sample has been divided into eight fractions.

Figure 1.   HPLC Chromatogram for A Light-Duty Diesel Particulate Extract

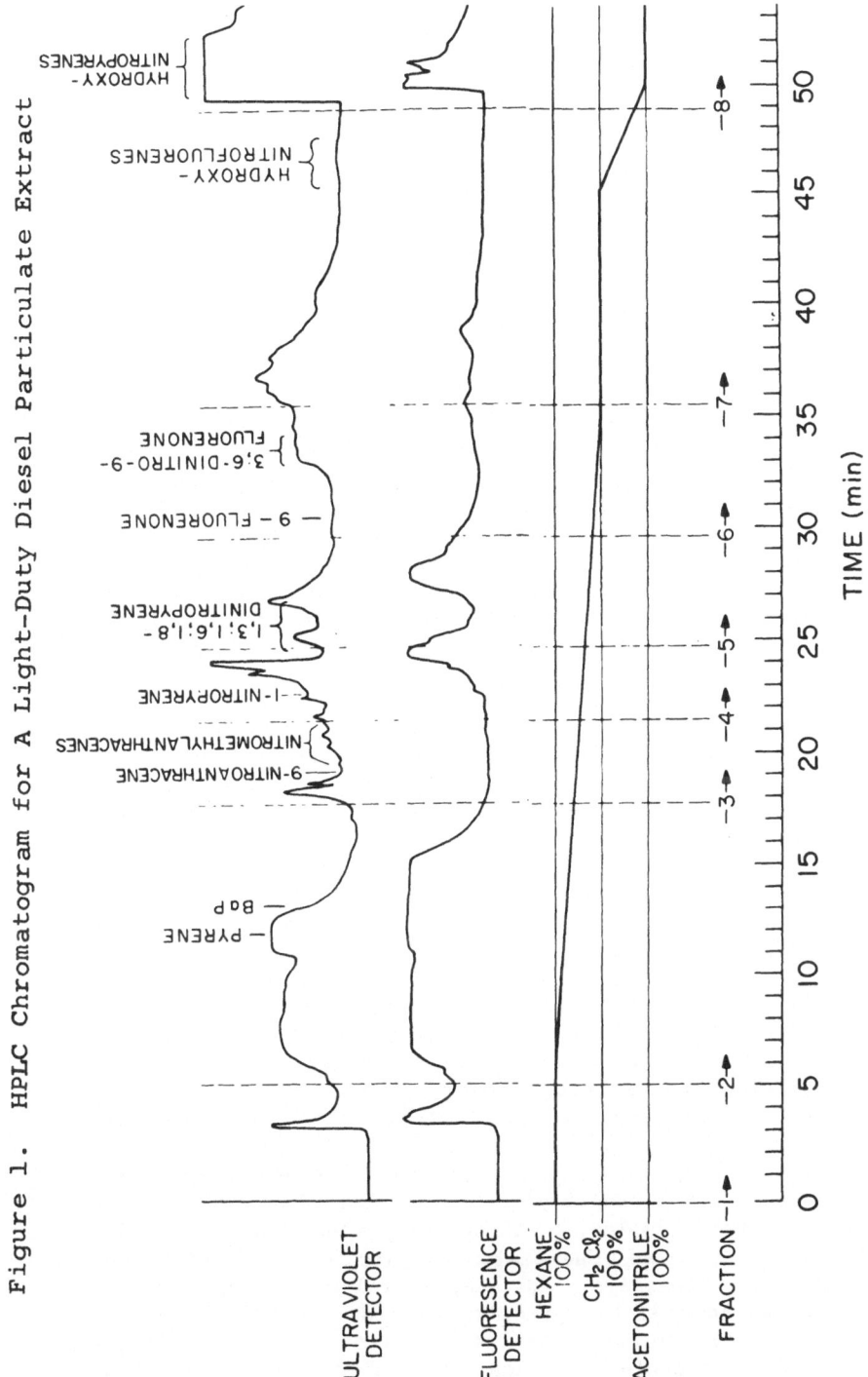

in which fractions 1 and 2 are designated as containing non-polar compounds, fractions 3-7 as containing moderately-polar compounds, and fraction 8 as containing polar compounds.

Mutagenicity Testing – Direct-acting Ames assay activities (TA98) for the total extract and individual fractions were measured using procedures previously described (3,14). The moderately polar fractions accounted for about 10% of the total extract mass and 70% of the total recovered extract mutagenicity. Fractions 3, 4 and 5 accounted for approximately 50% of the total recovered extract mutagenicity.

Chemical Composition of the Moderately-Polar Fractions – We have recently reported that the oxy-PAH account for most of the mass contained in the moderately-polar fractions for a light-duty diesel particulate extract(9). We have undertaken a similar analysis for the sample described in this work (figure 1) and found that it is qualitively similar but quantitatively different than that previously described for emissions collected from another light duty diesel engine(9). It can be concluded from these results that:

1. The moderately polar fraction of diesel particulates contain several hundred compounds, most of which are partially oxidized PAH and N- and S-containing heterocyclics.

2. There are order of magnitude differences in the concentration of oxy-PAH generated from different engines.

3. It is impossible to qualitatively and quantitatively determine the presence of each compound since standards for most compounds are not available and there is difficulty resolving isomers, even when state-of-the-art capillary column GC separation techniques are used.

During the past three years several procedures have been developed for simplifying the analysis of complex environmental samples. These techniques have included: "Bioassay Directed Chemical Analysis," (15) which uses short-term bioassays for determining which chemical fractions should be prioritized for detailed chemical analysis and "Survey Chemical Analysis", which utilizes analytical techniques such as high resolution mass

spectrometry (HRMS) (3, 7, 10) and mass spectrometry/mass
spectrometry (MS/MS)(10) for the analysis of groups of
isomers [e.g. nitro (pyrenes and fluoranthenes)].

We have used the Ames Salmonella typhimurium test (strain
TA98) to determine the direct acting activity for a select
number of PAH quinones, aldehydes, ketones and alchohols
which have been found to be major components in the
moderately polar fractions of several light-duty diesel
particulate extracts. We have not detected any significant
direct-acting activity for PAH containing these functional
groups(16). However, we have identified several nitro-PAH
which are responsible for up to 40% of the direct-acting
mutagenicity for three samples (9). In this work, capillary
column GC analysis was undertaken using a nitrogen selective
detector(ND) to further identify the presence of nitro-PAH
in the moderately-polar fractions.

Identification of Nitro-PAH — Figure 2 shows GC^2/ND
chromatograms for the analysis of three moderately polar
fractions of the diesel particulate extract shown
fractionated in Figure 1. These three fractions represent
approximately 50% of the total recovered extract
mutagenicity. Table 1 lists compounds which have been
identified in these fractions. Identification of specific
isomers was confirmed using the following criteria:

1. Elution of standards in the correct HPLC fraction.

2. The retention times of the standards and unknowns agree
   to better than $\pm 0.2\%$ when using the GC^2-ND
   techniques.

3. The retention times of the standards and unknowns agree
   to better than $\pm 0.4\%$ when using the GC^2/MS technique.
   The MS spectral pattern is similar for standards and
   unknowns.

4. The mass of the molecular ions for the unknowns are
   measured to an accuracy of beter than 10 ppm.

The GC^2/MS and GC^2-ND data was obtained using similar
conditions. GC^2/MS confirmation was not possible for the
low abundance peaks since the GC^2/MS sensitivity (5-10 ppm)
was an order of magnitude lower than that of the GC^2-ND.
Quantitation of 1-nitropyrene, 1,3-dinitropyrene,
1,6-ninitropyrene and 1,8 dinitropyrene was undertaken by
GC^2/MS using deuterated analogs of these compounds as
described previously (21). The data for these quantitative

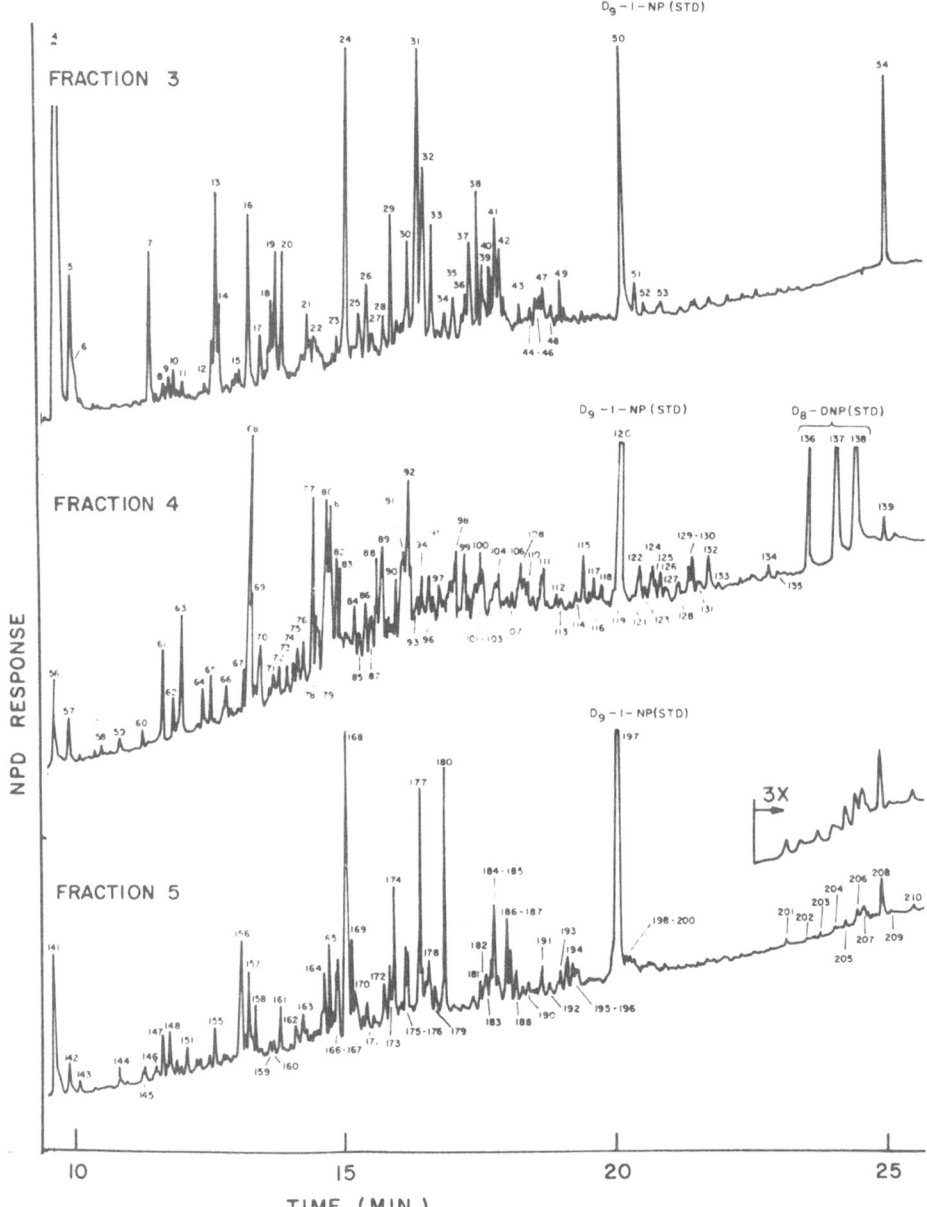

Figure 2. Gas Chromatograms for Fractions 3, 4 and 5 Obtained Using a 30m x 0.25mm DB-5 Fused Silica Column with A Nitrogen Detector.

TABLE 1-GC2/ND RETENTION TIME DATA AND GC2/MS SPECTRAL DATA
USED FOR IDENTIFICATION OF NITRO-PAH IN A DIESEL
PARTICULATE EXTRACTa

Peak	Compoundb	MASS	RRT(INP=1.00)c STD	Sample	GC2/MS Confirm	HPLC Fraction
2	1-Nitronaphthalene	173.048	0.433	0.424	nd	3
nd	2-Nitrobiphenyl	199.063	0.449	nd	nd	3
3	1-Nitro-2-methylnaph-					
	thalene	187.063	0.453	0.454	yes	3
nd	2-Nitronaphthalene	173.048	0.459	nd	nd	3
6	9H-Carbazole	167.074	0.504	0.502	yes	3
8	3-Nitrobiphenyl	199.063	0.566	0.566	nd	3
10	4-Nitrobiphenyl	199.063	0.586	0.584	yes	3
11	9-Methylcarbazole	181.089	0.593	0.589	yes	3
23	2-Nitrofluorene	211.063	0.722	0.722	yes	3
24	9-Nitroanthracene	223.063	0.741	0.741	yes	3
25	x-Nitro-Benzo-	231.105	sna	0.759	yes	3
	carbazole					
26	x-Nitro-Benzo-	231.105	sna	0.767	yes	3
	carbazole					
27	x-Nitro-Benzo-	231.105	sna	0.771	yes	3
	carbazole					
28	x-Nitrophenanthrene	223.063	sna	0.781	yes	3
29	x-Nitrophenanthrene	223.063	sna	0.788	yes	3
31	10-Nitro-9-methylan-					
	thracene	237.079	0.809	0.811	yes	3
32	9-Nitro-1-methylan-					
	thracene	237.079	0.818	0.817	yes	3
nr	2-Nitrophenanthrene	223.063	0.824	nr	yes	3
33	x-Nitro-y-methylan-					
	thracene	237.079	sna	0.826	yes	3
nr	2-Nitroanthracene	223.063	0.836	nr	yes	3
34	x-Nitro-y-methylan-					
	thracene	237.079	sna	0.838	yes	3
35	x-Nitro-y-methylan-					
	thracene	237.079	sna	0.846	yes	3
36	x-Nitro-y,z-dimethy-					
	lanthracene	251.095	sna	0.857	yes	3+4
37,99	x-Nitro-y,z-dimethy-					
	lanthracene	251.095	sna	0.861	yes	3+4
39,100	x-Nitro-y,z-dimethy-					
	lanthracene	251.095	sna	0.873	yes	3+4
40	x-Nitro-y,z-dimethy-					
	lanthracene	251.095	sna	0.879	yes	3+4
41,104	x-Nitro-y,z-dimethy-					
	lanthracene	251.095	sna	0.885	yes	3+4
43	x-Nitro-y,z,z'-tri-					
	methylanthracene	265.111	sna	0.907	yes	3+4
44	x-Nitro-y,z,z'-tri-					
	methylanthracene	265.111	sna	0.917	yes	3+4
45	x-Nitro-y,z,z'-tri-					
	methylanthracene	265.111	sna	0.921	yes	3+4
46	x-Nitro-y,z,z'-tri-					
	methylanthracene	265.111	sna	0.928	yes	3+4
47,111	x-Nitro-y,z,z'-tri-					
	methylanthracene	265.111	sna	0.936	yes	3+4
48	x-Nitro-y,z,z'-tri-					
	methylanthracene	265.111	sna	0.944	yes	3+4
54,139	x-Nitroterphenyl	275.095	sna	1.243	yes	3+4

TABLE 1 (Continued)

Peak	Compound[b]	MASS	RRT(INP=1.00)[c] STD	Sample	GC2/MS Confirm	HPLC Fraction
64	1,5 Dinitronaph-thalene	218.033	0.619	0.617	nd	4
86	1,8 Dinitronaph-thalene	218.033	0.768	0.767	nd	4
113	7-Nitrofluoran-thene	247.063	0.950	0.950	nd	4
nd	2,5-Dinitrofluorene	256.048	0.960	nd	nd	4
114	2-Nitrofluoran-thene	247.063	0.964	0.964	nd	4
nd	1-Nitrofluoran-thene	247.063	0.968	nd	nd	
115	3-Nitrofluoran-thene	247.063	0.971	0.970	nd	4
117	8-Nitrofluoran-thene	247.063	0.980	0.980	nd	4
nd	4-Nitropyrene	247.063	0.996	nd	nd [d]	4
120	1-Nitropyrene	247.063	1.000	1.000	75 ppm	4
nd	2-Nitropyrene	247.063	1.006	nd	nd	
124	2,7-Dinitrofluorene	256.048	1.031	1.029	no	4
131	3-Nitro-1-methyl-pyrene	261.079	1.075	1.074	yes	4
132	6-Nitro-1-methyl-pyrene	261.079	1.081	1.081	yes	4
nr	8-Nitro-1-methyl-pyrene	261.079	1.084	nr	yes	4
nd	7-Nitrobenzo(a)anthracene	273.079	1.099	nd	nd	4
134	1-Nitrochrysene	273.079	1.141	1.136	nd	4
135	6-Nitrocyclopenta-(c,d) pyrene	271.063	1.147	1.144	nd	4
200	2,5-Dinitro-9-fluorenone	270.028	1.041	1.043	nd	5
202	1,3-Dinitropyrene	292.048	1.178	1.174	0.30 ppm [e]	5
204	1,6-Dinitropyrene	292.048	1.204	1.202	0.40 ppm [e]	5
206	1,8-Dinitropyrene	292.048	1.222	1.226	0.53 ppm [e]	5
207	3-Nitrobenzo(a)pyrene	297.079	1.233	1.232	nd	5
208	6-Nitrobenzo(a)pyrene	297.079	1.243	1.245	yes	5
209	3-Nitroperylene?	297.079	sna	1.253	yes	5

(a) Abbreviations are as follows:

sna: standard not available-tentative identification by GC2/MS.
ne: not eluted from GC column:
nd: not detected is less than approximately 0.5 ppm by GC2/NPD and less than approximately 5 ppm by GC2/MS.
nr: not resolved from neighboring peaks.

(b) x,y,z notations refer to isomer position unknown.

(c) relative retention times with respect to 1-nitropyrene (INP = 1.00)

(d) 1-nitropyrene quantified by GC2/MS using d9-1-nitropyrene as an internal standard. See reference 19 for procedural details.

(e) 1,3-,1,6- and 1,8-dinitropyrenes quantified by GC2/MS using d8-1,3-,1,6- and 1,8-dinitropyrenes.

analyses are given in Table I.   Quantitative estimates for
the other nitro-PAH can be determined from relative peak
heights since the nitrogen detector has been found to
respond relative to the number of nitrogens present in the
compound (dinitropyrenes give twice the response for that of
nitropyrene).

It was found that most of the compounds responding to the
nitrogen detector were nitro-PAH.   A few N-heterocylic
compounds were also identified. The most notable finding in
these analyses is the large number of nitro-PAH which were
found. Approximately 200 nitro-PAH compounds are present in
these three fractions.   The complexity of these fractions
underscore the difficulty in obtaining definitive analysis
of individual nitro-PAH isomers.   The need for HPLC
fractionation   is   demonstrated   by   overlaying   the
chromatograms from fractions 3, 4 and 5 which results in
many peaks which would be unresolved. Even with high
resolution   capillary   column   GC   analysis,   HPLC
prefractionation is necessary for simplifying the mixture.
The difficulty in resolving isomers is illustrated for the
case of the nitro(fluoranthene and pyrene) isomer group.
Figure 3 shows the possible isomer positions available for
nitration.   Table I gives the retention times of these
isomers.

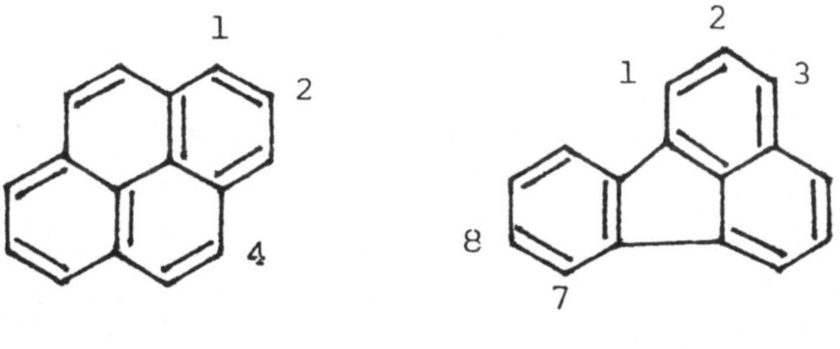

NITROPYRENES   NITROFLUORANTHENES

Figure 3.   Possible Isomers for Nitro (Fluoranthenes and
Pyrenes).

The GC^2/ND technique showed the possible presence of the
1-NF, 3-NF, 7-NF and 8-NF but the level was below 1-2 ppm
and not detected by GC^2/MS. 1-NP is the only nitropyrene
isomer found in significant concentration.

Reactions of Nitro-PAH - There are several possible sources of
nitro-PAH found in primary emission samples. These
compounds may be formed during the combustion process, in
the tail pipe, during the filter collection procedure, or a
combination of any of these processes.

We have previously shown that some 1-nitropyrene may be
formed during sampling, for which the degree of conversion
is dependent upon sampling conditions. This degree of
conversion can be estimated kinetically from a "global rate
constant" (11) as given by the equation below:

$$d(nitro\text{-}PAH)/dt = k^2(PAH)(NO_2)$$

Rate constants for the formation of 1-nitropyrene from
pyrene were determined experimentally using this expression
and found to give $1.4 \pm 0.3 \times 10^{-2}$ $h^{-1}ppm^{-1}$ for three
independent experiments. It can be concluded from these
results that:

1. Nitro-PAH are formed during sampling and may account
   for 10-20% of the 1-nitropyrene present in light-duty
   diesel particulates collected for 23 min. at 10-15/1
   dilution, in the presence of 3 ppm NO_2 and at 40°C.

2. Significant conversion of PAH can occur during filter
   sampling of emissions for long periods of time, at
   elevated temperatures and with undiluted samples
   containing high levels of NO_2. The kinetic expression
   given by equation 1 can be used to estimate the level
   of conversion during sampling.

The Effect of Engine Operating Conditions On Nitro-PAH
Emissions - We have operated a heavy-duty diesel engine
under a variety of load and speed conditions (11). Figure 4
shows the GC^2/ND chromatograms for a composite sample of
fractions 3-7 as defined in Figure 1. Each chromatogram was
obtained for this engine run under the three different
operating conditions. The components numbered in the
chromatograms are identified as follows: 1).
2-nitrofluorene, 2). 3-nitro-9-fluorenone, 3). 2-nitro-
9-fluorenone, 4). 9-nitroanthracene, 5).
nitrodibenzothiophene, 6). 9-nitro-1-methylanthracere, 7).
3-nitro-1,8-naphthalic acid anhydride, 8).
2,5-dinitrofluorene, 9). 1-nitro pyrene, 10). unknown, 11).
2,7-dinitrofluorene, 12). 2,5-dinitro-9-fluorenone, 13).
2,4,7-trinitro-9-fluorenone, 14). ?, 15). 1,3-dinitropyrene,
16). 1,6-dinitropyrene, 17). 1,8-dinitropyrene, 18).
3-nitro(BaP), 19). 6-nitro(BaP), 20). 3-nitroperylene?. A
detailed analysis of these chromatograms shows that there

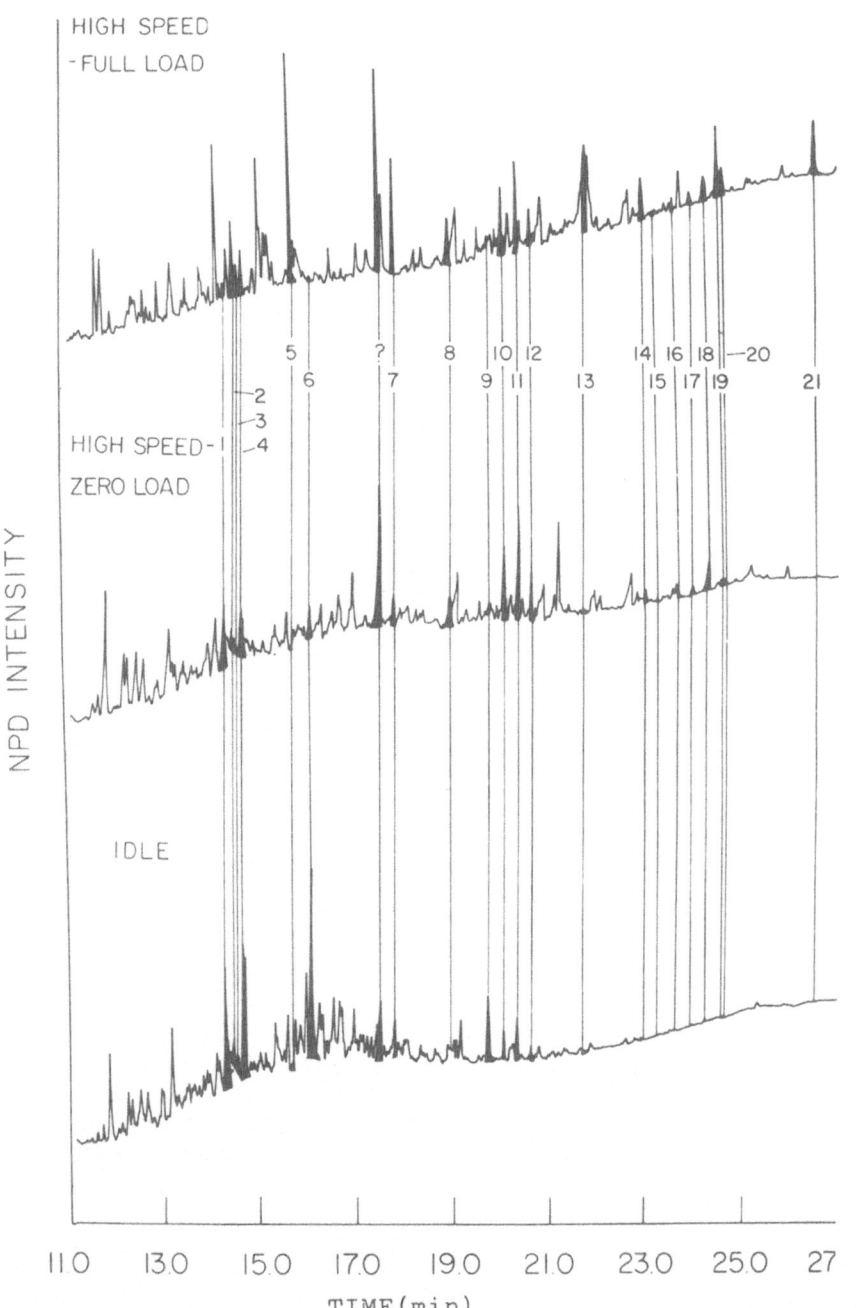

Figure 4. A Comparison of GC^2/Nitrogen Detector Chromatograms for A Heavy-Duty Diesel Engine Run Under Different Operating Conditions.

are dramatic changes in nitro-PAH emissions, depending upon engine operating conditions. Quantitative results were obtained for several of the nitro-PAH and the results are presented in Table 2.

TABLE 2-THE EFFECT OF ENGINE OPERATING CONDITIONS ON THE
CONCENTRATION OF NITRO-PAH IN HEAVY-DUTY DIESEL PARTICULATES

		PPM Concentration in Particulates[a]		
Peak	Compound	A	B	C
		Idle	High Speed Zero Load	High Speed Full Load
1	2-nitrofluorene	84(164)	62(134)	1.9(15)
2	3-nitro-9-fluorenone	18(35)	7.9(17)	8.0(63)
3	2-nitro-9-flurenone	10(19)	4.8(10)	3.7(29)
4	9-nitroanthracene	94(184)	16(35)	5.1(40)
6	9-nitro-1-methylanthracene	129(252)	13(28)	0.2(1.6)
7	3nitro-1,8-napthalic acid anhydride	23(46)	10(22)	22(174)
9	1-nitropyrene	14(28)	3.0(6.5)	0.13(1.0)
11	2,7-dinitrofluorene	15(30)	18(39)	3.9(31)
12	2,4-dinitro-9-fluorenone	5.5(11)	8.0(17)	2.1(17)
13	2,4,7-trinitro-9-fluorenone	(<1.0)(<2.0)	0.4(0.9)	NR
15	1,3-dinitropyrene	(<0.8)(<1.6)	0.6(1.3)	0.4(3.1)
16	1,6-dinitropyrene	(<0.8)(<1.6)	1.2(2.6)	0.8(6.3)
17	1,8dinitropyrene	(<0.8)(<1.6)	1.2(2.6)	0.8(6.3)
19	6-nitrobenzo (a) pyrene	(<3.2)(<6.5)	1.6(3.5)	0.3(2.4)

[a] Number in parentheses is ppm concentration in extract. Extraction of samples A, B and C gave 51.0, 45.9 and 12.7% extractables, respectively.
NR=not resolved

The following conclusions can be made from these experiments:

1. PAH and nitro-PAH emissions from diesels are significantly reduced under conditions of increased load and speed.

2. Some partially oxidized nitro-PAH emissions are increased under conditions of increased load and speed.

3. The dinitropyrene species show a slight increase in concentration under conditions of increased load and speed.

CONCLUSIONS

It can be concluded from these results that the emission of
nitro-PAH and possibly other oxy-PAH species are very
dependent upon source conditions, and in some cases
dependent upon the conditions of sampling.   General
conclusions regarding emissions of these compounds cannot be
made unless emissions are collected from a number of
representative engines which have been run under a variety
of operating conditions.

REFERENCES

1. Huisingh et al., "Application of Bioassay to The
    Characterization of Heavy Duty Diesel Particulate
    Emissions," papers presented at the Symposium on
    application of Short-term Bioassays in the Fractionation
    and Analysis of Complex Environmental Mixtures,
    Williamsburg, Va. (1978).

2. Wang, Y., Rappaport, S. M., Sawyer, R. F., Talcott, R.
    C., and Wei, E. T., Cancer Letters, 1978, 5, 39.

3. Schuetzle, D., Lee, F. S.-C., Prater, T. J. and Tejada,
    S. B., Proceedings of the 10th Annual Symposium on the
    Anal. Chem. of Pollutants, Gordon and Breach Science
    Publishers, New York, May 28-30, 1980, pp. 193-244; also
    published in Int.'l J. Environ. Anal. Chem., 1981, 9,
    93.

4. Pederson, T. C. and Siak, J. S., "The Role of
    Nitroaromatic Compounds in The Direct-Acting
    Mutagenicity of Diesel Particulate Extracts," J. of
    Applied Toxicology, 1981, 1, 54.

5. Yu, M., and Hites, R. A., 1981, Anal. Chem., 1981, 3,
    951.

6. Newton, D. L., Erickson, M. D., Tomer, K. B.,
    Pellizzari, and Pamela Gentry, Environ. Sci. Technol.,
    1982, 16, 206.

7. Xu, X. B., Nachtman, J. P., Jin, Z. L., Wei, E. T.,
    Rappaport, S. M. and Burlingame, A. L., Analyt. Chem
    Acta, 1982, 136, 163.

8. Pederson, T. C. and Siak, J. S., J. of Applied
    Toxicology, 1981, 1, 61.

9. Schuetzle, D., Sampling of Vehicle Emissions for
    Chemical Analysis and Biological Testing." Environmental
    Health Perspectives Journal, Jan., 1983.

10. Schuetzle, D., Riley, T. L., Prater, T. J., Harvey, T.
    M. and Hunt, D. F., "The Analysis of Nitrated
    Polynuclear Aromatic Hydrocarbons in Diesel
    Particulates," Anal. Chem., 1982, 54, 265.

11. Schuetzle, D. and Perez, J. M.,   "Formation of
    Nitrated Polynuclear Aromatic Hydrocarbons During
    Sampling of Diesel Emissions," J. of Air Pollution
    Control Assoc., Submitted, 11/82.

12. Schuetzle, D. and Perez, J. M., "A CRC Cooperative
    Comparison of Extraction and HPLC Techniques for Diesel
    Particulate Emissions," Proceedings of the 74th Annual
    Meeting of the Air Pollution Control Association, Paper
    81-56.4, Philadelphia, PA. (June 16-21, 1981).

13. Paputa, M., Hampton, C., Marano, R., Schuetzle, D.,
    Riley, T. L., Prater, T. J. and Skewes, L. M., "Analysis
    of Nitrated Polynuclear Aromatic Hydrocarbons in
    Particulates Using Capillary Column GC/Nitrogen
    Selective Detection", Anal. Chem., Submitted, Nov.,
    1982.

14. Salmeen, I., Durisin, A. M., Prater, T. J., Riley, T. L.
    and Schuetzle, D., Mutation Research, 1982, 104, 17.

15. Schuetzle, D., Riley, T. L., Prater, T. J., Salmeen, I.,
    "The Identification of Mutagenic Chemical Species in Air
    Particulate Samples", Proceedings of the 2nd
    International Congress on Analytical Techniques in
    Environmental Chemistry, Barcelona, Spain (11/23/81) in
    "Analytical Techniques in Environmental Chemistry-II",
    editor, J. Albaiges), Pergamon Press, Oxford, England,
    1982, pp. 259-280.

16. Salmeen, I., Pero, a., Riley, T., Hampton, C., Prater,
    T., Gorse, R. and Schuetzle, D., "Identification and
    Rigorous Quantitation of Direct-Acting Ames Assay
    Mutagens in Diesel Particulate Extracts", to be
    presented at the Sixth International Symposium on
    Polynuclear Aromatic Hydrocarbons, Columbus, Ohio (Oct.
    27-29, 1982).

TRENDS IN TRANSPORTATION FUELS

J M Tims

Esso Petroleum Company, Abingdon, England

SUMMARY

Forecasts for the next decade show a decline in total oil demand, with a proportional increase in demand for transportation fuels. This change in demand pattern will be met by conversion of residual fuel into light components. Conversion products are rich in olefins and, especially, aromatics, and this will be reflected in changes in the composition and combustion quality of gasoline and diesel fuels. Another factor which will have an effect on fuel composition is the removal of lead for health protection reasons, but the extent of the changes, will depend on a number of as yet undecided factors. Foreseen changes in Europe are similar to those which have already taken place in the USA, and today's situation in the USA can give a good picture of the European position in ten years' time. Assuming that the appropriate technology is transferred to Europe we can expect little impact on emissions of polynuclear aromatics from vehicles.

1 INTRODUCTION

The past decade has been one of unparalleled change for the oil industry in response to world political and economic events, and to environmental concerns originating in the United States and now receiving much attention in Europe and other parts of the world. These influences together with technical developments in vehicle design, largely in the interests of improved fuel economy, have led to major changes in the way petroleum refiners manufacture fuels, and consequently, in the

313

D. Rondia et al. (eds.), Mobile Source Emissions Including Polycyclic Organic Species, 313–326.
© *1983 by D. Reidel Publishing Company.*

composition of those fuels. The period of change is not over, and it is to be expected that further significant developments will occur during the next decade.

The question arises as to what the effect of these changes will be on the composition of exhaust gas from vehicles, which plays such a large part in determining the quality of air in our cities. Much work has been done to demonstrate the relatively small effect of fuel quality and composition on the gaseous exhaust components currently subject to legislation, namely carbon monoxide, hydrocarbons and nitrogen oxides, but the effect on other components of exhaust has been less well characterized.

This paper reviews some of the factors which will effect processes for fuel manufacture in European refineries, forecasts likely trends in composition and discusses the likely effect on emissions of polynuclear aromatics from vehicles using those fuels.

2 SUPPLY AND DEMAND

2.1 The Global Scene

The dramatic escalation in crude oil prices since 1973, and the realisation that reserves of crude oil cannot continue indefinitely to supply the fraction of world energy they have traditionally supplied, have set the stage for major changes in the supply/demand outlook. Price, availability and security of supply considerations have led to both extensive energy conservation measures and to progressive replacements of oil-based fuels by coal, gas and nuclear-based electricity. In 1980, demand for petroleum products in Europe was less than 1% higher than in 1970. Forecasts for 1985 are 1% lower than for 1980, and a further 1% drop is expected by the year 2000. The surplus primary distillation capacity existing in European refineries has been widely discussed in the media and a number of refineries will have to be closed. In the context of a declining market, fears expressed during the early 1970s., that crude oil discoveries were not matching anticipated expansion in demand are no longer valid. The view of the oil industry at the present time is that supply and demand for crude oil will be broadly in balance. Furthermore, crude prices are unlikely to be high in the short term enough to justify massive investment in processes such as coal liquefaction and extraction of heavy oils from tar sands and shale. This has been illustrated by the cancellation or postponement of a number of such projects in the USA, Canada, Australia and elsewhere. During the period under review crude oil will continue to be the primary source liquid fuels.

Declining demand has not been spread evenly across all
fuels products, and there is a major and continuing shift in
the relative demand for the different products. First, world
economic recession has impacted more heavily on industrial
activity than on transportation, and has reduced demand
for heavy industrial fuels proportionally more than lighter
transporation fuels. Secondly, while it is relatively easy to
convert steam raising and process heat equipment to burn coal
or gas and hence to replace heavy industrial oil fuels, there
is no substitute in prospect for the liquid fuels used in
road and air transport. Industrial users will continue to
turn to coal and to

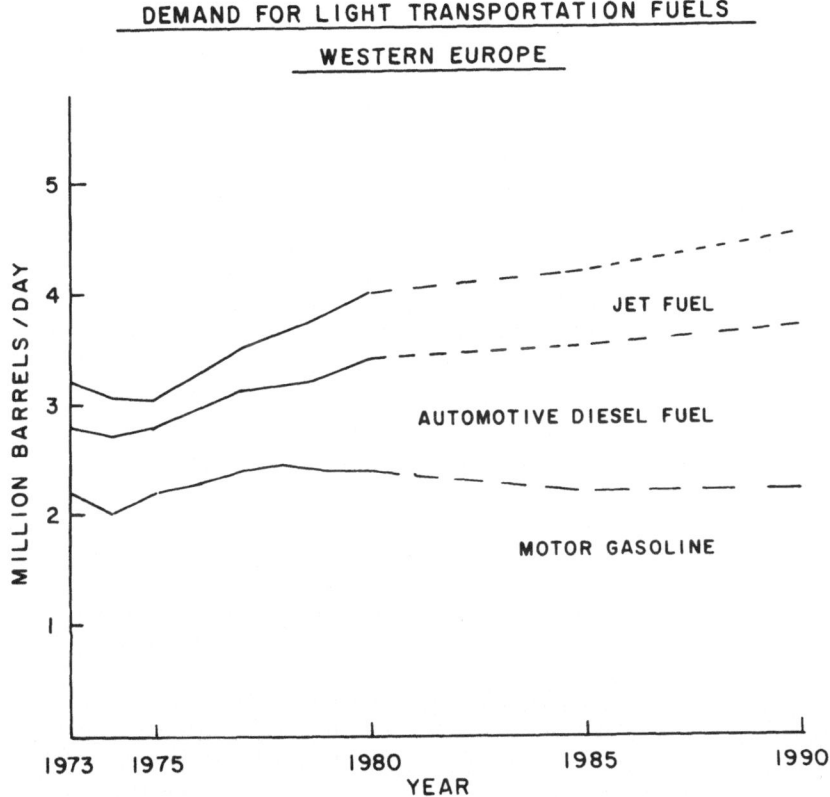

Fig.1

some extent gas in response to economic pressures and the
encouragement offered by governments in many industrialized
countries, but aviation and motor fuels will continue to be
provided from petroleum sources this century.

2.2   The Changing Barrel

These different outlook for transportation and industrial
fuels are illustrated by Fig 1 and 2, which show the historical
demand pattern for different fuel products since 1973 and
projected to 1990 for Western Europe. Fig 1 relating to
transportation fuels, shows an overall modestly rising
trend following the trough which came immediately after the
problems associated with the Arab-Israeli war in 1973. Within
this pattern, motor gasoline demand is expected to decrease
somewhat in volume terms as a result of improvements in fuel
economy achieved in new motor car designs, improved driving
habits and the growth in diesel engines as power units
for both commercial and private vehicles. Today's motor cars
are substantially more fuel efficient than those of a decade
ago, and it is widely accepted that by the end of the century,
the average European car will go at least 30% further on a
litre of gasoline than does the average car today. Penetration
of the diesel engine into the passenger car sector reached 4%
in 1980 and is likely to achieve 12% by 2000. Demand for
automotive diesel fuel is forecast to grow accordingly.
Jet fuel growth will be slow but steady, with the increase in
air traffic somewhat offset by higher efficiency new aircraft
and by a decline in the use of kerosene (the same product) in
the household sector.

The outlook for domestic and industrial gas, oil and
residual fuels (Fig 2) is one of steady decline under pressure
from alternative fuels at attractive prices. This is particularly
marked in the heavy residual fuel sector, where coal is the
major competitor, but is also apparent in the domestic heating
market as a result of efficiency improvements and competition
from gas and electricity.

This changing pattern of demand means that from each
barrel of crude the refinery must produce a steadily increasing
proposition of the lighter hydrocarbons needed for transportation
fuels and a smaller and smaller proportion of heavier molecules.
Relative demand for various products is the recent past and
out to 1990 is shown in Fig 3.

It can be seen from Figure 3, that the relative demand
for distillate and residual products changes from 62%:38% in
1973, to 70%:30% in 1980 and 75%:25% in 1990. The trend is
likely to continue after this date, and the time may come when

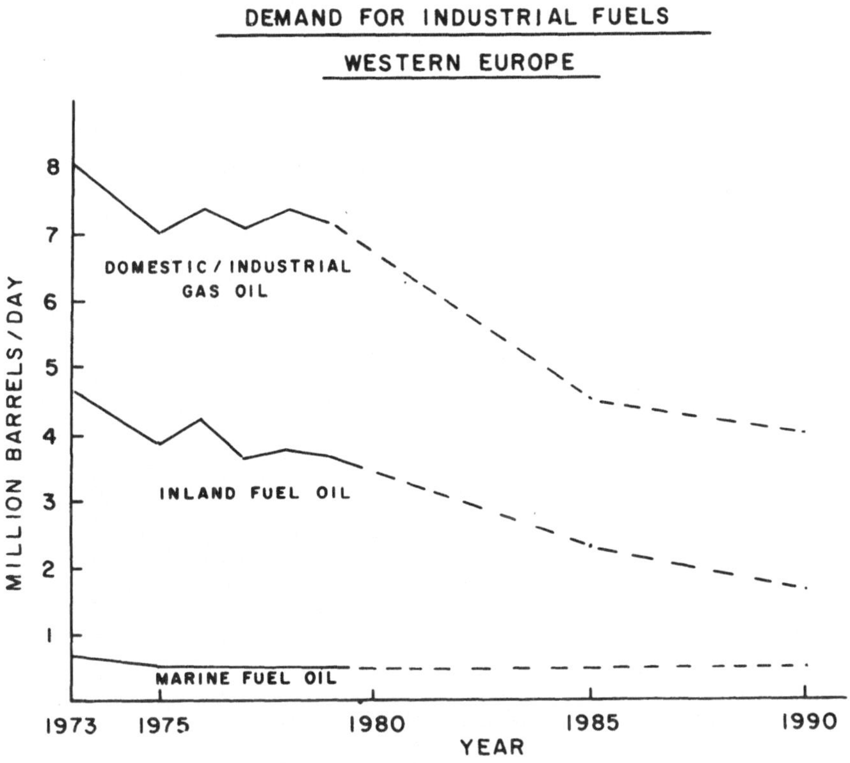
DEMAND FOR INDUSTRIAL FUELS
WESTERN EUROPE
MILLION BARRELS/DAY
8
7
6
5
4
3
2
1
DOMESTIC/INDUSTRIAL
GAS OIL
INLAND FUEL OIL
MARINE FUEL OIL
1973
1975
1980
1985
1990
YEAR

Fig.2

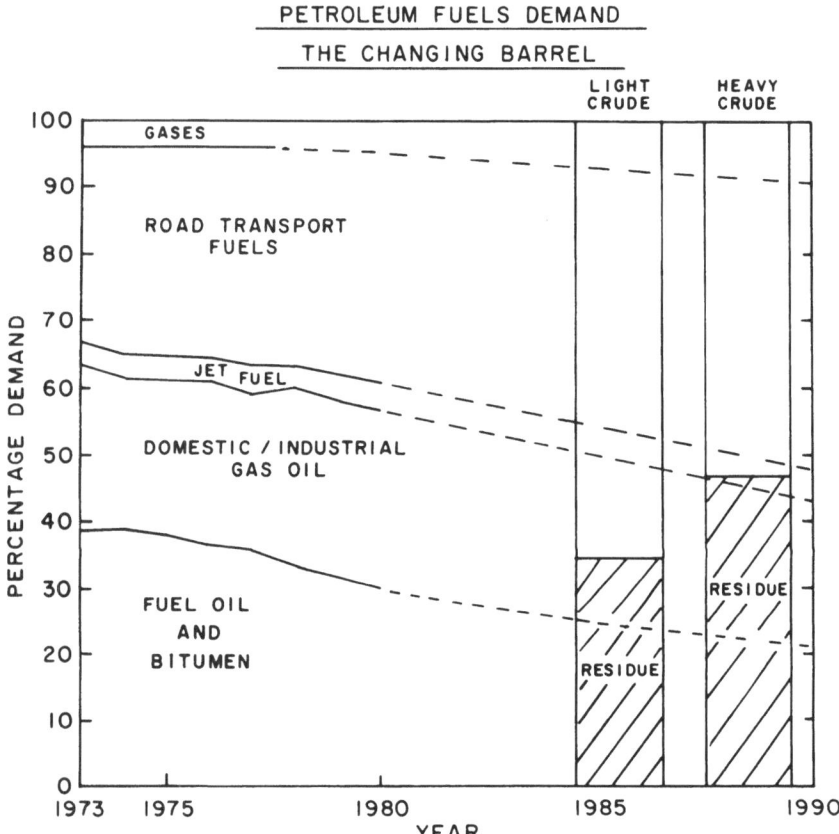

Fig.3

only small volumes of residual fuel are required for marine bunkers and special industrial applications. Also shown in Fig 3 is the percentage of residue naturally in light and heavy crudes and obtained in simple atmospheric distillation. The imbalance between residue demand and that occurring naturally is apparent even for light crudes such as North Sea and North African; for some of the heavy crudes from the Middle East and Mexico, which make up a growing part of our European crude slate, the situation is even worse. the imbalance must be corrected in the refinery by conversion of heavy residues into lighter products. It is unlikely that light components of non-petroleum origin will play a large role in correcting the situation during the next decade. Oxygenated materials, such as methanol from coal or natural gas are being, and will continue to be, used in motor gasoline on a localized basis, but as octane supplements partially replacing lead rather than for volume. Within the time frame considered here, their cost, limited availability and potential for problems in distribution and use must limit their applicability.

2.3 Residue Conversion

The need to convert heavy residues into light products to meet demand is not new. In the United States demand has for many years been skewed towards light products, and the trend is already well-established in Europe. Accordingly, refinery processes are already in operation which can bring about this conversion. They are essentially thermal cracking processes which take place in the presence or absence of catalysts. The feeds for these processes have in the past been mainly heavy distillates obtained by vacuum distillation of residues. However, technology is now available which makes it possible to crack residues themselves in such processes as visbreaking, residue cat cracking and the various types of coking.

Conventional Cracking processes give products rich in aromatics and olefins the predominant hydrocarbon types depending on boiling ranges. The following table gives some typical values in comparison with products in the same boiling range from straight distillation.

Addition of increasing amounts of cracked components to gasoline and diesel fuel to meet the volume needed in the market place will therefore make these products more and more aromatic and olefinic. The consequences of this are discussed later.

Table

	Gasoline Range, %		Diesel Range, %	
	Aromatics	Olefins	Aromatics	Olefins
Light catalytic	15	31	70	<5
Heavy catalytic	47	15	70	
Steam cracked	58	18	-	
Visbroken	9	39	25	<5
Straight Run	4	1	20	<5

3   ENVIRONMENTAL FACTORS

In the early days of exhaust emissions control, it was
established that, within the likely commercial range, motor
gasoline and diesel fuel composition have relatively little
effect on the composition of vehicle exhaust in terms of the
regulated components carbon monoxide, unburnt hydrocarbons and
nitrogen oxides. Legislation relating these exhaust components
has therefore had little direct effect on the way in which
fuels are refined and, hence, their composition. The only
exception to this is where emissions controls are so severe as
to demand catalytic exhaust control and hence unleaded gasoline.

3.1 Motor Gasoline

In the case of motor gasoline, legislation on lead
reductions, either for hygiene or for catalyst protection
reasons, has a very major effect on gasoline composition and
the potential for PNA emissions. Lead content of motor
gasoline is now limited to 0.4g/1 throughout Western Europe,
and in a number of countries legislation has either already
taken effect or is in place to reduce lead to 0.15g/1. These
reductions have not removed the pressures for a total ban on
lead, and it is possible that unleaded gasoline will appear
in Europe within the next decade. The effect of these moves
on motor gasoline composition have to be considered in two
stages: first, the reduction to 0.15g/1, and then to unleaded.

So far, in those countries which have reduced lead to
0.15g/1 there has been no commensurate drop in octane quality,
and in order to maintain octane quality it has been necessary
to increase aromatics content to compensate for lead. For

example, a survey carried out by Associated Octel in 1979 showed the average aromatics content in West German gasoline (at 0.15g/1 lead) to be close to 40% compared with an average 30% for France, Benelux and United Kingdom at 0.4g/1. The refinery process employed almost exclusively for octane improvement in motor gasoline in a reducing lead context is catalytic reforming. A typical catalytic reformate contains as much as 70-75% aromatics, of which 10% is benzene, 45% C_7-C_8 and 17% C_9-C_{10} compounds. The PNA content of reformates depends on the precise boiling range but is much higher than that of other gasoline blending components. Typically it is in the range 500-1000 microgram/l.

Those countries which adopt 0.15g/1 lead without any change in octane quality can, then, expect to see more highly aromatic motor gasoline in the market place.

In the case of unleaded gasoline the composition of the gasoline depends on the octane quality associated with the unleaded grade. During 1979, a study was published (Rational Use of Fuel Private Transport) which had been carried out jointly by the European oil and motor industries under the auspices of EEC. This showed that for each lead level there was an optimum octane number in terms of the balance between energy used in the refinery in manufacturing gasoline and the energy recovered in cars on the road. The optimum associated with different lead levels are:

Lead Level g/1	Research Octane Number
0.4	97
0.15	95
unleaded	93

In reducing lead from 0.4 to 0.15g/1 without a commensurate drop in octane quality, governments have pursued a non-optimum course in energy terms. This can be expresssed in terms of additional crude oil necessary to produce a given mass of gasoline. It is now broadly accepted that in moving to an unleaded grade there can be no question of maintaining octane quality since, quite apart from the massive wastage of crude oil, the industry would be quite unable to supply the volumes required. Recognising this, it is expected that any country demanding unleaded gasoline will follow the precedent of USA and Japan and do so at the "Regular" grade quality of 92-93 RON. Figure 4 shows the general relationship between aromatics content and octane quality for unleaded gasoline.

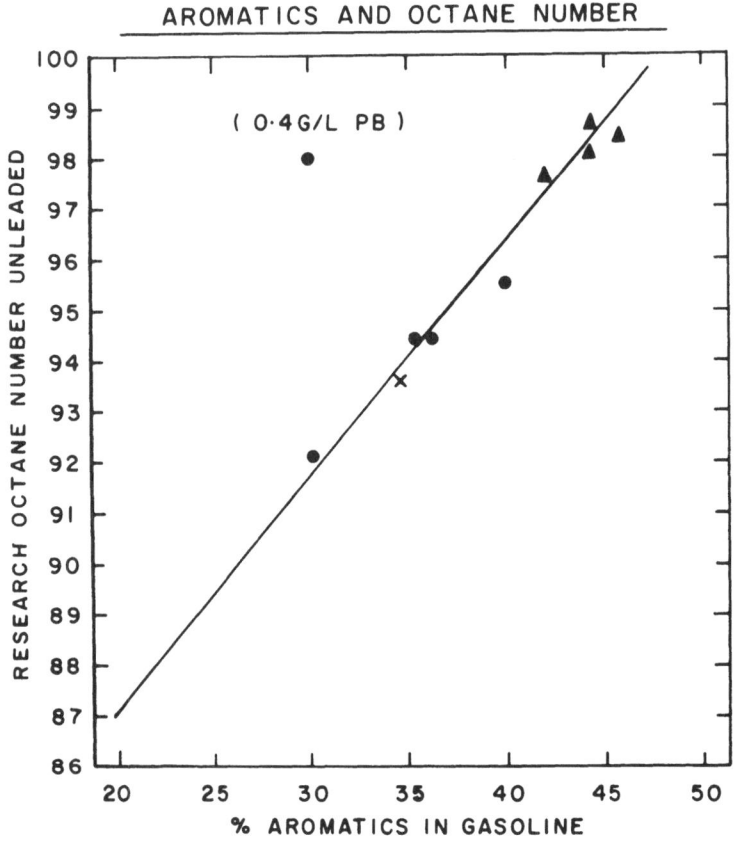

Fig.4

The Figure also shows a typical aromatic level for a
current Premium grade motor gasoline at 0.4g/1. It is clear
from this that in terms of aromatics we might expect a future
unleaded Regular grade to be about the same as today's leaded
Premium.

The other environmental concern related to motor gasoline
which might at first sight seem to affect composition is
benzene. Exposure of workers and consumers to benzene from
evaporation of gasoline during manufacture, distribution and
use has received a good deal of attention, and some countries
actually limit the maximum allowable benzene content of motor
gasoline. However, it is now clear that controlling the
benzene content of the gasoline is a relatively ineffective
way of protecting the population at large, since benzene
is created in the engine during combustion and is emitted from
exhaust even in fuels very low in benzene. The only way of
controlling this source of benzene is through exhaust catalysts.

3.2 Diesel Fuel

Environmental factors are perhaps less important than supply and demand considerations in their influence on diesel fuel. Sulphur content is subject to legislative control, but this has not affected composition in a major way.

Control of particulates from diesels is a matter of concern in the USA but is not yet considered for Europe. Although some workers have reported a relationship between diesel fuel volatility and aromaticity, and particulates it is clear that engine design plays the major role. Legislation on particulates is therefore unlikely to bring to bear a major influence on fuel quality.

4   FUELS OF THE FUTURE

Previous sections have outlined some of the factors which will affect the way in which transportation fuels are refined in the future. It now remains to forecast the characteristics of these fuels and to consider how they might change the potential for PNA emissions in the operating conditions under which they will be used.

The published literature is far from unanimous on the question of the exact inter-relationships between fuel composition and PNA in exhaust emissions, but certain principles do seem to be clear. First, some of the PNA in the fuel appears to pass through the engine unchanged, at least in the case of the gasoline engine. Combustion chamber deposits seem to play some role in storage and subsequent emission of these compounds. Second, some workers report a relationship between total aromatics in the fuel and PNA emissions, others identify correlations only with the heavier aromatics ($>C_9$). In principle then, it can be expected that a change in fuel composition may lead to changes in PNA emissions. On the other hand, vehicle design and operating conditions also have a big effect on PNA emissions, and it is difficult to predict how future vehicle populations will behave. If anything we should expect PNA emissions from individual vehicles to reduce in the future in line with the trend to leaner air fuel ratios. The one unanimous conclusion drawn by all workers in the field is that advanced emissions control systems are able to reduce PNAs in exhaust by up to 99% of the uncontrolled level.

4.1 Motor Gasoline

As a consequence of relative supply and demand for light
and heavy fuels motor gasoline will contain more cracked
components. This means that it will tend to be relatively
high in olefins (up to say 25%) and marginally poorer in Motor
Octane number (important under severe conditions). In some
locations oxygenates may constitute up to 10% of the composition
of the gasoline. The author is not aware of any evidence
linking either olefins or oxygenates with PNA in exhaust.

Measures to reduce lead are likely to have a two stage
effect. In the short term, reduction from present levels to
0.15g/l with no change in octane quality will be compensated
by increased use of catalytic reformate and higher aromatics
contents (up to 10% more). Assuming that there is a valid
relationship between total aromatics content (and fuel PNA
content) and PNA emissions there is the potential for some
increase in PNA emissions in the short and medium term.
Studies by the Danish Environmental Protection Agency have
suggested that a 10% increase total aromatics can give around
20% more PNA in exhaust. When unleaded gasoline is introduced
it is expected to be at 92-93 RON and as such will not require
supplementary aromatics. With the advent of unleaded gasoline
there will be no bar to the use of the exhaust catalysts
already in use for some years in the United States. Introduction of these devices in Europe would bring this region
into line with the USA and permit effective control of both
benzene and PNA in vehicle exhaust, as well as substantially
reduced gaseous emissions. From this point of view the
gasoline powered vehicle would no longer be a significant
source of PNA. It should be stressed that, as yet, no exhaust
catalyst has been shown capable of extended operation on
gasoline containing lead.

4.2 Diesel Fuel

As a consequence of supply - demand considerations diesel
fuel will contain increasing proportions of components made by
conversion of residues by thermal and catalytic cracking
processes. These components are highly aromatic (70-80% for
catalytically cracked gas oil), low in cetane number and high
in density. Traditionally, high cetane number virgin distillates
have been reserved for diesel fuel and the lower quality
cracked components absorbed into domestic heating oil. This
outlet for lower quality components is diminishing under
commercial pressure from other fuels and more cracked material
must appear in diesel fuel. At the present time the potential
for placing cracked material in diesel fuel is limited by the

cetane number specification contained in various national standards. It is anticipated that changes will be made in the standards of those countries demanding high quality to bring them into line, first with other European countries at about 45 cetane, and then with the USA at 42 cetane number. There is a strong correlation between cetane number and total aromatics content of a diesel fuel.(Fig 5).

It is readily apparent that a reduction in cetane number from 50 to 42 permits an increase in aromatics content from say 27% up to 38%. The increase in aromatics will be in those components boiling in the range 180-320°C, that is, in the $C_{12}-C_{20}$ group.

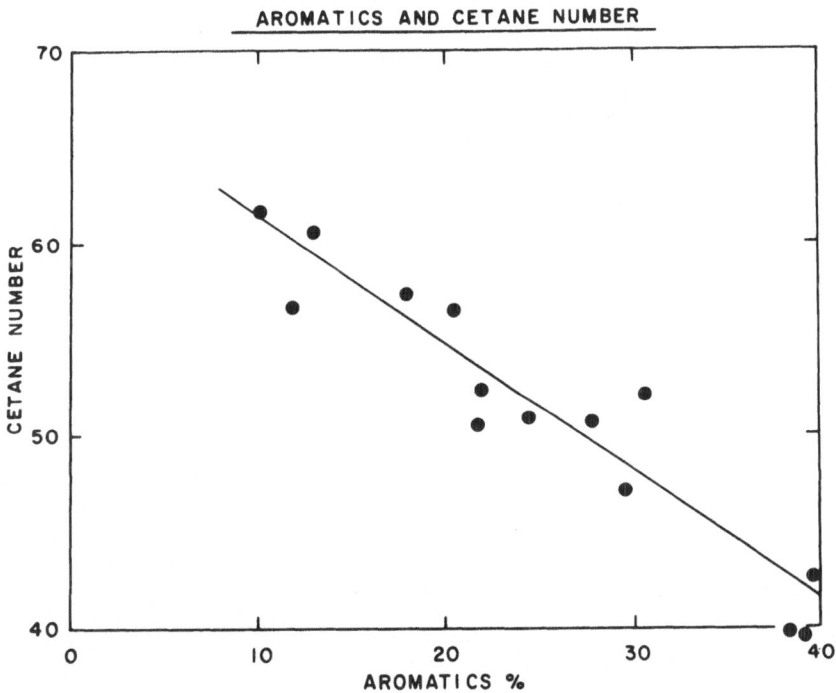

Fig.5

Diesel fuel may also become heavier in the sense that higher final boiling points may appear (up to 400°C). Should this be realised there is unlikely to be any marked effect on PNA emissions. The heavy material is unlikely to contain much aromatics, since aromatics boiling above 300°C are solid. However, in 1981 Beltzer and Bouffard reported a correlation between percent boiling above 340°C and particle bound organic material in exhaust which needs to be considered.

A major difference between diesel engines and gasoline powered engines is that the potential for controlling hydrocarbons in diesel exhaust through exhaust catalysts is very poor. The most that can be expected is some form of trap to contain diesel particulates. This may also be effective in controlling PNA, since a substantial part of the PNA emission from diesels is associated with the solid material.

In conclusion it may be expected that although there will be considerable changes in the composition and properties of European motor fuels during the next decade there is every hope that the effect on PNA emissions may be small. Technology to achieve this is available for the gasoline engine; which will still make up more than 80% of our car population in the year 2000, and there is no reason to expect that the diesel will cause major concern.

MODERN PAH-ANALYSIS AND FATE OF PAH IN AIR

K.Van Cauwenberghe and L.Van Vaeck

Chemistry Dept. University of Antwerp (UIA)
B-2610 Wilrijk Belgium

Current analytical methods for the determination of PAH in environmental samples require the separation of individual compounds, either by thin layer or high performance liquid chromatography with fluorescence detection, or better, by gas chromatography on capillary columns, coupled to a flame ionisation detector or a mass spectrometer. Modern thermostable capillary columns will allow the elution of PAH in the molecular weight range above coronene. Prior isolation of PAH as a compound class by liquid-liquid partition and column chromatography is necessary before most of these techniques can be applied.

Recently, alternate approaches to PAH-analysis have been proposed, which could possibly eliminate tedious separations in the future. These luminescence - based techniques make use of the Shpol'skii effect, matrix isolation or fluorescence line narrowing spectrometry, combined with site selective laser excitation. Also, the refined mass spectrometric identification technique MS/MS was applied successfully to the analysis of PAH in synthetic fuels.

Next to the limited analytical information available from studies of PAH-metabolism, there is a surprising lack in the current literature of research data on analytical methods for polar PAH-derivatives in environmental samples, e.g. epoxides, phenols, nitroarenes, which may be of great toxicological significance.

The direct mutagenic activity observed in applying the Ames reversion assay to both emission and ambient particulate samples can, at least in part, be explained by the presence of polar PAH-derivatives, e.g. nitroarenes in diesel exhaust.

D. Rondia et al. (eds.), Mobile Source Emissions Including Polycyclic Organic Species, 327–347.
© *1983 by D. Reidel Publishing Company.*

The possibilities for chemical transformations of PAH by gas-particle interactions in emission plumes, exhaust systems or during atmospheric transport are many. However, such interactions may also occur upon prolonged Hi-Vol sampling of particulate matter and thus create a chemical sampling artifact.

Photochemical transformations of PAH, adsorbed on various substrates are well documented in the literature: they may involve radical initiated auto-oxidation, singlet molecular oxygen or hydroxyl/hydroperoxyl attack. In model systems, several PAH coated on filters or simulant particles have been shown to react fast with ozone or nitrogen dioxide in the dark and to result in the formation of direct mutagens. Any future policy aimed at the reduction of PAH emission levels should therefore also be assessed in terms of possible conversion of PAH into other hazardous products.

The proper evaluation of the importance of atmospheric transformation of PAH will require a better understanding of the physical and chemical factors, which influence the reactivity of aerosol constituents during gas-particle interactions, e.g. particle size and specific surface, catalysis of the particle matrix. Thus, more elaborated kinetic and product identification studies are needed. These should not be limited to simplified model systems, but also include the direct exposure of combustion related particles in a reaction chamber to ambient levels of gaseous pollutants.

I. MODERN PAH-ANALYSIS

Our major interest in polycyclic aromatic hydrocarbons (PAH) lies in the fact that they constitute a class of environmental pollutants, primarily related to the combustion of fossil fuels and hence of anthropogenic origin, of which several members exhibit carcinogenic properties in experimental animals. Therefore, analytical procedures for the determination of PAH were already developed decades ago, but several new methodologies are continuously being explored for application to PAH-analysis.

Recently, the application of short term bio-assays to a variety of emission and ambient particulate samples has demonstrated that other compounds than PAH, most likely PAH-derivatives, can be responsable for a major part of the observed biological activity of the extracts. These results have forced us to widen our interest to the field of particulate polycyclic organic matter (POM), which besides hetero-atomic PAH (polycyclic aromatic compounds, PAC, containing nitrogen or sulfur) may also contain oxidized, nitrated, or sulfonated derivatives of PAH, formed in exhausts, emission stacks, plumes or during transport through the atmosphere.

Sampling of exhaust or ambient particulates for PAH-analysis

may represent a problem from two different aspects. The incorporation of PAH into particulate matter is the result of a condensation
/adsorption process from the gas phase and thus, some temperature
dependent partition of PAH between gas and particulate phase will
remain. Furthermore, various possibilities for transformations
through gas-particle interactions will exist. These physical or
chemical interactions may not only occur in the atmosphere, but also during the act of sampling.

The physical loss of PAH upon prolonged sampling has been demonstrated experimentally and reported in the literature for twenty years (Commins, Rondia, Pupp et al., Tomingas and Voltmer, Peters and Seifert (1-5)). For the lower PAH-analogs, a non-negligi-
ble part of the gas phase collected downstream of the particulate
filter originates from evaporation (blow-off) during the often long
sampling times, required to collect sufficient sample for PAH-analysis. This can represent a serious problem when sampling ambient
aerosols in background areas, even with the nanogram sensitivity
available in current analytical techniques. Impregnation of filters
with glyceroltricaprylate has been used to reduce this blow-off artifact (Brockhaus (6), König et al. (7)).

Recently, the actual gas phase-particulate matter distribution
of PAH has been measured in ambient samples (Van Vaeck and Van Cauwenberghe (8), Thrane and Mikalsen (9)). A general criterion has
been proposed to check the importance of sampling artifacts, due to
either blow-off or reactions: it requires the repeatibility of the
mass of PAH collected per unit volume as well as the invariability
of PAH-profiles with increasing sampling time at constant air flow
rate (Grimmer (10)). Sampling of exhaust gases for PAH must always
involve filter trapping in combination with gas phase enrichment
by adsorbent or cryogenic techniques. The presence of reactive gases in the exhaust, e.g. NO_x (or SO_2 in stacks) may result in fast
conversion of some PAH into new compounds (Hughes et al. (11)).
Therefore, dilution tunnels are used frequently to effectively mix
automobile exhaust with ambient air and thus to simulate the changes in composition occurring upon immediate dilution in the atmosphere as a result of physical and chemical gas-particle interactions.

The extraction of PAH from particulate matter is most effective with the solvents benzene or toluene and normally recovery decreases with increasing molecular weight, both for diesel particulates (Grimmer (12)) and fly-ash (Griest et al. (13)). The spiking
of a sample with labeled PAH is often used to verify the quantitative character of PAH-recovery. However, the coating of particles
with additional labeled compounds under laboratory conditions is
not to be compared with the condensation of PAH in the combustion
process and therefore absolute results of such studies are questionable. Vacuum sublimation of PAH from carbonaceous material has also
been shown to be efficient, but time-consuming for the higher ana-

logs (Stenberg and Alsberg (14)).

   Analytical procedures for the determination of PAH can be divided into two classes: whenever a non-selective detection method
is used, e.g. flame ionisation, several enrichment and separation
steps are preceeding the actual identification and quantitation;
in combination with a highly specific detection method, e.g. single
ion monitoring of PAH molecular masses, a much more simplified
work-up of the extract will suffice.

   In the older literature, PAH are first isolated as a compound
class by column chromatography on alumina and subsequently separated by thin layer or paper chromatography (Sawicki et al. (15)).
Acetylated cellulose has been recommended as an excellent adsorbent
for the reversed phase separation of PAH (Pierce and Katz (16)).
Identification and quantitation will usually require the recovery
of the individuals spots in order to measure them by spectrofluori-
metry. Two kinds of problems may arise: low or irreproducible recovery from the plates and possible photochemically induced oxidation of compounds.

   Modern procedures prefer the use of gel filtration on Sephadex LH 20 to isolate the class of PAH.(Wilk et al. (17), Oelert (18)
Retention volumes do not depend on water content as with deactivated alumuna and column capacity is considerably increased. The gas
chromatographic separation of PAH on glass capillary columns is
the method of choice today, and is only limited by considerations
of vapor pressure, thermolability or polarity (important for PAH-
derivatives). GC^2 provides superior resolution of isomeric pairs
such as BeP and BaP with different carcinogenic properties. Stationary phases commonly used are SE-52 and OV-17 (Carugno and Rossi (19)
Grimmer and Böhnke (20)). Better control of the column surface characteristics (acid leaching) greatly improved the deactivation and
separation efficiency since (Lee et al. (21)). The recently developed fused silica capillaries have shown to be excellent material
for the production of well deactivated high temperature columns.
(Dandeneau and Zerenner (22)) Only minor selectivity improvements
are observed with more polar phases than SE-52 (5% phenyl silicon)
(Cantuti et al. (23), Borwitzky and Schomburg (24)).

   A number of attempts to extend the usual temperature range of
capillary columns in order to elute PAH above coronene have been
successful. Grob (25) was able to chromatograph stable PAH-compounds
from coronene up to rubrene (molecular weight 300 to 532) on a short
OV-101 column. Romanowski et al. (26) used a 15 meter SE-54 fused
silica column to separate PAH in the molecular weight range from
300 up to 402, in combination with FID and MS detection. A recent
review on the performance of capillary columns in PAH-analysis has
been published by Lee and Wright (27).

Besides the universal FID-detector, the selective electron capture (ECD), nitrogen-selective thermionic (NPD) and flame photometric (FPD) detectors have been used for the measurement of PAH, and nitrogen or sulfur containing heterocyclic analogs respectively (Wakeham (28), Willey et al. (29)).

The application of GC-MS to PAH-analysis will allow considerable simplification of the work-up procedure. As far as solubility and dynamic range considerations will permit, the major PAH in airborne particulate matter samples can be accurately measured in the electron impact single or multiple ion monitoring mode, using the molecular ions of the respective PAH as specific ions, in the presence of much larger amounts of aliphatic hydrocarbons and carboxylic acids. Polar and ionic material is first removed from the combined benzene-methanol extract by liquid-liquid partition in water-diethylether. After drying, the addition of diazomethane results in derivatisation of the acidic components and the sample can be injected onto the column (Van Vaeck and Van Cauwenberghe (30)).

The separation of PAH-fractions can also be performed by high pressure liquid chromatography. HPLC is particularly suited for the analysis of PAH of high molecular weight, whose low vapor pressure or thermolability would preclude elution in GC. However, HP LC will provide considerably lower chromatographic resolution (despite the increasing number of possible isomers of higher mass) and lower column capacity (decreasing solubility of higher PAH). The retention of twelve PAH on columns packed with octadecyl bonded phases from different manufacturers have been compared (Ogan and Katz (31)), using acetonitrile-water (80:20) as the mobile phase. It was shown that not only capacity ratios (k'-values) of the PAH may vary significantly, but also that selectivity factors were different. These data did not correlate with carbon content or surface concentration of alkyl groups. The resolution of reversed phase HPLC columns is much inferior to that of capillary GC and the separation of isomeric pairs is often incomplete. The use of a variable wavelength fluorescence detector can provide a means for the selective monitoring of some PAH emission bands and thus enable a deconvolution of overlapping peaks in certain cases (Dong and Locke (32)). About fifty PAH with molecular weight above 300 could be positively identified or be given tentative structural assignments in a carbon black extract (Peaden et al. (33)). Only limited information is available on reference compounds containing six or more aromatic rings (Pierce and Katz (34)).

Considerable efforts have been made in recent years to develop an efficient LC-MS interface (Arpino and Guiochon (35)). Whereas present coupling techniques, involving conventional HPLC columns , suffer from a lack of sensitivity, due to the high splitting ratio of the elute (a consequence of the limited pumping capacity,

available on commercial mass spectrometers), the use of micro-HPLC
columns with flow rates in the range of microliters per minute,
combined with mass spectrometric detection in the chemical ionisation mode, using the acetonitrile-water solvent mixture as a reagent gas, seems very promising for PAH or POM analysis (Schäfer
and Levsen (36)).

Some polar bonded-phase sorbents have also been explored for
the HPLC separation of PAH. A silica-diamine packing was proposed
as an alternative to the C_{18}-packing, using methylenechloride-hep-
tane as the eluant (Chmielowiec and George (37)). The major advantages of this system would be a better ring size selectivity for
alkylated PAH and the absence of water which eliminates splitting
off for LC-MS analysis and hence increases sensitivity.

All methodologies discussed so far are based on first isolating PAH as a compound class, followed by their separation into
individual components and quantitation. These procedures are not
only time-consuming, but also complete resolution of structural
isomers is rarely achieved. Therefore, there is a continuing interest in alternate approaches that would not require or at least
considerably simplify the enrichment steps of the protocol.

Ambient temperature luminescence spectrometry has been applied
to the characterisation of complex PAH-mixtures with only limited
success because of excessive overlap of excitation or emission
spectra. However, considerable sharpening of PAH absorption bands
is observed, when these molecules are incorporated into appropriate
matrices and solidified at low temperatures. Generally, three different techniques to achieve fluorescence line narrowing have been
attempted, which only differ by the nature and the preparation of
the host matrix. Methods to perform fluorescence line narrowing
spectrometry (FLNS) are based on the Shpol'skii effect, matrix isolation or the use of organic glasses.

The selectivity of the Shpol'skii effect lies in the inherent
sharp absorption bandwidths (10 cm^{-1}), exhibited by PAH in a fro-
zen n-alkane host. However, different orientations of the PAH guest
molecules lead to non-equivalent crystal field effects and hence to
multiple site spectra. As a consequence Shpol'skii effect spectra
of PAH mixtures are still quite complex. These narrow absorption
bandwidths allow the site specific excitation of a given PAH through
the utilisation of a tunable dye laser (Laser Excited Shpol'skii
Spectrometry: LESS, Yang et al. (38)). Thus, application of LESS to
synthetic PAH-mixtures looks very promising. Some limitations on
Shpol'skii effect spectrometry may be non-reproducibility of relative site populations and restricted linear working range because
of aggregate formation. According to the authors, proper dilution
of the sample, reproducible cooling regimes, reduction of laser power or the choice of an alternate excitation line would eliminate

these drawbacks.

In matrix isolation fluorescence spectrometry (MIF), a liquid
or solid sample is vaporised, mixed with a large excess of diluent
gas (nitrogen or argon) and deposited on a cold surface. This results in narrower bandwidths in the fluorescence spectra. However,
when the matrix gas is a Shpol'skii solvent, sharp line spectra are
observed for PAH (Maple et al. (39)), presumably because of the existence of vacancies in the alkane host lattice, which are comparable in size to the PAH guest molecules. Again, tunable dye laser
excitation will result in the production of site-selective fluorescence spectra. The linear dynamic range of MIF is much extended in
comparison with LESS and quantitative precision is high. Artifacts
from quenching and solute aggregation are totally eliminated. Some
PAH-fractions from a coking plant waste water sample were analysed
by the authors.

Fluorescence line narrowing spectrometry of PAH can also be
performed in organic glasses. A glycerol-water glass doped with a
mixture of PAH and cooled to about 4° K can be irradiated with a
tunable dye laser and only a small subset of solvent-matrix confi-
guration sites, whose absorption overlaps the excitation line will
fluoresce (Brown et al. (40)). Mixtures of fourteen PAH and a standard solvent refined coal sample have been successfully analysed
by this technique. A well-reproducible optical quality of the glasses has the advantage to allow quantitation without recourse to the
method of standard additions or internal standards. Although in
Shpol'skii matrices, inhomogeneous absorption line broadening is
reduced much further and hence will allow for a greater degree of
selective excitation than in organic glasses, Shpol'skii matrices
are polycrystalline, opaque and an effective light scatterer. Thus,
quantitation is more complex.

Mass spectrometry-mass spectrometry (MS-MS) is rapidly emer-
ging as a very efficient technique for the identification and quantitation of trace organics in complex mixtures. The basic idea behind it, is to ionize the sample obtained after only minimal preparation (either in EI or CI), then focus a selected molecular ion
(either positive or negative) and study its further fragmentation
upon collisional activation. The technique, originally called MIKES,
has been applied to the analysis of some isomeric nitrogen-contain-
ing PAH in coal liquid samples (Zakett et al. (41)). The use of
charge stripping reactions and ammonia as the CI reagent gas provided enhanced selectivity for N-PAH. More recently, PAH were measured in untreated solvent refined coal by negative chemical ionisation
charge inversion MS-MS (Zakett et al. (42)). MS-MS studies were also performed in order to distinguish between isobaric amino and aza-
PAH in coal liquids (Wilson (43)).

The presence of nitrated PAH-derivatives in diesel particula-

tes has been investigated by means of a special MS-MS scanning procedure (Schuetzle et al. (44)). Using a triple stage quadrupole,
operated in the constant neutral loss mode, a rapid qualitative
screening of nitro-PAH analogs was developed. Positive ion chemical ionisation conditions were used to form the nitro-PAH pseudo
molecular ions. During collisionally activated dissociation, these
predominantly eliminated the neutral fragment OH in the second quadrupole. The first and third quadrupoles were scanned simultaneously with a mass differential of 17 amu.

In contrast to the variety of analytical methodologies discussed above, which have been applied to the determination of PAH
and their hetero-atom analogs, there is only limited information
available on the identification and quantitation of PAH-derivatives. Unlike other carcinogens, that are polar and can bind directly to biological molecules, PAH require metabolic activation. By
enzymatic reactions in vitro they can be transformed to a variety
of polar derivatives such as epoxides, phenols, dihydrodiols, quinones etc., among which some structures are probable ultimate carcinogens, e.g. BaP-dihydrodiolepoxides. Also, there is now substantial evidence for the chemical transformation of several PAH under
thermal or photochemical conditions in simulated atmospheres, which
modify the pattern of mutagenicity of these compounds. However,
apart from studies of the metabolism of single PAH (Jerina et al.
(45)), major efforts to develop analytical procedures for the detection of polar derivatives of PAH in environmental samples are
just starting.

Recently, considerable work has been performed on the identification of PAH-derivatives in diesel particulate extracts using
a combination of preparative HPLC separation and GC-MS or HR-MS
analysis (Schuetzle et al. (46)). Whereas the composition of the
non-polar fraction in terms of aliphatic and polyaromatic hydrocarbons has been well characterised, the compounds present in the moderately to highly polar fractions had not been identified before.
The transition fraction has been shown to contain mostly hydroxy-,
dihydroxy-, ketone, quinone, carboxaldehyde and acid anhydride derivatives of PAH. Also, some nitro-PAH were detected in this fraction. Their contribution to the direct mutagenicity of diesel extract is believed to be important. The easy conversion of hydroxy-
substituted PAH to the corresponding quinones was observed during
the analytical procedures, even in HPLC separation and direct probe
MS-analysis. At present, it is unknown whether these oxy-derivati-
ves were formed in the engine combustion process, the exhaust system or during sample collection on filters. Polycyclic ketones
(mostly 9-fluorenone homologs) and quinones have been identified
in diesel particulates (Choudhury (47)) by capillary GC-MS, but
their analysis seems restricted to the lower analogs (3 to 5 rings).
Higher molecular weight quinones have been isolated from airborne
particulate matter by using thin layer separation procedures.

(Pierce and Katz (34), Sawicki (48))

Hydroxylated metabolites of PAH (Mc Clusky (49)), amino-deri-
vatives of PAH (Wilson (43)) in solvent refined coal and nitro-
PAH (Schuetzle et al. (44)) in fractions of diesel soot extract
have all been identified and sometimes quantified by MS-MS procedures.

Modern analysis of PAH-derivatives in environmental samples
is still in a qualitative stage. Complex separations by off-line
HPLC are usually required and identification by GC-MS has only limited success because of the low volatility and/or labile character of certain compounds. Direct LC-MS coupling or MS-MS seem to
hold much promise for future work. The possibility for achieving
separation efficiencies comparable to that of capillary GC-columns
by using capillary LC-columns is being explored. Their interfacing
to a mass spectrometer should be splitless and hence very sensitive
(Ishii and Takeuchi (50)). As to MS-MS, its possibilities in using
linked scans to detect several characteristic neutral losses in the
mass spectra of functionalised PAH molecules offer a new means of
specific detection, probably comparable to the breakthrough of multiple ion monitoring in the past.

II.   FATE OF PAH IN THE ATMOSPHERE

The general application of the Ames' Salmonella microbiologi-
cal reversion assay (Ames et al. (51)) to environmental samples
has demonstrated the mutagenicity of the organic extracts of ambient particulate matter on several occasions (Pitts et al. (52-53),
Tokiwa   et al. (54), Talcott and Wei (55), Teranishi et al. (56),
Möller and Alfheim (57)). The mutagenicity of the frame shift type
detected in those samples (TA 98 was the most sensitive strain) was
often not enhanced by the addition of liver microsomes, thus indicating the presence of mutagens, which do not require metabolic activation and therefore could not be simple PAH or hetero-atom analogs.
Direct acting mutagens were shown to be present in the exhaust of
spark ignition and diesel engines (Wang et al. (58), Barth and Blacker (59)). Fly-ash and soot extracts also exhibit mutagenic activity (Chrisp et al. (60), Kaden and Thilly (61)). Recently, hetero-
atomic PAH have also been investigated: some sulfur heterocyclics,
typical of combustion soots, are moderately active (Pelroy et al.
(62)), amino-PAH isolated from coal liquefaction samples and nitro-
PAH are generally stronger mutagens than the corresponding aza-com-
pounds (Ho et al. (63)). Among the oxygen containing PAH-derivatives, phenols and quinones can be considered as moderate mutagens.
(Dean (64), Brown (65))

The direct acting mutagenicity of ambient particulates is associated predominantly with the respirable particles (<2μm, Talcott

and Harger (66)), on which organic compounds accumulate through
condensation processes at the emission source. Whereas the chemical composition of the neutral and basic fractions of ambient aerosols can explain the mutagenicity, observed on metabolic activation, by the presence of a variety of PAH and aza-analogs, the direct biological activity of the moderately polar neutral and acidic fractions has not been accounted for so far (Cautreels and
Van Cauwenberghe (67)). Indeed, our knowledge of oxygenated compounds with possible biological activity in ambient aerosols is
very limited. Sawicki (48) has identified several carbonyl compounds derived from PAH, e.g. 7H-benz(de)anthracene-7-one and phe-
nalene-9-one in ambient dust using a combination of TLC separation
and fluorescence detection. Some higher molecular weight quinones
were discovered later in ambient air by Pierce and Katz (34).

HPLC has been successfully applied in a recent study to obtain structural information on the oxygenated PAH-derivatives in
the moderately polar neutral fraction of diesel exhaust, which ac-
counts for the highest degree of direct mutagenic activity (Schuetzle et al. (46)). The presence of nitroarenes could explain this
effect. Also, the pyrene 3,4-dicarboxylic acid anhydride isolated
by Rappaport et al. (68) from diesel extract could be one of a new
class of weak direct mutagens in diesel emissions. The mutagenicity of other major compound classes, e.g. the hydroxy- or carboxaldehyde derivatives of PAH is presently under investigation. A similar effort of identification on ambient particulate matter extracts has, to our knowledge, not been attempted so far. The collection of sufficiently large quantities of atmospheric dust may
be prohibitive in such a study.

The formation of direct acting mutagens has been observed in
laboratory exposure studies of single PAH to gaseous pollutants
under simulated atmospheric conditions and the compound(s) responsable have been isolated (Pitts et al. (69-70)). Thus, nitro-ben-
zo(a)pyrenes or benzo(a)pyrene-4,5-epoxide formed upon reaction
with NO_2 or O_3 respectively are direct mutagens. These results
support the idea that, contrary to some statements (Berry and Leh-
mann (71)), PAH as primary pollutants on particulate matter can be
subject to chemical transformation through gas-particle interactions, either in emission plumes, exhaust systems, during atmospheric transport or, to some extent during filter sampling. The
latter possibility has large implications on the validity of mutagenicity data, all of which have been gathered for Hi-Vol filtration samples. Polar PAH-derivatives will be more hydrophilic in
nature, their increased water solubility will probably facilitate
removal from the air, but, when they are inhaled, it is also likely to promote resorption in the lung tissue. The presence of polar functionalities in PAH is essential for their binding to the
DNA, which, similarly to the binding of certain PAH-metabolites,
is assumed to provide a proper molecular basis for the induction

of frame shift mutations. Indeed, from several studies of the metabolism of PAH in mammalian cells (Jerina et al. (72)), it followed that the enzymatically formed epoxides, phenols or dihydrodiol-
epoxide of benzo(a)pyrene (the ultimate carcinogen) are direct
mutagens. By analogy, it can be expected that some PAH-derivatives
formed by electrophilic attack of chemical reagents will also exhibit direct mutagenic activity.

A major problem in this hypothesis is the possibility of transformation of parent PAH upon Hi-Vol sampling, especially the artifactual formation of direct-acting mutagens during prolonged gas-
particle contact. Its importance is presently difficult to assess
and will depend on the age of the collected particles and their
whole previous history of gas-particle interactions during atmospheric transport. When sampling on filters, the effective contact
area between the adsorbed PAH and the air flow will be reduced, because of the penetration of particles between the fibers. In contrast however, gaseous pollutants have been known to adsorb selec-
tively to filters or particle surfaces and to undergo filter catalyzed reactions (Burton et al. (73)).

In the older literature, substantial evidence is found for
the reactivity of PAH in solution, especially involving electrophilic substitution. Also, photochemical transformation of PAH,
adsorbed on a variety of supports or particles has been observed.
The National Academy of Science document (74) on particulate organic matter has stressed the importance of this reactivity for gas-
particle interactions. Several common gaseous pollutants, e.g. O_3,
NO_2, ground state or excited singlet molecular oxygen (in the presence of light), or hydroxyl radicals all can react as electrophi-
les with the carbon positions or double bonds of highest electron
density of PAH.

While it is generally accepted that ground state molecular
oxygen does not react with PAH in the dark (losses of BaP up to 10
% in 24 h are the result of blow-off under Hi-Vol sampling conditions, Peters and Seifert (5)), upon irradiation, photooxidation
of PAH seems easy in the adsorbed state. Thus, Inscoe (75) deposited fifteen PAH on four different adsorbents (silicagel, alumina,
cellulose and acetylated cellulose) and exposed them to UV and room
light in ambient air. Eleven PAH underwent pronounced changes. On
the less polar substrates, conversion occurred much more slowly.
These reactions may be interpreted as radical initiated autooxida-
tions: Inomata and Nagata (76) suggest the formation of a PAH radical cation on polar substrates based on EPR measurements. Oxidation
by singlet molecular oxygen, formed by energy transfer from an irradiated sensitizer (Schaap (77)) is another possible mechanism. Most
tetracyclic and pentacyclic PAH absorb visible light strongly and
have high intersystem crossing quantum yields to the triplet state.
Therefore, they can not only generate singlet oxygen, but also sub-

sequently react with it. Geacintov (78) coated solid polystyrene
fluffs with twenty PAH and irradiated them in the presence of oxygen. Although no photodegradation products were isolated, efficient
energy transfer from PAH to O_2 was observed. Grossman (79) performed the singlet oxygen oxidation of several PAH coated on diatoma-
ceous particles in an irradiated fluidised bed under oxygen, using
dye-coated particles as sensitizer. Anthracene-like PAH reacted by
cycloaddition to form endoperoxides and quinones, which were isolated and identified by MS.

The earliest study of PAH photochemistry was conducted by
Falk et al. (80). A striking result in these measurements was the
higher reactivity of pure PAH, compared to that for PAH adsorbed
on soot. However, Tebbens et al. (81) studied the chemical transformation of benzo(a)pyrene and perylene in smoke, passing through
a flow reactor. Upon irradiation, he found that 35-65% of the original content in PAH had disappeared or had been transformed. Subsequently, Thomas et al. (82) employed a similar flow system to
measure the reaction of BaP on soot particles and found a 58% decrease at the exit of the chamber upon irradiation.

Ozone can react as an electrophile with PAH and some of their
aza-analogs according to two different mechanisms: one step attack
on the most electron-dense double bond will yield a primary ozon-
ide, followed by ring opening and further oxidation (Criegee (83));
two   step attack starting at the most negative carbon atom will
yield disubstituted oxidation products (diphenols, quinones). Both
types of reaction occur readily in solution with BaP (Moriconi et
al. (84)) and require no irradiation. Participation of the solvent
in the reaction mechanism can lead to complex ozonolysis products.
(Kemps and Van Cauwenberghe (85), Chen et al. (86) Degradation
studies of PAH in solution are not relevant to the determination
of their half life on particles. Thus, based on kinetic measurements in solution and extrapolation to the gas phase, Radding et
al. (87) reported a half life for BaP of 870 h, while Lane and
Katz (88) measured about 0.6 h at nearly ambient ozone concentration at the gas-solid interface.

Free radicals such as the hydroxyl or hydroperoxyl radicals
are expected to react with PAH by analogy to their reactivity with
simple aromatics in the gas phase: e.g. toluene is converted into
cresols (Atkinson et al. (89)). The rate constant of hydroxyl radical attack on anthracene, pyrene and benzo(a)pyrene has been measured recently in aquous solution (Chekulaev and Shevchuk (90)).
It exceeded the diffusion rate constant for BaP, which is explained by the possible formation of hydroperoxyl radicals in the system.

Especially the larger PAH are sensitive to electrophilic substitution as well as oxidation. Thus, nitration or sulfonation of

PAH proceeds readily in solutions of nitric or sulfuric acid.
(Dewar et al. (91), Vollman et al. (92)) The half lives of twenty-
one PAH in diluted solutions of nitric/nitrous acid (10:1) have
been measured recently by Nielsen (93).

These last few years, several experiments have been performed
in which PAH adsorbed onto substrates or particles were exposed to
nearly ambient levels of gaseous pollutants. Both kinetic and structural information resulted from these experiments, which will be
shortly reviewed here.

The kinetics of the dark reaction of ozone toward several PAH
was shown to be very fast at nearly ambient ozone levels. Lane and
Katz (88) reported a half life of 0.62 h for BaP exposed in petri-
dishes to an ozone level of 190 ppb. Irradiation did not seem to
affect this reactivity significantly. PAH, containing five-member-
ed rings such as benzo(k)fluoranthene were by far more resistant
to oxidation. Pitts et al. (70) used Hi-Vol glass fiber filters
as the substrate to study the kinetics and the products of the
reaction of BaP with O_3 in the dark. Conversion yields of 47% were
observed after a 1 h exposure to 110 ppb of ozone. A maximal and
nearly constant yield of 57% was obtained for longer exposures,
which could be an indication for a surface limited reaction of BaP
on glass fibers (Van Cauwenberghe and Van Vaeck (94)). Other almost
equally reactive PAH toward ozone are benz(a)anthracene and pyrene.

By HPLC the major reaction products of the BaP-O_3 system were
separated and tentative chemical structures were assigned to the
peaks, based on the interpretation of the low and high resolution
MS-data as well as on comparison with reference spectra. The reaction mixture consisted mainly of ring-opened compounds such as di-
aldehyde and dicarboxylic acid, but also contained disubstitution
products such as diphenols and quinones. The benzo(de)anthracene-
7-one 3,4-dicarboxylic acid found in the reaction mixture is probably the precursor to the benzo(de)anthracene-7-one, detected by
Sawicki (48) in the extract of ambient aerosols. The benzo(a)pyrene
quinones were discovered in ambient air by Pierce and Katz (34).
The direct mutagenicity of the reaction mixture was shown to be
caused by the presence of the benzo(a)pyrene-4,5-epoxide in small
yield (Pitts et al. (70)). This compound is a photosensitive intermediate of the ozonolysis and is known as a DNA-binding metabolite
in biological systems (Grover and Sims (95a).

Recently, Peters and Seifert (5) used ^{14}C-labeled BaP-impreg-
nated glass fiber filters to evaluate the BaP loss under Hi-Vol
sampling conditions. While only losses up to 10% were observed for
[14]C in 24 h exposures to ambient air, TLC determinations of BaP indicated that chemical reactions could account for losses up to 90%.
A correlation of these results with ambient ozone levels (about 30
ppb) was suggested, but photodecomposition of BaP was also shown

to be important (estimated half life of 2 h). According to these
authors, the kinetics of BaP photodecomposition on glass fiber is
not affected significantly by the presence of atmospheric particulate matter, since similar decay curves were obtained on dust-coa-
ted and uncoated filters.

In exposure experiments of benzo(a)pyrene and perylene on filters to 1 ppm of nitrogen dioxide in the dark (Pitts et al. (69));
significant conversion into nitro-derivatives was observed (60% in
8 h). Traces of nitric acid (at ppb level) had to be present for
the reaction to proceed. The nitration products were direct mutagens in the Ames' test. Nitroarenes have been detected in one ambient air sampling location (Jaeger (95b) and in diesel particulate matter (Schuetzle et al. (44),(46)). Indirect evidence that the
direct mutagenicity of airborne particulates might originate from
nitroaromatics or nitroheterocyclics was suggested by the decreased activity of the extracts of ambient aerosols in a nitroreductase deficient bacterial strain derived from TA 98 (Wang et al.(96)
). However, the chemical structure of these compounds in ambient
air appears to be more complicated than that of the simple mono-
and dinitroarenes produced in laboratory exposures.

A better simulation of the actual exposure conditions of PAH
to gaseous pollutants in the atmosphere can be performed, when aerosol particles collected from real emission sources or generated
artificially by combustion processes are used to study the possible
reactions of PAH adsorbed at their surface. Recently, some experiments with particles have been described in the literature.

Jaeger and Hanus (97) studied the interaction of gaseous nitrogen dioxide (1.3 ppm in air) with four PAH adsorbed on coal fly-ash,
carbon deposits, silicagel and alumina in a glass reactor. The
rate of nitration decreased in the order silicagel, fly-ash, alumina, carbon, irrespective of the adsorbed PAH, thus illustrating the
importance of the carrier for the quantitative aspect of the conversion. Earlier, Jaeger and Rakovic (98) studied the reaction of
pyrene and benzo(a)pyrene, adsorbed on fly-ash and alumina, with
10% sulfur dioxide in air, and isolated many sulfur containing compounds including sulfonic acids. However, under typical ambient
conditions, no significant conversion of benzo(a)pyrene, exposed
to 1 ppm of SO_2 in air was observed on a glass fiber filter.

Hughes et al. (11) exposed coal fly-ash, enriched with PAH
through vapor phase adsorption (Miguel et al. (99)), to gaseous
concentrations of 100 ppm of NO_2, SO_2 and SO_3 (levels about 10 times higher than in a power plant emission plume) and detected several nitration, sulfonation and oxidation products of PAH by HPLC.
the reaction also occurred on other types of particles, such as
alumina, silica and activated charcoal. In photochemical experiments however, PAH adsorbed onto coal fly-ash seemed to resist pho-

todegradation when irradiated in air with actinic UV-light. Yet,
the same compounds (anthracene, benzo(a)pyrene) photolyse efficiently when they are adsorbed on activated alumina or in solution.
(Korfmacher et al. (100)) The reason for this stabilizing effect
of fly-ash is not clear.

Atmospheric particulate matter, enriched with anthracene by
sublimation underwent efficient photodecomposition in air upon exposure to sunlight (Fox and Olive (101)). Some of the reaction products identified provided evidence for the participation of singlet
molecular oxygen generated at the particle surface as in the laboratory experiments of Grossman (79).

Butler and Crossley (102) studied the reactivity of PAH, naturally present on soot particles, generated in a flame from an eth-
ylene-air burner, in a reaction chamber with ambient air containing 5-10 ppm of sulfur dioxide or a mixture of nitrogen oxides.
While exposure of up to three months to SO_2 did not yield any significant loss of PAH, degradation by NO_x occurred readily and resulted in half lives varying between 7 and 30 days (for benzo(a)-
pyrene and phenanthrene respectively),in correspondence with the
order of reactivity for electrophilic substitution of aromatic
systems.

An interesting experiment which is particularly relevant to
the problem of chemical sampling artifacts as well as to the reactivity of PAH with gaseous pollutants has been performed by Löfroth et al. (103). During Hi-Vol sampling of urban particulate matter, the effect of adding traces of ozone or nitrogen dioxide was
investigated in parallel sampling systems. At an added ozone level
of 200 ppb no significant degradation of PAH seemed to occur, but
upon exposure to 960 ppb of NO_2, degradation of pyrene, benz(a)an-
thracene and benzo(a)pyrene in the range of 20, 40 and 60% respectively was observed over a 24 h period. Furthermore, a three to
fourfold enhancement of direct mutagenic activity both in nitroreductase proficient and deficient strains was detected. Thus, both
the chemical composition and the biological activity of the sample
exposed to NO_2 were seriously affected by artifactual reactions on
the filter.

All the experiments described above give strong indications
for the atmospheric reactivity of several PAH with gaseous pollutants or with oxygen in the presence of light. However, in these
simulation experiments, both the particle material and the exposure conditions used are often chosen as a function of the available experimental facilities. Thus, simplifications are introduced
(choice of carrier, use of single PAH, etc.) which are certainly
appropriate for fundamental studies, but make it difficult to ex-
trapolate kinetic data obtained in laboratory exposures to the actual gas-aerosol chemistry in the atmosphere.

In ambient aerosols, organic compounds released in combustion processes will be incorporated into the particles through adsorption or condensation. As a result, organics are found predominantly in the accumulation mode of the aerosol, i.e. the respirable particles (<2μm) which have a large specific surface. Each aerosol particle should therefore be considered as a matrix of various shape and composition (e.g. fly-ash: spherical particles built from metal oxides; diesel exhaust: conglomerates of carbon) to which one or more surface layers of PAH are adsorbed. The chemical reactivity of PAH will therefore be affected by two new parameters, depending on the particle matrix.

When a reactive compound is finely divided over a surface, the access of a gas molecule will be largely determined by particle size and specific surface area. While this parameter is probably more or less constant for glass fiber filters, impregnated with PAH from solutions, it can vary immensely in the exposure experiments with particles. Thus, it follows that experiments designed to evaluate the reactivity on particles in the atmosphere should be performed with particles of known small size and narrow size range. In this respect, the use of coarse silica, alumina and fly-ash particles (usually > 50μm) as carriers to study PAH transformation should not be encouraged. Experiments involving irradiation are even more complex: PAH deposited inside porous particles or inner layers of PAH can easily be shielded from the incident light.
The distribution of PAH into only one monolayer or multilayers on the particle will affect heterogenous reactivity in general: only the outer layer directly exposed to the gas phase may react, while the inner layers are being protected from further attack by this surface layer of products. In that respect, the coating of aerosol particles with a single PAH ('spiking') is often difficult to control and may result in an uneven multilayer distribution, atypical of the original particles.

Furthermore, the adsorption of organics onto the aerosol matrix can modify their reactivity through catalytic effects, induced by constituents of that matrix. The suppression of photochemistry of PAH, adsorbed onto coal fly-ash (Korfmacher et al. (100)), has been tentatively explained by the role of transition metal ions at the surface in quenching excited states formed upon irradiation.

From this discussion, it follows that characterisation of the physical and chemical environment of the reactive sites of a particle on a molecular level will be required to better understand its fate in the atmosphere. Exposure experiments should be continued, not only using simplified model systems, but also using natural combustion related aerosols of various particle size and matrix composition. Filter exposures of particulate matter should yield more detailed information on sampling artifacts. Exposure of aerosol particles to gaseous compounds should be tried in a static or dynamic reaction chamber in order to closely resemble atmospheric conditions of gas-particle interaction. However, the (re)suspension of amounts of particulate matter, sufficient for detailed chemical analysis, in a chamber for exposures to gaseous pollutants over a period of several hours still presents serious experimental problems and needs further investigation. In the meantime, more simplified approaches with simulant particles can be useful. The solution chemistry of PAH should also be further explored. In view of the surprising affinity of airborne particles for water (De Wiest and Brull (104)), the role of relative humidity on the rates and mechanisms of PAH conversion should not be underestimated.

REFERENCES

1) Commins B.T.: 1962, National Cancer Institute Monograph 9,pp 225
2) Rondia D.: 1965, Intern. J. Water Air Poll. 9, pp 113
3) Pupp C., Lao R.C., Murray J.J. and Pottie R.F.: 1974, Atmos. Environ. 8, pp 915
4) Tomingas R. and Voltmer G.: 1978, Staub Reinhalt. Luft 38,pp 216
5) Peters J. and Seifert B.: 1980, Atmos. Environ. 14, pp 117
6) Brockhaus A.: 1974, Atmos. Environ. 8, pp 521
7) König J., Funcke W., Balfanz E., Grosch B. and Pott F.: 1980, Atmos. Environ. 14, pp 609-613
8) Van Vaeck L., Broddin G. and Van Cauwenberghe K.: 1980, Biomed. Mass Spectrom. 7, pp 473-483
9) Thrane K.E. and Mikalsen A.: 1981, Atmos. Environ. 15,pp 909-918
10) Grimmer G.: 1981, Workshop on Polycyclic Aromatic Hydrocarbons, OECD Paris, Report to the Air Management Policy Group, Appendix V
11) Hughes M.M., Natusch D.F.S., Taylor D.R. and Zeller M.V.: 1980, 'Chemical Transformations of Particulate Polycyclic Organic Matter' in Polynuclear Aromatic Hydrocarbons, eds. Björseth A. and Dennis A.J., Battelle Press, Columbus Ohio.
12) Grimmer G.: 1981, Workshop on Polycyclic Aromatic Hydrocarbons, OECD Paris,'Analysis, Data Reporting and Profile Concept' background paper.
13) Griest W.H., Yeatts Jr L.B. and Caton J.E.: 1980, Anal. Chem. 52, pp 199-201
14) Stenberg U.R. and Alsberg T.E.: 1981, Anal. Chem. 53,pp 2067-2072
15) Sawicki E.: 1964, Chemist Analyst 53, pp 24, 56, 88
16) Pierce R.C. and Katz M.: 1975, Anal. Chem. 47, pp 1743
17) Wilk M., Rochlitz J. and Bende H.: 1966, J. Chromatog. 24, pp 414-416
18) Oelert H.H.: 1969, Fresenius Z. Anal. Chem. 224, pp 91
19) Carugno N. and Rossi S.: 1967, J. Gas Chromatog. 5, pp 103
20) Grimmer G. and Böhnke H.: 1972, Z. Anal. Chem. 261, pp 310
21) Lee M.L., Bartle K.D. and Novotny M.V.: 1975, Anal. Chem. 47, pp 540
22) Dandeneau R.D. and Zerenner E.H.: 1979, J. High Resoln Chromatog. Chromatog. Commun. 2, pp 351

23) Cantuti V., Cartoni G.P., Liberti A. and Torri A.G.: 1965, J. Chromatog. 17, pp 60

24) Borwitzky H. and Schomburg G.: 1979, J. Chromatog. 170, pp 99

25) Grob K.: 1974, Chromatographia 7, pp 94

26) Romanowski T., Funcke W., König J. and Balfanz E.: 1981, J. High Resoln Chromatog./Chromatog. Commun. 4, pp 209-214

27) Lee M.L. and Wright B.W.: 1980, J. Chromatog. Sci. 18,pp 345-358

28) Wakeham S.G.: 1979, Environ. Sci. Technol. 13, pp 1119

29) Willey C., Lee M.L., Iwao M. and Castle R.N.: 1982, Anal. Chem.

30) Van Vaeck L. and Van Cauwenberghe K.: 1978, Atmos. Environ. 12, pp 2229-2237

31) Ogan K. and Katz E.: 1980, J. Chromatog. 188, pp 115-127

32) Dong M., Locke D. and Hoffmann D.: 1977, Environ. Sci. Technol. 11, pp 612-618

33) Peaden P.A., Lee M.L., Hirata Y. and Novotny M.: 1980, Anal. Chem. 52, pp 2268-2271

34) Pierce R.C. and Katz M.: 1976, Environ. Sci. Technol. 10,pp 45

35) Arpino P.J. and Guiochon G.: 1979, Anal. Chem. 51, pp 682A

36) Schäfer K.H. and Levsen K.: 1981, J. Chromatog. 206, pp 245-252

37) Chmielowiec J. and George A.E.: 1980, Anal. Chem. 52, pp 1154 -1157

38) Yang Y., D'Silva A.P. and Fassel V.A.: 1981, Anal. Chem. 53, pp 894-899

39) Maple J.R., Wehry E.L. and Mamantov G.: 1980, Anal. Chem. 52, pp 920-924

40) Brown J.C., Duncanson J.A. and Small G.J.: 1980, Anal. Chem. 52, pp 1711-1715

41) Zakett D., Shaddock M. and Cooks R.G.: 1979, Anal. Chem. 51, pp 1849-1852

42) Zakett D., Ciuper J.D. and Coöks R.G.: 1981, Anal. Chem. 53, pp 723-726

43) Wilson B.W., presented at the Second Symposium on Environmental Analytical Chemistry, Provo Utah 1980

44) Schuetzle D., Riley T.L., Prater T.J., Harvey T.M. and Hunt D.F. : 1982, Anal. Chem. 54, pp 265-271

45) Jerina D., Lehr R., Yagi H., Hernandez O., Dansette P., Wislocki P., Wood A., Chang R., Levin W. and Conney A.: 1976 in de Serres F., Fouts J., Bend J. and Philpot R. (eds)' In Vitro Metabolic Activation in Mutagenesis Testing, Elsevier, Amsterdam

46) Schuetzle D., Lee F.S.C., Prater T.J. and Tejada S.B.: 1981, Intern. J. Environ. Anal. Chem. 9, pp 93-144

47) Choudhury D.R.: 1982, Environ. Sci. Technol. 16, pp 102-106

48) Sawicki E.: 1967, Archives Environ. Health 14, pp 46

49) Mc Clusky G.A., Huang S.K., Moore C.J. and Selkirk J.K.: 1980, Annual Conference of the American Society for Mass Spectrometry New York Abstr. RAMOA5

50) Ishii D. and Takeuchi T.: 1980, J. Chromatog. Sci. 18, pp 462

51) Ames B.N., Mc Cann J. and Yamasaki E.: 1975, Mutation Res. 31, pp 347

52) Pitts Jr J.N., Doyle G.J., Lloyd A.C. and Winer A.M.: 1975,

'Chemical Transformations in Photochemical Smog and their Applications to Air Pollution Control Strategies', 2nd Annual Report
to NSF-RANN, Grant N° AEN 73-02904-A02

53) Pitts Jr J.N., Grosjean D., Mischke T.M., Simmon V.F. and Poole
D.: 1977, Toxicol. Lett. 1, pp 65-70

54) Tokiwa H., Tokeyoshi H., Morita K., Takahashi K., Soruta N.
and Ohnishi Y.: 1976, Mutation Res. 38, pp 351-359

55) Talcott R. and Wei E.: 1977, J. National Cancer Institute 58,
pp 449-451

56) Teranishi K., Hamada K. and Watanabe H.: 1978, Mutation Res.
56, pp 273-280

57) Möller M. and Alfheim I.: 1980, Atmos. Environ. 14, pp 83

58) Wang Y.Y., Talcott R.E., Sawyer R.F., Rappaport S.M. and Wei
B.T.: 1978, 'Mutagens in Automobile Exhaust' presented at the
Symposium on Application of Short Term Bioassays in the Frac-
tionation and Analysis of Complex Environmental Mixtures, U.S.
EPA, Williamsburg Virginia

59) Barth D.S. and Blacker S.M.: 1978, 'EPA 's Program to Assess
the Public Health Significance of Diesel Emissions' presented
at the 71st Air Pollution Control Association Annual Meeting,
Houston Texas

60) Chrisp C.E., Fisher G.L. and Lammert J.E.: 1978, Science 199,
pp 73

61) Kaden D.A. and Thilly W.G.: 1978, 'Genetic Toxicology of Kerosene Soot' in Proceedings of the Workshop on Unregulated Diesel Emissions and their Potential Health Effects, U.S. Dept.
of Transportation, Washington DC, pp 612-633

62) Pelroy R., Stewart D., Tominaga Y., Iwao M., Castle R. and
Lee M.: 1981, Mutation Res., in press

63) Ho C., Clark B., Guerin M., Barkenbus B., Rau T. and Epler J.:
1981, Mutation Res., in press

64) Dean B.J.: 1978, Mutation Res. 47, pp 75-97

65) Brown J. and Brown R.: 1976, Mutation Res. 40, pp 203-224

66) Talcott R. and Harger W.: 1980, Mutation Res. 79, pp 177

67) Cautreels W. and Van Cauwenberghe K.: 1976, Atmos. Environ.10,
pp 447

68) Rappaport S.M., Wang Y.Y., Wei E.T., Watkins B.E. and Rapoport
H.: 1980, Environ. Sci. Technol. 14, pp 1505

69) Pitts Jr J.N., Van Cauwenberghe K., Grosjean D., Schmid J.P.,
Fitz D.R., Belser W.L., Knudson G.B. and Hynds P.M.: 1978,
Science 202, pp 515

70) Pitts Jr J.N., Lokensgard D.M., Ripley P.S., Van Cauwenberghe
K., Van Vaeck L., Shaffer S.D., Thill A.J. and Belser W.L.:
1980, Science 210, pp 1347

71) Berry R.S. and Lehman P.A.: 1971, Ann. Rev. Phys. Chem. 22,pp
47-84

72) Jerina D., Yagi H., Hernandez O., Dansette P., Wood A., Levin
W., Chang R., Wislocki P. and Conney A.: 1976, 'Synthesis and
Biological Activity of Potential Benzo(a)pyrene Metabolites'
in Freudenthal R. and Jones P. (eds), Polynuclear Aromatic Hy-

drocarbons: Chemistry, Metabolism and Carcinogenesis, Raven
Press, New York

73) Burton R.M., Howard J.N., Penley R.L., Ramsay P.A. and Clark
T.A.: 1973, J. Air Pollut. Control Assoc. 23, pp 277

74) National Academy of Science: 1972, 'Particulate Polycyclic Organic Matter', Committee on Biological Effects of Atmospheric
Pollutants, Washington DC

75) Inscoe M.N.: 1964, Anal. Chem. 36, pp 2505

76) Inomata M. and Nagata C.: 1972, GANN 63, pp 119

77) Schaap P.A. (ed): 1976, 'Singlet Molecular Oxygen', Dowden,
Hutchinson and Ross Inc., Strondsburg PA

78) Geacintov N.E.: 1973, 'Reactivity of PAH with O_2 and NO in the
presence of light', U.S. EPA, Washington DC, Publ. N°650/1-74-010

79) Grossman B.: 1976, 'Photooxidation of PAH by Singlet Oxygen',
Ph.D. thesis, Universitaire Instelling Antwerpen

80) Falk H.L., Markul I. and Kotin P.: 1956, A.M.A. Arch. Ind.
Health 13, pp 13

81) Tebbens B.D., Thomas J.F. and Mukai M.: 1966, J. Amer. Ind.
Hyg. Assoc. 27, pp 415

82) Thomas J.F., Mukai M. and Tebbens B.D.: 1968, Environ. Sci.
Technol. 2, pp 33

83) Criegee R.: 1957, Rec. Chem. Progress 18, pp 111

84) Moriconi E., Rackoczy B. and O'Connor W.: 1961, J. Am. Chem.
Soc. 83, pp 4618

85) Kemps M. and Van Cauwenberghe K.: unpublished results

86) Chen P.N., Junk G.A. and Svec H.J.: 1979, Environ. Sci. Technol. 13, pp 451

87) Radding S.B., Mill T., Could C.W., Liu D.H., Johnson H.L.,
Bomberger D.C. and Tojo C.J.: 1976,'The Environmental Fate of
Selected PAH', Stanford Research Institute EPA N° 68-01-2681

88) Lane D.A. and Katz M.: 1977,'The Photomodification of Benzo(a)
pyrene and Benzo(b)- and (k)fluoranthenes under Simulated Atmospheric Conditions' in Fate of Pollutants in the Air and Water
Environment, Part 2, Suffet I.(ed), Wiley-Interscience, New York

89) Atkinson R.A., Darnall K.D.,Lloyd A.C., Winer A.M. and Pitts Jr
J.N.: 1979,'Kinetics and Mechanisms of the Reactions of the Hy-
droxyl Radical with Organic Compounds in the Gas Phase' in
Advances in Photochemistry 11, Pitts Jr J.N., Hammond G.S. and
Gollnick K.(eds), Wiley, New York

90) Chekulaev V.P. and Shevchuk I.M.: 1981, Eesti NSV Teod. Akad.
Toim. Keem. 30, pp 138-140

91) Dewar M.J., Mole T., Urch D.S. and Warford E.W.: 1956, J. Chem.
Soc. pp 3572

92) Vollman H., Becker H., Corell M., Streeck H. and Longbein G.:
1937, Ann. Chem. 531, pp 1459

93) Nielsen T.: 1981,'A Study of the Reactivity of PAH', Nordic
PAH project, Report N° 10, Central Institute for Industrial
Research, Oslo

94) Van Cauwenberghe K. and Van Vaeck L.: 1980,'Physical and Chemical Transformations of Organic Pollutants during Aerosol Samp-

ling' in Quayle A.(ed), Advances in Mass Spectrometry 8, pp 1499, Heyden and Son Ltd, London

95a) Grover P.L. and Sims P.: 1970, Biochem. Pharmacol. 19, pp 2251

95b) Jaeger J.: 1978, J. Chromatog. 152, pp 575

96) Wang C.Y., Lee M., King C.M. and Warner P.O.: 1980, Chemosphere 9, pp 83

97) Jaeger J. and Hanus V.: 1980, J. Hyg. Epidemiol. Microbiol. Immunol 24, pp 1

98) Jaeger J. and Rakovic M.: 1974, J. Hyg. Epidemiol. Microbiol. Immunol 18, pp 137

99) Miguel A.M., Korfmacher W.A., Wehry E.L., Mamantov G. and Natusch D.F.: 1979, Environ. Sci. Technol. 13, pp 1229

100) Korfmacher W.A., Wehry E.L., Mamantov G. and Natusch D.F.: 1980, Environ. Sci. Technol. 14, pp 1094

101) Fox M.A. and Olive S.: 1979, Science 205, pp 582

102) Butler J.D. and Crossley P.: 1981, Atmos. Environ. 15, pp 91

103) Löfroth G., Toftgard R., Carlstedt-Duke J., Gustafsson J.A., Brorström E., Grennfelt P. and Lindskog A.: 1981, 'Effects of Ozone and Nitrogen Dioxide present during Sampling of Genuine Particulate Matter as detected by Two Biological Test Systems and Analysis of PAH' presented at EPA 's Diesel Emissions Symposium, Raleigh North Carolina

104) De Wiest F. and Brull P.: 1980, 'Interfacial Physicochemical Characteristics of Airborne Soot Particles' in Proceedings of the 14th International Colloquium on Atmospheric Pollution Paris, Benarie M.(ed), Studies in Environmental Science 8, pp 227, Elsevier Scientific Publ. Co, Amsterdam

ACKNOWLEDGEMENT

Dr. Van Vaeck L. thanks the National Foundation for Scientific Research (N.F.W.O.) for a grant as associate research chemist. (aangesteld navorser)

THE RATIONAL UTILISATION OF FUELS IN PRIVATE TRANSPORT(RUFIT)-
EXTRAPOLATION TO UNLEADED GASOLINE CASE, REPORT 8/80

CONCAWE, Oil Companies International Study Group for
Conservation of Clean Air and Water.
Europe, Den Haag, Netherlands.

ABSTRACT

The concept of optimum octane number assumes a car population
matching the available motor gasoline quality. Reducing the
permissible gasoline lead content without measures being taken
to optimise the RON requirement of the car population may
have two consequences :

First, crude oil consumption in the refinery will be increased.
For example, providing the existing car population with an
unleaded gasoline pool of 96 RON could lead to energy
penalties in the order of 8-10% compared to providing the same
quantity of 96 RON gasoline containing 0.4 gram lead/litre.

Secondly, it is important to note that requiring pool octane
numbers above 96 RON at 0.15 gram lead/litre and above 93 RON
for unleaded fuels would touch on the limits of technology for
some existing refinery configurations.

Thus, although it may be possible to manufacture these gasolines,
they could not be produced in the quantities necessary to
satisfy market demand without the construction of additional
secondary processing capacity.

It may therefore be concluded that if further restrictions on
gasoline lead content are considered desirable for environmental reasons, their introduction should be phased in with the
introduction of engines designed to minimise the inevitable
increase in total energy consumption by having octane requirements in line with the optimum RON described in this report.

D. Rondia et al. (eds.), Mobile Source Emissions Including Polycyclic Organic Species, 349–363.
© 1983 CONCAWE

INTRODUCTION

Since private transport currently consumes a significant part
(around 11%) of the total energy used in Western Europe, it is a
field which may be capable of yielding worthwhile savings in
monetary and crude oil utilization terms.

One of the most effective ways of reducing fuel consumption in
the private car is through engine design changes, but these may
also affect gasoline quality requirements. Measures directed
towards protection of the environment, i.e. emission control,
which involve design changes in engines or associated equipment
or reductions in the lead content of gasolines, may lead to
higher fuel consumption in the engine and also to the use of
more crude oil in the manufacture of the appropriate gasoline.
There is, therefore, a complex inter-relationship between
total energy economy, gasoline quality and emission control.

The Energy Directorate of the European Economic Comission (EEC)
requested the petroleum and automobile manufacturing industries
to conduct a study into these relationships to obtain a rational planning basis for any future legislative actions dealing with vehicle exhaust emissions and energy conservation.

In response to this request the CCMC (Comité des Constructeurs
du Marché Commun) issued a report (1) on the relationships
between vehicle fuel consumption and compression ratio/octane
requirement for vehicles meeting the 1975 EEC emission regulations.
At the same time CONCAWE (the oil companies' international
study group for Conservation of Clean Air and Water - Europe)
conducted a study to evaluate the crude oil requirements and
refining costs of manufacturing gasolines at different octane
quality and lead additive levels (2). CEC, the Co-ordinating
European Council for the development of performance tests for
lubricants and motor fuels assisted these studies in providing
the necessary interphase between the two organizations and in
combining the results of both groups in order to reach general
conclusions (3).

The original CONCAWE study (2) dealt only with the impa ct on
crude requirements and refining costs of reducing lead from
0.6 g/L to 0.4 g/L to 0.15 g/L. However, EEC subsequently asked
CONCAWE to extend the study to include an estimate of the impact
of unleaded gasoline on the total energy consumption of road
transport. The purpose of the current report is to describe the
basic methodology of these joint studies (1,2,3), the main results
concerning gasolines containing 0.6-0.15 g Pb/L, and the best
estimates obtained by extrapolating the refining study (2) to an
unleaded gasoline situation.

CURRENT EUROPEAN GASOLINE PATTERN/POOL

There are considerable differences in gasoline specifications and
patterns of usage between the various European countries. Permissible lead contents range from 0.15 g/L maximum in West Germany
through 0.635 g/L in Italy to 0.84 g/L in Greece. The Premium
Grade (about 98 RON min.) share of the market varies from some
55% in West Germany to about 94% in Italy.

An analysis of data on gasolines marketed in the EEC countries in
the period 1976/77 indicated a "leaded gasoline pool" which could
be closely represented by the following characteristics :

Research Octane Number (RON)	96.0
Motor Octane Number (MON)	86.0
RON 100° Dist. min.	87.0
Lead Content	0.4 g/L
Evap. at 100° C % vol.	50-65
Vapour Pressure Reid. max.	0.7 bar
Final Boiling Point °C	205

This specification was used as the Base Case for the CONCAWE
study (2) into the rational use of fuel in transport.

THE OPTIMUM OCTANE NUMBER CONCEPT.

The purpose of the original RUFIT study (2) was to look at the
gasoline engined motor vehicle and European refineries as a
single integrated system with the Research Octane Number (RON)
as the link, taking into account a range of possible restrictions
on the use of lead additives.

The concept starts with a constant number of miles which need to
be covered by the vehicle population and investigates how the
quantity of crude oil consumed varies when vehicles are optimized
to run on a single grade of gasoline but at different octane number
levels. For the purpose of the study, this constant mileage is that
which correctly matched vehicles would cover using 1,000 tons of
gasoline with a specification equal to the current European pool
described in 2 above, i.e. 96 RON with 0.4 g Pb/L.

The next step is to consider the rate at which vehicle gasoline
consumption varies when compression ratio, and consequently
octane requirement, is raised or lowered. In the CONWAWE study
(2) this is taken to be the "change in gasoline consumption (% wt)/
unit change in vehicle octane requirement (RON)" and is termed
the Car Efficiency Parameter (CEP). This is assumed to remain
constant over the octane requirement range 90-100 RON.

This is illustrated in the upper part of Fig. 1 for CEPs of 0.8 and 1.2% wt/RON, e.g. an engine with a CEP of 0.8 will require an additional 32 tons of gasoline to travel the constant number of miles if the compression ratio is lowered to reduce the octane requirement from 96 to 92 RON.

The next step is to consider the amount of crude oil required in the refinery to manufacture the gasoline using a lead level of 0.4 g/L. Two factors must be taken into account:

- the energy (crude oil) consumption of the various octane improvement processes such as catalytic reforming and cracking, isomerisation, alkylation, etc. necessary to achieve the required quality;

- the amount of gasoline required.

The lower pair of curves in Fig. 1 shows the variations in refinery fuel consumption, for manufacturing the quantities and qualities of gasoline required to provide constant mileage as shown in the upper set of curves. Finally, the sum of these two elements (vehicle and refinery fuel consumption) will give the total energy consumption in terms of crude oil for driving over the constant number of miles at each octane level. Since the two elements are moving in opposing directions they will generally show a minimum at some point, as can be seen in the central curve in Fig. 1. This minimum is termed the Optimum Octane Number.

Using this concept, companies participating in CONCAWE provided data utilizing models of typical European refineries, to calculate the total energy requirements, relative to the base case, over the range 90-98 RON for CEPs between 0.4 and 1.6% wt/RON.

Data were calculated for different types of refineries all having primary (crude oil) distilling capacities in the order of 5 million tonnes/annum, followed by various configurations of secondary up-grading facilities (reformer, catalytic cracker, alkylation, etc.); gasoline production was about 700,000 tons/annum in each case.

Although hydroskimming type refineries with only atmospheric distillation and catalytic reforming have been a significant factor in the European refining pattern in the pas, the trend towards a "lighter barrel" is expected to cause a progressive change with installation of more conversion facilities in the future.

In the case it was assumed that the existing refinery facilities were fully utilized. Crude slates and product packages were those actually existing in each individual refinery. In the variable cases the output of all products was kept constant

Fig. 1 Variation in Total Energy Consumption for Constant Mileage

Gasoline Lead Constant 0.4 g/L
Change in Gasoline Consumption 1.2 % wt/RON   ------
Change in Gasoline Consumption 0.8 % wt/RON   ———

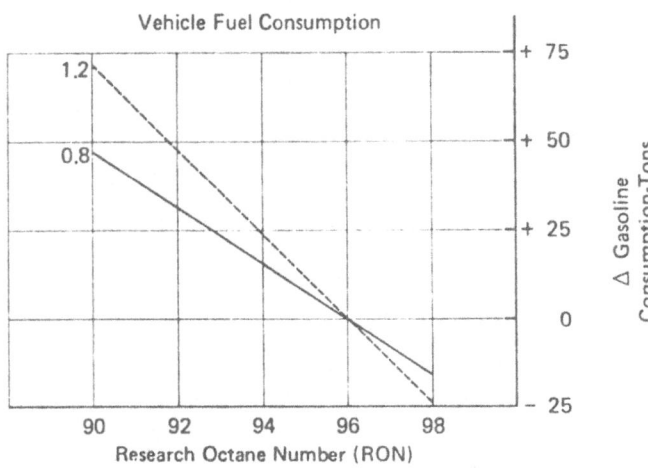

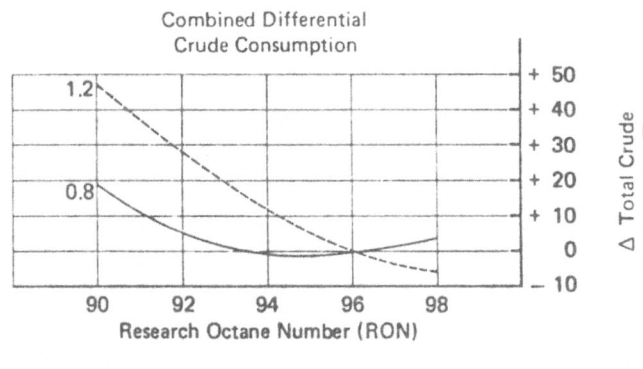

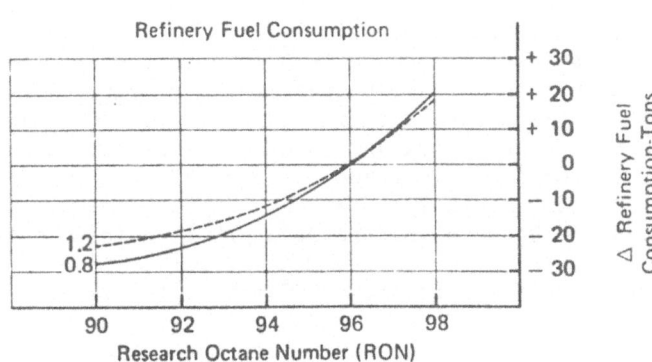

with the exception of motor gasoline production which could
vary to match the requirement for constant mileage at each
RON level.

Computer models were allowed to utilize "non-existing" primary
or secondary processing facilities if required to meet the specified product demand (both quantitative and qualitative). Any
incremental crude required by the models to meet the specified
output was introduced as Arabian light crude to enable an
easier comparison.

An annual capital charge of 25% of capital cost was specified
for any new process plant taken up in the calculation. All
costs were calculated in terms of 1976 U.S. dollars.

It should be noted that each model represented a specific set
of circumstances and that the changes in processing severity/
complexity were left to the discretion of the refiner concerned.
Consequently, the reproted results represented a range of real
circumstances and indicated that some refineries would be
better placed than others to make the necessary changes.

OPTIMUM OCTANE NUMBERS

Leaded cases

Table 1 shows the variations in optimum octane number with
change in vehicle fuel consumption/RON for CEPs between 0.4
and 1.6% wt/RON for gasolines with lead levels of 0.6,0.4 and
0.15 g/L.

This table shows that both the CEP and the permissible lead
content have a marked effect on the value of the optimum octane
number.

Following publication of the CCMC report (1) it was possible
for CEC to narrow the range of CEPs applicable to the bulk of
current European vehicles (see Table 2). Their analysis of
CCMC data (3) showed that the bulk of the engines tested had
CEPs (or in CEC terminology, Engine Response Factors) in a
band 0.7-1.2% wt/RON with a median value of 1.0. CEC also
reassessed the CONCAWE data and concluded that the optimum
octane number for a single grade of gasoline which minimizes
crude consumption is likely to be between 95.3 and 97.8 at
0.4 g/L lead and 94.6 to 96.5 at 0.15 g/L lead (see Table 2)

The following quotation from the CEC report (3) is worthy of
note :

Table 1   Variations in Optimum Octane Number with Change in Vehicle Fuel Consumption/RON (Ref. 3)

Assumed Change in Vehicle Gasoline Comsumption		Gasoline Lead Content gram/litre		
(CEP)		0.6	0.4	0.15
% wt/RON*	% vol/RON**	Optimum Gasoline RON for minimum crude oil consumption		
0.4	(0.67)	95.5	94.0	93.5
0.6	(0.90)	96.2	94.7	94.0
0.8	(1.10)	96.9	95.4	94.5
1.0	(1.32)	97.6	96.1	95.0
1.2	(1.55)	>98.0	96.8	95.5
1.4	(1.76)	>98.0	97.5	96.0
1.6	(1.98)	>98.0	>98.0	96.5

* % wt/RON: change in gasoline consumption (% wt)/unit change in vehicle octane requirement (RON)

** % vol/RON: change in gasoline consumption (% vol)/unit change in vehicle octane requirement (RON)

Table 2   CEC Analysis of CCMC/CONCAWE Data (Ref. 3)
Optimum Octane Numbers for Minimum Overall Crude Use

Engine Response Factor (CEP) (% wt/RON)	Lead Level (g/L)	
	0.4	0.15
0.7	95.3 RON	94.6 RON
1.2	97.8 RON	96.5 RON

"Understanding the meaning of engine response factors is
extremely important to the concept of this study. High engine
response factors do not necessarily mean high efficiencies
in the gasoline engine. All the response factor tells us is
the way in which increasing compression ratio affects both
fuel consumption and octane requirement for a particular engine
design. The point of the CCMC study was to measure this
parameter for 1976 cars tuned to meet ECE 15.01 standards.
Cars tuned to meet different emission standards may not show
the same response factor and hence their appreciation of high
octane gasolines may be different. Thus, engine response work
should be updated in the light of changing emissions standards"

UNLEADED CASE

All the refining data discussed so far were obtained form the
refinery models.

To determine the consequences of reducing the lead content from
0.15 g/L to zero the results of optimization at 0.15 g Pb/L were
extrapolated to an optimum octane number for unleaded gasoline.
The results of this extrapolation are as follows :

Fig.2 shows the lead response, i.e. the increase in Research
and Motor Octane Numbers, obtained when different concentrations
of either TEL or TLM are added to various typical gasoline
components.

For the weighted average, the lead response of the components
comprising the European motor gasoline pool is about 3 RON for
0.15 g Pb/L.

Fig. 3 shows the effect on energy consumption of reducing the
lead content of gasoline whilst maintaining constant gasoline
consumption/production (i.e. 1,000 tons). This corresponds to a
situation where permissible gasoline lead levels are restricted
without any measures to reduce the octane requirements of the
vehicle population. The curves for 0.60 , 0.40 and 0.15 Pb/L are
taken from the original CONCAWE report (2). If the lead is not
added to the latter case, than the curve will be displaced to
the left by 3 RON.

It should be noted that the 0.15 g/L and zero lead curves are
dotted above 96 and 93 RON respectively. This is to indicate that
certain refinery configurations, particularly hydroskimmers, may
be either incapable of reaching the required quality or can
achieve it only at a reduced output; it therefore indicates a
potential supply problem.

Based on the above assumption, reducing lead from 0.15 g/L to

Fig. 2   Lead Response of Typical Gasoline Components

FRSC : Full-Range Steam Cracked Gasoline
FRCC : Full-Range Catalytically Cracked Gasoline
FRCR : Full-Range Catalytic Reformate
SRG   : Straight Run Gasoline
SRB   : Straight Run Benzine

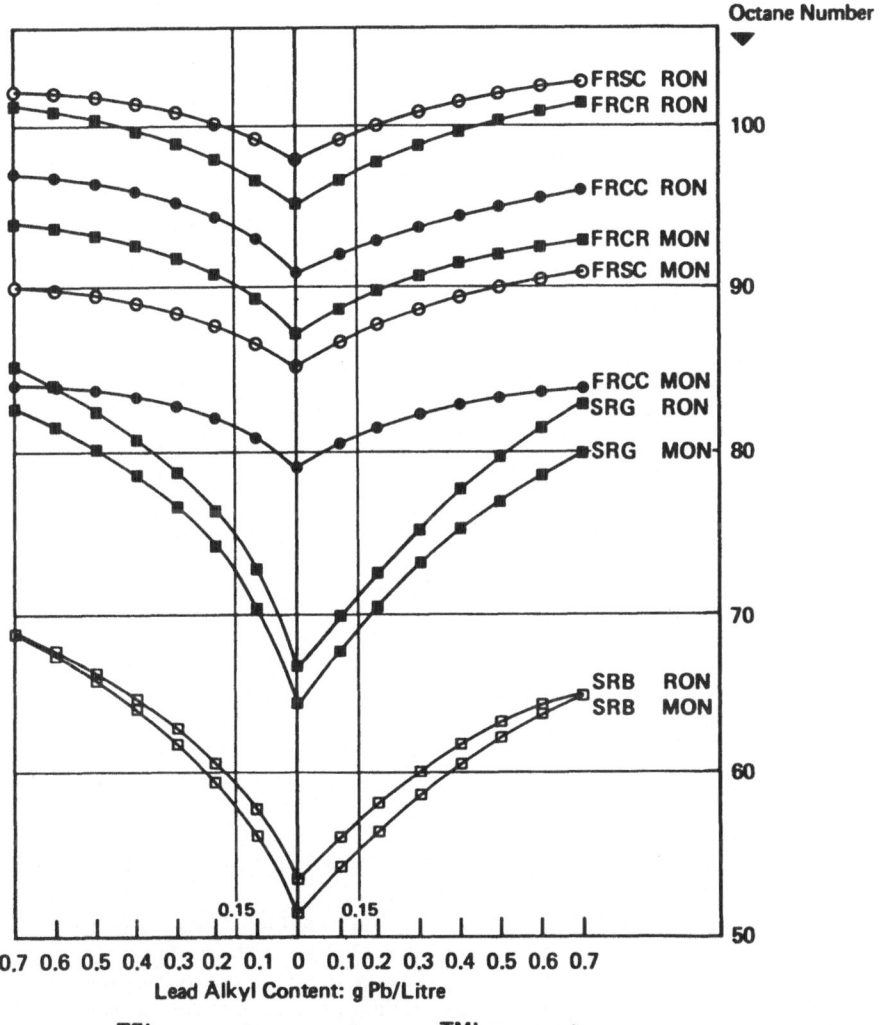

Fig. 3 Conversion Refineries/Refinery Fuel Consumption

Incremental Crude vs. RON: Constant Gasoline Consumption
(Related to Base Case: 1000 ton, 96 RON at 0.4 g Pb/L)

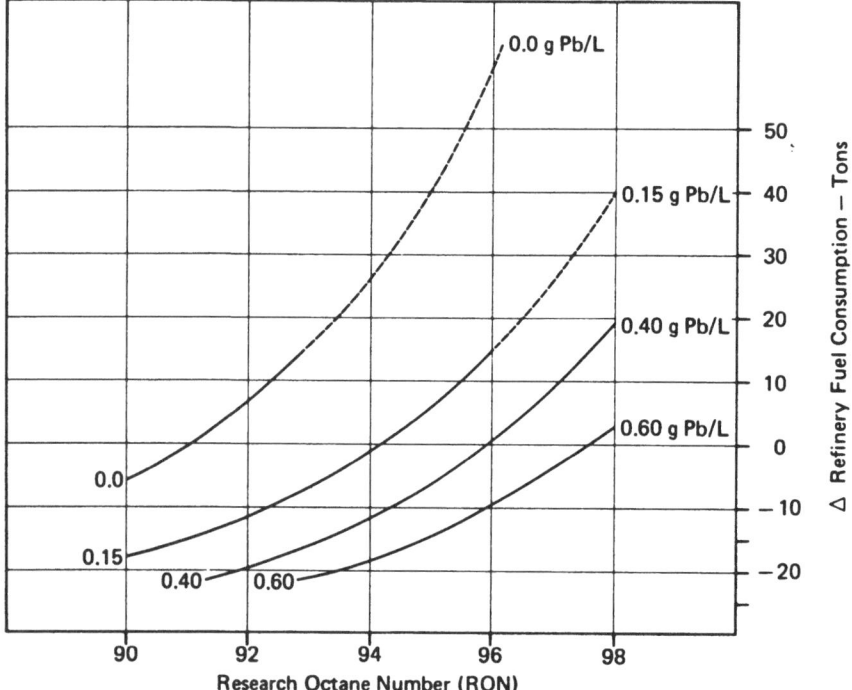

zero results in either a loss of 3 RON or an energy debit, in terms of refinery fuel, equal to between 1 and 5% of the gasoline produced, depending on the level of RON. However, this does not take account of the fuel consumption characteristics of the vehicles and, as has been shown, the amount of gasoline required to cover the constant mileage at a given octane level depends on the CEP.

Fig. 4 illustrates the complete evaluation of the unleaded case.

The (black) curve designated 0.15 g Pb/L is the total energy consumption for both the refinery and the vehicle – the vehicle having a CEP of 1.0% wt/RON, i.e. the median value estimated by CEC for current European vehicles. The optimum octane number determined by CONCAWE (2) at 0.15 g Pb/L is about 95 RON and without the lead this falls to 92 RON (dotted curve):

Fig. 4 Developments of Optimum Octane Numbers for Unleaded Gasoline

All Data Relative to Base Case: 1000 ton Gasoline, 96 RON, 0.4 g Pb/L
Car Efficiency Parameter: 1.0

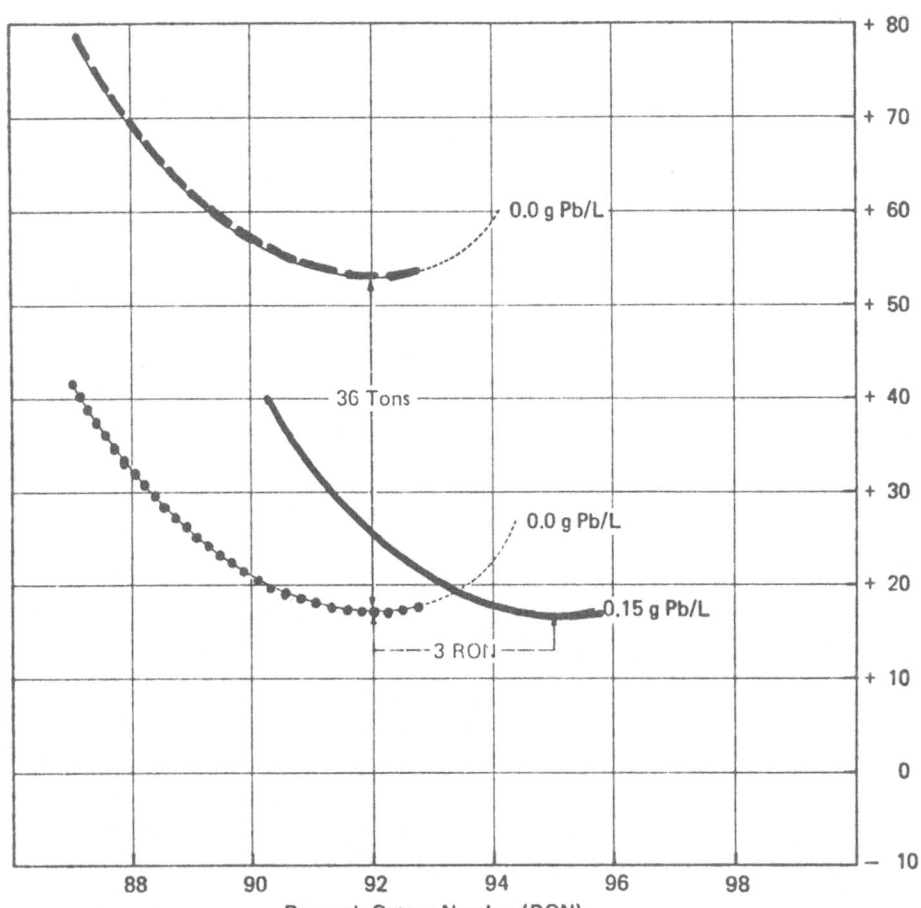

Research Octane Number (RON)

- However, a vehicle which operates just knock-free at 95 RON
  and which has a CEP = 1.0, will require an additional 30 tons
  of gasoline to cover the same mileage when tuned to operate
  knock-free at 92 RON; and,

- to manufacture this additional quantity of gasoline, the
  refinery will also consume additional energy. This will vary
  depending on the octane number and the type of refinery. From
  the CEC report (3) it is indicated that a factor of 1.2 is a
  reasonable average value for calculating the energy demand
  for manufacturing motor gasoline. That is, 1.2 tons of equivalent crude oil are required to produce 1 ton of gasoline.

Applying this factor to the case in <u>Fig.4</u> (dashed curve) shows
that even under optimized conditions, the total energy consumption for constant mileage increases by 36 tons if we move from
an optimized situation at 0.15 g Pb/L to an unleaded gasoline
situation.

Fig. 5   The Effect of Lead Content On Crude Consumption/Optimum Octane Number

Base Case: 1000 ton Gasoline, 96 RON, 0.4 g Pb/L
Car Efficiency Parameter: 1.0

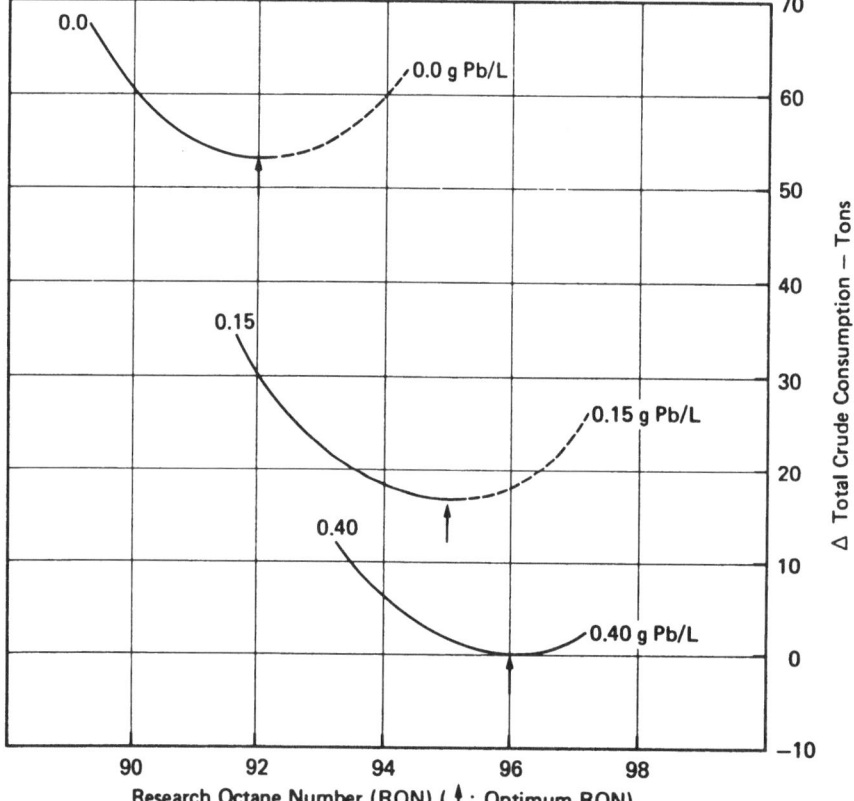

In <u>Fig.5</u> this extrapolation has been applied to the complete
results of the CONCAWE study (2). The figure represents the
total crude oil consumption in typical European conversion
refineries to produce gasolines of various RONs and lead contents for constant mileage in vehicles with a CEP = 1.0

For this CEP, the optimum octane number at a lead level of 0.4
g/L is about 96 RON. Reducing the lead level to 0.15 g/L lowers

the optimum RON to 95 and incurs an energy debit of nearly 2%
relative to the gasoline consumed in the base case. Again it is
indicated that requiring pool octane numbers of above 96 RON for
the 0.15 g Pb/L case would touch on the limits of refinery technology for a number of refinery configurations. This is not to
say that it is impossible to manufacture such gasoline at all
but that it is not possible to manufacture such gasoline in the
assumed quantities without substantial investments.

In this unleaded case, the optimum Research Octane Number for
conversion refineries would drop to about 92 RON and, compared
to the base case, an energy penalty of more thatn 5% would be
involved.

ENERGY

It has been shown that the total energy consumption in terms of
crude oil of motoring can be influenced by vehicle characteristics and the permissible lead content of gasoline.

Analysis of the data (assuming a CEP of 1.0) shows increased
energy requirements in terms of crude oil equivalent, <u>at</u>
<u>optimum RONs</u>, as follows :

   <u>at 0.15 g Pb/L</u>, an increase of more than 1.6%
                       relative to the base case.

   <u>at zero lead</u> content, an increase of over 5.0%
                       relative to the base case.

The effects on additional crude oil consumption incurred by
providing the quantities/qualities of gasoline required by
vehicle populations designed to operate efficiently at octane
levels other than the optima defined by this study can be
deduced from the increasing steepness of the curves in <u>Fig.5</u>
For example, providing an unleaded gasoline pool of 96 <u>RON</u> for
the existing car population could lead to energy penalties in
the order of 8-10%. Even more important would be the inability
of the refining industry to provide the unleaded gasoline in
the necessary quantities.

Compared to lead reductions, the effect of CEP on energy consumption, in the range identified by CEC as representing the bulk
of present European vehicles, is relatively small.

However, it should be noted that no information is available at
present about the CEP of engines operating on unleaded gasolines
or at more severe exhaust emission limits.

COST

The CONCAWE RUFIT study (2) did include the determination of
additional investment requirements and operating costs. The
figures were not reported by they were found to be more or
less the same, after inflation allowances, to those previously
published by CONCAWE (4,5).

One important fact derived from the RUFIT study (2), however,
is that the optimum octane number for the minimum cost of motoring is 1-2 RON below the minimum for energy conservation. This
would indicate that minimum driving cost would be achieved at
about 91 RON.

APPLICATION

In applying the findings of these studies, several additional
factors should be taken into account.

Severe reductions in the lead content of gasolines for use in
existing vehicles can lead only to considerable debits in the
total energy consumption of road transport.

The supply of low lead/low octane gasolines at optimized levels
should be phased progressively to match the requirements of
new vehicles able to use them efficiently.

In selecting optimum gasoline specifications at low lead levels
it should be borne in mind that the results described above
assume a constant difference between RON and MON (Motor Octane
Number), namely, asensitivity of 10. Recent trends in the
refining industry towards increased conversion make this
sensitivity more and more difficult to achieve. At the same time
the relative importance of the MON for satisfactory vehicle
performance is increasing. It is recommended that a value of
10 for sensitivity should be considered a minimum for unleaded
gasolines to avoid an unnecessary wastage of crude oil. Further,
an adequate RON 100°C value for the gasoline is considered
essential to avoid knock under accelerating conditions and this
characteristic is also more difficult to achieve for an unleaded
gasoline. These points indicate that the optimum unleaded octane
number shown by the CONCAWE/CEC study (2,3) may be optimistic.
It would be prudent to select an unleaded octane number in the
order of 91-92 RON to provide some leeway for a number of
factors which have not yet been evaluated.

Finally, all results of this study have been developed assuming
a single gasoline grade. If engine manufacturers are to match a
given octane number and to utilize fully the potential offered
by this octane number, then they must design their engines so

that all of them require this defined octane number. Experience has shown that this is not possible. On the one hand, production tolerances lead to a spread of octane requirements for individual cars of a given model. On the other hand, during the first few thousand kilometres of use, octane requirement increases until it stabilizes at a level which can vary considerably for individual cars.

Under these conditions a number of cars will have lower octane requirements and, therefore, not justify the energy used to manufacture a higher RON gasoline. Greater energy economies may be achieved by an additional grade of lower octane quality.

REFERENCES

1. Committee of Common Market Automobile Constructors (CCMC) : Study on the Rational Use of Energy "Engine-Fuel Matching". Report on the Experimental Programme, Phase I (final results), March 1977

2. CONCAWE Report N° 6/78 : "A study on the rational utilization of fuels in private transport (RUFIT)"

3. CEC Report : "The rational use of energy in transportation : interpretation and integration of the findings of CCMC and CONCAWE"

4. CONCAWE Report N° 2/72 (with Supplement) : "The problem of gasoline engine exhaust control"

5. CONCAWE Report N° 10/77 : "Automotive emission regulations and their impact on refinery operations".

SUMMARY CHAPTER - NATO ADVANCED RESEARCH WORKSHOP - MOBILE
SOURCE EMISSIONS INCLUDING POLYCYCLIC ORGANIC SPECIES

D. Rondia, M. Cooke[*], and R. Haroz[**]

Université de Liège, Laboratoire de Toxicologie
Industrielle, B-4000 Liège, Belgium
*Battelle Memorial Institute, 505 King Avenue,
Columbus, Ohio 43201, USA
**Battelle Memorial Institute, Centres de Recherche
de Genève, 7 Route de Drize, 1227 Carouge (Genève),
Switzerland

INTRODUCTION

The publications in this book and the discussions which were
an important part of their presentation, represent documentation
of an international meeting on mobile source emissions held
August 30 to September 2, 1982, in Liège, Belgium. The meeting
was an Advanced Research Workshop sponsored by the NATO Scientific
Affairs Office, the Belgian Regional Ministry for the Environment,
and the Battelle Memorial Institute. The 37 participants arrived
from 13 countries and represented many of the current research
programs relating mobile source emissions to engine type and
operation. Several participants were also studying the effects
of fuel and relating fuel type, including gasoline and diesel
(with and without additives and blending agents), to engine
performance.

This summary chapter is designed to review the principal
ideas discussed and give the specific recommendations that came
from the technical discussions at the meeting. This chapter is
organized into four sections: an Introduction; a Summary which
includes a technical review of the formal papers; Specific
Recommendations which were developed at the meeting and reviewed
in written form by the participants; and, an Overview which
summarizes the public health effects of the data presented.

D. Rondia et al. (eds.), Mobile Source Emissions Including Polycyclic Organic Species, 365–379.
© 1983 by D. Reidel Publishing Company.

SUMMARY OF PAPERS PRESENTED AT THE MEETING

   The summary presents a review of the technical data presented
at the meeting. This summary is organized into six separate
sections representing the topical sessions held during the meeting:
Basic Studies, Health Effects, Analytical Methods, Data Bases,
Engineering Studies, and International Programs. The sequence of
these sessions led from fundamental research, through discussions
of occurrence and effects of chemical species, to a discussion of
engineering approaches toward optimizing engine performance. This
sequence served well to carry the discussions of formal papers in
an ordered manner and easily led to the development of general
recommendations on mobile source emission research.

Basic Studies

   This set of papers dealt with the fundamental processes by
which polycyclic organic species and soot products are produced
during the internal combustion processes and the transformations
that occur in discharges to the atmosphere. A mass balance for
polycyclic aromatic hydrocarbons (PAH) was presented which showed
the relative impacts of PAH species from several sources. Polycyclic organic matter as an atmospheric pollutant results from
such a wide variety of emissions that the calculation of emission
factors per process, per fuel, or per country present many
difficulties and is generally the result of large extrapolations.
Consequently, most figures have a relative rather than absolute
value. Keeping this in mind, it was estimated that in the general
urban environment, polycyclic aromatic hydrocarbons from automobile and road transport sources represent 4 to 20 percent of the
total amount of PAH emitted.

   Most reported data are based on total polycyclic aromatic
hydrocarbons, not benzo(a)pyrene alone, and usually neglect the
bulk of accompanying substances, and the soot emitted. The
importance of the biological properties of the whole has, until
now, been nearly neglected.

   In summing the relative contributions of various mobile and
stationary sources to the total atmospheric burden of PAH, one
group of researchers emphasized the importance of combustion
temperature, reaction time, and degree of turbulence in the
combustion processes studied. These factors can all affect the
amount of PAH produced. For instance, aluminum smelters are an
important stationary source of PAH due to an optimum temperature
for production of PAH, reducing conditions in the vicinity of
carbon electrodes refining the metal, and the contained nature
of the reactions involved. PAH balances were given for several
countries. Transportation sources accounted for a much as 20%
of the PAH burden. It was stressed that the changes occurring

in new engine designs, displacement, fuel composition, and
driving habits, cause these estimates of relative PAH contribution to be subject to periodic reappraisal. Exact estimates of
PAH burdens are often confounded by difficulties in comparing
data from diverse studies. This research group made a strong
plea for standardization of reporting units so that accurate
global budgets can be made.

In this session, world-wide, long-term decreases in average
ambient PAH levels, decreases in average sulfur oxide concentration,
and decreases in anthropogenic particulate levels were noted. This
trend is not followed by average ambient levels of nitrogen oxides,
which seem to track the increased energy consumption found in
most industrial countries. These general effects were seen to
develop quickly in the Netherlands during the 1970s when the
discovery of low cost natural gas led to a rapid changeover to
this clean fuel. A sharp drop in PAH, sulfur oxides, and ambient
particulate levels was observed in a short time frame. Holland
thus has an interesting test population for data accumulation.
In fact, several studies are being performed in the Netherlands
to track indicator species such as benzo(a)pyrene with specific
correlation to exposure of local populations. Belgium has
performed extensive studies of population exposure on a regional,
and source specific basis, using benzo(a)pyrene as the indicator
compound. These data are very difficult to interpret in public
health terms, however, because synergistic human exposures
cannot be separated. People are exposed through several
environmental media. Although the use of indicator species is
helpful, several species are present in the exhaust gases of
mobile sources which pose a spectrum of health risks. The
epidemiological data base is very limited and a great deal of
additional data on populations, chemical species, and combinations
of species will be necessary before definitive comparisons can
be made.

Emissions of PAH have been extensively studied and show, in
many places of the world, a continuous decrease over the last
fifteen years as a result of better fuel burning and cleaner
burning fuels (oil and gas instead of coal, for example). However,
wood burning, which is becoming current practice in several wood-
rich developed countries, has generated a net increase in the PAH
burdens in these areas.

Basic studies on flame chemistry revealed that PAH species
are produced very rapidly (millisecond processes). A basic
formation mechanism was proposed in which benzene molecules are
attacked by small, unsaturated radical species. The necessary
conditions for PAH production are the presence of benzene and a
high concentration of the radical intermediates which then form
stable adducts. Multiple ring systems are autocatalytic and

promote further ring condensation. The data presented were taken
from studies on aromatic fuels in low pressure, laminar flames.
The same processes occur in aliphatic fuels, but the rate determining step induces production of the first benzene ring rather
than a multiple ring system. Thus, PAH production is generally
slower with aliphatic fuels. Since PAH species occur just before
sooting in the systems studied, the processes of condensation
and inclusion within soot particles, work to reduce the free
concentrations of PAH species present in post-flame gas-phase
discharges.

The knowledge of flame kinetics has made much progress
recently thanks to the elaborate analytical methods now available.
Kinetics of aromatic fuel combustion were compared to that of
aliphatic fuels. Aromatic fuels will normally promote the formation of PAH. Formation of polycyclics in flames is an intermediate
step in soot formation. PAH are normally burned out or transformed into soot. On the other hand, demethylation of higher
benzene homologues in the internal combustion motor gives rise
to high concentrations of benzene in the exhaust. Much of the
benzene in residential air could originate either in such
demethylations, in synthesis from engine reactions, or through
discharge of unburned fuel.

The formation of other gaseous compounds (e.g., formaldehyde,
acrolein, aromatic aldehydes) in the motor and their concentration
in air has not been discussed, even after demonstration of the
mutagenic activity of the gas phase of engine exhaust. Also, the
biological activity of gaseous products is seldom studied (except
in inhalation studies where results are not yet known) and the
interpretation of such data differs among the reporting authors.

Very little consideration has been given until now to the
fate of PAH in the air. The nature and concentration of their
transformation derivatives is practically unknown. Oxidation
derivatives of PAH are a complex mixture of different compounds
and probably generate toxification as well as detoxification
processes. PAH emission levels, if regulated in the future,
should be assessed in terms of possible hazardous conversion
products.

Research presented in this session also noted that once PAH
are discharged from the exhaust system, the compounds are subject
to a number of chemical reactions which alter the composition and
biological activity of the species emitted. These transformations
are especially difficult to characterize because they can occur
during sampling either with filtration or on solid sorbent samplers.
Artifact formation and substrate degradation can lead to false

high and low results in chemical analyses. Ambient sunlight induced photochemical reactions, ozonolysis, nitration, and sulfonation are a few examples of the possible reactions that occur with ambient samples.

Information presented also indicated that several catalytic reactions are possible on particulate matter, and that future studies should include better data on size and chemical characteristics of particulate matter. In fact, mobile discharges have a high moisture content, reactive acid anhydrides (e.g., NO_2) catalytic-adsorptive particulate matter, and contain reactive PAH species. Future research must deal with this reactive matrix and better identify the products of combustion and the reactions induced by sampling the emission stream.

Health Effects

This meeting brought together individuals from a variety of organizations having an interest in mobile and stationary source emissions. Various aspects of the generation, processing, and assessment of engine emission exhausts were presented and debated. It was agreed that any potential health risks, as assessed by experimental approaches available today, must be considered not only by assessing the biological activity of the final sample but also by acknowledging the effects attributable to generation, collection, extraction, and storage methods.

The generation of the sample is dependent on the engine type, operating conditions, applied load, and the test cycle. Any comparative judgements must be drawn from standardized tests. Changes in the fuel composition, e.g., lead reduction and increased aromatics content, may have resultant health, environmental and economic consequences. Catalysts also change the character of emissions, generally by a reduction of many components. However, these new substances could introduce other problems which need to be assessed. The evolution of engine types and modified fuels should be assessed biologically in order to minimize the possibility of introducing new health hazards.

Reactions in the exhaust system can easily modify the biological properties of emissions. Changes in sampling methods and conditions can also result in changed biological properties. In addition, subsequent biological effects can be modified by conditions of collection, including flow rates employed, the duration of collection, the type of filter or other techniques (e.g., electrostatic precipitation) used, and by gases flowing through material trapped on the filter. Also, it was noted that the method of extraction and particularly the choice of solvents could affect biological activity of the final sample. The

specific activity of the final sample is usually based on extractable organics and it was clear that the quantity and chemical
makeup of extracted material was dependent on the polarity of the
solvent or solvents employed.

Analytical Methods

    As a majority of the participants in the workshop were
familiar with analytical work in the PAH field, this topic brought
significant conclusions to the meeting. Some analytical problems
were so demanding that a subcommittee on standardization of
analytical methods and units was formed.

    The primary conclusion of the session on PAH analysis was
that the results of older and analytical methods are generally
confirmed by the present advanced analytical techniques. The
older sampling methods, however, were recognized as deficient
techniques in many aspects.

    Most of the new analytical methods, like low temperature
fluorescence (Shpolskii effect), or MS/MS have been developed in
research laboratories but are mainly used at the qualitative
stage, especially for the study of oxygenated or transient
derivatives of PAH and for the identification of isomers. As
quantitative tools, some progress is still necessary before many
of these new methods can be made available and usable by survey
laboratories.

    Each separation method has benefits and drawbacks. HPLC
looks promising for the separation of high molecular weight
substances and metabolites, but suffers a lack of resolution.
GLC is affected by the thermolability of some PAH and their
derivatives. LC/MS interfaces still present many difficulties
but micro-HPLC/MS looks promising. TLC was used in some work
presented at this meeting but incomplete recovery and possible
oxidation hampers use of this method in quantitative work.

    The extraction of PAH and their derivatives from soot by
analytical solvents was noted as a fundamental concern. The
need exists to gain a better knowledge of the elution by natural
solvents like blood, plasma, albumin, and lipoprotein solutions.

    Other discussions concerned the artifacts produced by sampling
methods, especially the nitration of PAH from diesel exhaust on
filter media or in the exhaust muffler. Reaction of Tenax or
XAD-2 adsorbents with inorganic pollutants like SO_2 or NO_x gives
rise to other problems by increasing the chemical background
through polymer decomposition. In long-term open air sampling,
bias often results from slow equilibration between gas-phase and
already-sampled particulates especially in filter only sampling

Another important point underlined by participants is the need for measuring a large number of PAH rather than just benzo(a)-pyrene (BaP) alone, to characterize an emission or a situation. The reactivity of BaP to light and oxygen makes this compound a poor standard. The sum of 15 to 20 PAH would be a better profile, but incomplete still, as has been shown by biological activity (through the Ames test). A very strong statement was made for agreement on which compounds to study. A suitable reference list would promote comparability of results and correlation of biological effects.

New methodologies are continuously being applied to PAH analysis, yielding increased sensitivity and specificity. At the same time, biological assessment by short-term assays is experiencing a parallel increase in sensitivity and useful applications. Bioassay gives a meaningful complement to chemical analysis.

Data Bases

Presentations and discussion in this session fully recognized that data bases on PAH either do not exist or do not offer sufficient quality. Calculation of emission factors was difficult because of a lack of homogeneous data. Risk assessment of the use of diesel fuel in transportation is nearly impossible in the U.S. given the paucity of epidemiological data related to PAH and their derivatives. In the same way, the basic question formulated at the beginning of the workshop; whether polycyclics increased in exhaust as a function of the concentration of aromatics in the fuel, remains controversial. Standardization of sampling and analysis data and units is necessary before data bases can be built.

Much of the inadequacy in data bases is related to the conditions of the assays on motors burning modified fuels or synthetic fuels without adequate tuning of the motor, yielding a totally unrealistic model for the use of such fuels. The role of soot deposits in the cylinder seems very important, hence the necessity of very careful conditioning of the motor and vehicle after each sequence of fuels.

Engineering Research Programs

A primary objective of the automobile industry is to relate emissions to classical air pollutants and optimize motor performance while insuring the most efficient or economic use of the fuel used. Most investigations of unregulated pollutants like PAH were designed with experiments that evaluated motor performance. A fundamental difference in perspective exists between the automobile industry and the classical analyst or biochemist. For example, the high efficiency of the noble metal catalysts, even in the presence of lead, cannot be evaluated by only examining differences in results obtained with and without fuel components, but must be accompanied by a careful system balance due to possible back-pressure in the muffler. Thus, automotive engineers made a very valuable contribution to the meeting by adding a perspective on the total system.

Some PAH problems arising from diesel cars in the U.S. could be caused by a lack of homogeneity or specifications of diesel fuel and the associated cetane number. This situation is also likely to be important in the future since the quality and origin of crudes could vary widely as diverse sources are used to meet near-term demands. Distillation schemes will be adapted according to specific demands, but the cracking or conversion of residual fuel will lead to products richer in aromatics and olefins which may influence PAH, hydrocarbon, and soot emissions.

Despite present predictions for the number of diesel-powered cars in Europe to increase markedly, discussions on the engineering approach to the control of automobiel emissions have centered on the gasoline motor. Cylinder wall temperature, air-fuel ratios, exhaust gas catalysts, and lead traps were presented and discussed as remedies. Blended fuel, for instance gasohol or synthetic fuels, were not considered because they are used on a local basis or during transient, difficult conditions. The contribution of lubricating oil in PAH emissions exists but was not considered an important issue, and was not discussed in the meeting.

Sampling methods and driving cycle procedures were examined Two methods of sampling (cryogenic condensation and dilution tunnel) and three driving cycles were used as reference methods for comparability of results. Concern was expressed on the choice of appropriate test vehicles, and motor tuning when using experimental, or laboratory-modified fuels. It is necessary to copy, as closely as possible, the normal use of such fuels in the automobile. The test cycle and vehicle tuning parameters are of primary importance and a need exists for engineering standardization, as well as for analytical method and sampling standardization. This could lead to the definition of a standard fuel.

It was concluded that any remedy to regulated automobile pollutants (hydrocarbons, carbon monoxide) would decrease the polynyclear aromatic hydrocarbons.

International Research Programs

The last session of the meeting discussed the need for international research programs in the field of PAH. Besides the stated need for sampling, analysis and engineering procedures standardization, no special topic was estimated an emergency in the fields surveyed. We need more data before answering questions on the dangers related to a possible increase in POM (as an unavoidable consequence of decreased lead antiknock additives) and on the use of diesel fuel. Presently, no standard exists for PAH in ambient air in any country. There is a TLV for the benzene soluble fraction of coke oven smoke in factories, and a standard exists for PAH in drinking water, but these conditions are not relevant to the case of automobile exhaust.

The World Health Organization through the International Agency for Research on Cancer evaluated POM-risk in 1973 and will hold at least three meetings in 1983 on health effects of polynuclear aromatic hydrocarbons. These meetings will not be concerned specifically with mobile source emissions. The European Community has not issued regulations and is not considering issuing any in the near term on PAH-POM from vehicle exhausts. The EEC has recommended a continuous survey of lead body burden in the populations of member countries.

SPECIFIC RECOMMENDATIONS FOR FURTHER RESEARCH

1. A great need exists for PAH-POM sampling and analysis standardization and validation. Progress is continuing in this field and should be encouraged. Composition of the exhaust gas phase and examination of the role of the particulate phase needs much investigation.

2. The role of transportation sources in the emissions of PAH-POM is not known with accuracy, especially as biologically active POM is concerned. One participating research group cites mobile sources in Scandinavia as contributing 6 to 9% of the total atmospheric PAH, and around 24% in the U.S., although the data base used for the U.S. was not corrected for the use of catalytic convertors. Such estimates should be measured for the local urban environment rather than a national or continental scale which is subject to great variances.

3. The addition of lead additives to a given gasoline does not appear in itself to affect PAH emissions, providing the hydrocarbon composition does not change.

Generally, aromatic blending agents are
used as a replacement for lead additives
which can influence PAH emissions. The
point remains controversial but it was
generally accepted that high levels of
aromatics in the fuel increases PAH at
the tailpipe.

4. Biological assessment of the direct and
    indirect mutagenicity of PAH is well
    characterized and is a valuable tool
    for studying emissions. The mutagenicity
    of POM however, is much more complex and
    could involve other enzymes and other
    metabolites. Divergent results in this
    field show the need for more knowledge
    on POM composition and POM-metabolic
    activation, deactivation or chemical
    decomposition processes.

5. In estimating the biological properties
    of POM, two fields are nearly completely
    neglected and need emphasis; i.e., the
    biological fate of PAH adsorbed on inhaled
    airborne particulates, and the fate of PAH-
    POM in the chemically and photochemically
    reactive atmospheric environment.

6. After having improved our knowledge in the
    above points, we will need to include the
    magnitude of emission impacts due to
    increase (or decrease) of mobile sources,
    types (gasoline and diesel powered
    vehicles, exhaust catalysts), fuel
    availability and composition, and to
    develop source impact models capable
    of evaluating health impacts on target
    populations.

OVERVIEW

Hated, tolerated or adored, the automobile has become an
important component of our daily lives. The amount and severity
of mobile source exhaust components on urban air pollution is
difficult to measure with accuracy for significant changes in the
last ten or fifteen years have occurred as a result of various
factors such as motor types, fuels, and catalysts.

In the last five years, catalysts have been used in the U.S.,
Canada, and Japan, and have resulted in significant decreases in
hydrocarbon and carbon monoxide emissions. Use of exhaust catalysts
requires unleaded gasoline which greatly reduced lead emissions
in these three countries. In Europe, a lowering of lead emissions
was obtained by a regulatory decrease of lead antiknocking agents
in gasoline, from levels of 0.70 g/l or sometimes more, to maximum
levels of 0.40 g/l imposed by most Common Market countries.

Decreasing the concentration of antiknocking agents induces
changes in fuel composition in order to keep a suitable octane
number. Such changes lead to a modified exhaust composition.
Very often the decrease in lead-alkyls has been compensated by
an increase of the aromatic fraction in the fuel, resulting
possibly in an increase in the emission of polycyclic organic
matter, either in gaseous or particulate phases.

A third modification of exhaust gases composition is a
consequence of the energy or petroleum crisis, fuel prices, and
its increasing importance in family or state budgets. These
new circumstances caused a slow but already marked increase in
the use of diesel fuel in road transportation, especially in
the personal car. The consequences of this trend on the composition of mobile source exhaust are multiple. Emissions of soot,
nitrogen oxides, polycyclic organic matter and odorous substances
are significant. It has already been suggested that diesel and
jet exhaust in Los Angeles, from vehicles burning 5% of the total
fuel, are responsible for 62% of the elemental carbon present in
atmospheric suspended matter although contributing only 25% of
the total carbon (CASS et al., 1980). The problem of air pollution
from urban traffic could thus present a very different pattern in
a few years.

A last factor which may lead to a change in the nature or
the amount of exhaust gases from transportation could arise from
the modifications of fuels, as imposed by the types and quantities
of crude oil available on the world market, or as a consequence
of the replacement of a portion of gasoline by other fuels offered
by the chemical industry, e.g., alcohols or liquefied coal derivatives. Some data presented at the meeting suggested that alcohol
blending would reduce harmful emissions. Synthetic fuel emissions
are very different than existing conventional fuel emissions
supplies and require extensive study. The automobile will remain
as an indispensable need in our society but the characteristics
of emissions are subject to slow changes due to a number of
factors that affect engine design, fuel, and control technology.

In fact, not only are the emissions changing but also our
knowledge about their chemistry, their toxicity, and their effects
in general. But as in many environmental problems we have to

deal with two opposite aspects of knowledge: on one side, the scientific knowledge that prompted us to gather in Liège and discuss laboratory results or views on the situation. On the other side, we find public pressure and appraisal of the same problem. The power of the public can be of tremendous weight as seen in public decisions, and in the generation of new regulations. The Bureau of the European Consumers Union (BECU) together with some national committees (like CLEAR in Great Britain, Campaign for Lead Free Air), with the help of the media, are presently launching a campaign for running all automobiles with lead-free gasoline as soon as possible and in any case from 1985 on.

On the other side, we have the work of SINN (1981) in Frankfurt which shows that in a few years, the decrease of respirable lead has reached 75 percent in the street studied, while the blood lead among persons living in the same streets decreased about 15 percent in the same interval. Another study, partly sponsored by the EED in Italy, evaluated the relationship between blood lead and leaded fuel. This three year study was based on the isotopic ratios of lead antiknocking ingredients. The program attempted to measure the portion of atmospheric lead in the total body burden. The conclusions agreed with the Frankfurt study, i.e., a limited importance of atmospheric lead in the total body burden. Another study in France (SERVANT, 1981) based on lead isotopic ratios yielded a similar conclusion.

Decreasing lead antiknock additives could possibly create increases in the emission of polycyclic aromatic hydrocarbons. Conversion to diesel motors for automobile transportation could also create in our environment a new hazard due to the possible mutagenic/carcinogenic properties of the aromatic combustion products from this type of engine, e.g., nitroaromatics.

We have to be cautious and critical in studying the available data on these problems, both in the interpretation and drawing conclusions. If we come to the conclusion that aromatics in gasoline or diesel fuel will increase the amount of mutagens or carcinogens in the air we breathe, we should be able to quantify this type of population exposure relative to other known harmful bronchial agents, like cigarette smoke, and adjust the warning in the same mode. Such epidemiological or biological links are difficult to draw from the existing data base.

This book documents formal, written papers from a NATO Symposium on mobile source emissions, a meeting in which scientists and engineers from several disciplines came together to discuss engine performance. The common tie was that these individuals were all using exhaust gas as a probe to evaluate clean and efficient operation of motor driven vehicles. The participants identified many new ideas and their discussions raised as many questions needing scientific investigation.

The participants expected to bring diverse research results together and reach a consensus by meshing disparate pieces into a uniform picture of engine performance. Such was not to be the case. The meeting was often plagued with the inability to correlate one study with similar investigations because several test parameters were handled differently. The participants soon came to a consensus, mobile source investigations like many other technical problems require standardization of test procedures before intelligent comparisons can be made. The attendees attempted to deal with this problem by making specific recommendations for standardization in this field. Standardization will give comparisons, a tool to gauge new data.

This is the reason why we have deliberately used the structure expressed in the NATO Advanced Research Workshop rules asking us to convenue a "limited number" of scientific experts from member countries of NATO, from different scientific fields and different sectors (university, government, industry), for a short period to achieve one or more of the following objectives:

(a) exchange of thought at the frontiers of knowledge or at the frontiers of the different fields

(b) review and critical assessment of the state-of-the-art

(c) formulation of recommendations for future research directions

(d) formulation of plans for large international scientific experiments.

This meeting was not a discussion of the toxicity of lead, nor a congress in analytical chemistry, or lung cancer epidemiology, nor a confrontation of the possible or needed refinements of the Ames test. Other recent meetings have been organized on these subjects and we are happy that several of the participants actively participated to those discussions, for example:

● The meeting of the Air Management Policy Group of OECD on "Polycyclic Aromatic Hydrocarbons" held in Paris, October, 1981.

● The meeting on "Short Term Bioassays in Mutagenicity Testing" sponsored by the Karolinska Institute of Hygiene, Stockholm, in March, 1982.

- The EPA Symposium on "Toxicological
  Effects of Emissions from Diesel Engines"
  held in Raleigh, North Carolina, USA,
  in October, 1981.

- The NATO-ASI on "Soot in Combustion
  Systems and Its Toxic Properties" held
  in Le Bischenberg, France, August, 1981.

- The annual Battelle Symposium on Polycyclic
  Hydrocarbons.

Members of the CONCAWE (oil companies international study
group for Conservation of Clean Air and Water - Europe) in The
Hague could not attend this meeting but made available the
report they issued in 1980 titled "The Rational Utilization of
Fuels in Private Transport-Extrapolation to Unleaded Gasoline
Case" which is included in this proceedings.

We are well aware that most of us come from different
disciplines. It is perhaps the first time that biologists and
analytical chemists face automobile or petroleum engineers to
delineate automobile exhaust gases and investigate possible
scenarios for the future. We are however confident that such
a composition for the discussion panel contributed to the kind.
of "objective knowledge" that we most urgently need.

At the same time this conference was being held, another
congress was meeting in Geneva, brought together by the United
Nations to deal with the proliferation of geographical names on
world maps. It seems each country selects its own suitable list
of names to identify regions and localities around the world
which leads in turn to massive confusion and a great redundancy
in geographical terminology. The group of people occupying a
location (termed the "donor") has the right to name its own
territory and the proper title that is given should be accepted
by all other peoples in the world (termed the "receivers"). This
is not the case. National groups name places by their own sets
of identifiers, causing a world-wide multiplication in names for
the same place. The science of geographic naming, toponymy, is
suffering much the same confusion as studies on engine behavior;
the lack of common, accepted reference points. The Geneva
conferees returned to 60 nations having left the job of standardization in toponymy undone. Alas so did the NATO participants,
returning to 13 nations with the job of standardizing mobile
source measurements also undone. This is perhaps the single
problem most stressed in this meeting, standardized test procedures.
Until this problem is addressed, much fine research will have to
stand alone without a reference for comparison.

REFERENCES

1. Cass, G. R., Boone, P. M., Macias, E. S.: 1981, Emissions
    and Air Quality Relationships for Atmospheric Carbon
    Particles in Los Angeles, In: Particulate Carbon Life
    Cycle, G. T. Wolff and R. L. Klimish, editors, Plenum Press,
    New York.
2. Servant, J., Delapart, M.: 1981, Blood Lead and Lead 210
    Origins in Residents of Toulouse, Health Phys., 41,
    pp. 483-487.
3. Sinn, W.: 1981, Relationship Between Lead Concentration in
    the Air and Blood Lead Levels of People Living and Working
    in the Center of a City (Frankfurt Blood Lead Study),
    Int. Arch. Occup. Environ. Health, 48, pp. 1-23.

AUTHOR INDEX